MIGRATION AND DISPLACEMENT IN A CHANGING CLIMATE

This book provides insight into the impact of climate change on human mobility – including both migration and displacement – by synthesizing key concepts, research, methodology, policy, and emerging issues surrounding the topic. It illuminates the connections between climate change and its implications for voluntary migration, involuntary displacement, and immobility by providing examples from around the world. The chapters use the latest findings from the natural and social sciences to identify key interactions shaping current climate-related migration, displacement, and immobility; predict future changes in those patterns and methods used to model them; summarize key policy and governance instruments available to us to manage the movements of people in a changing climate; and offer directions for future research and opportunities. This book will be valuable for students, researchers, and policymakers of geography, environmental science, climate and sustainability studies, demography, sociology, public policy, and political science.

KELSEA BEST is an Assistant Professor of Civil, Environmental and Geodetic Engineering and City and Regional Planning at the Ohio State University where she studies equity in climate impacts and adaptations, including climate-related mobility. She has consulted for the US Department of State and the National Geospatial Intelligence Agency. She is an active member of the Association of American Geographers and the Society for Risk Analysis.

KAYLY OBER is an adjunct professor at Georgetown University. She has worked on climate, migration, and conflict for 15 years, including as a lead author of the World Bank report "Groundswell: Preparing for Internal Climate Migration." She is currently a senior advisor for the Bureau of Conflict and Stabilization Operations at the US Department of State.

ROBERT A. McLEMAN is Professor of Geography and Environmental Studies at Wilfrid Laurier University in Waterloo, Canada. He has researched climate-related migration and displacement for more than two decades, advised many governments and international agencies on related issues, and served as coordinating lead author for the Sixth Assessment Report of the Intergovernmental Panel on Climate Change and has been a consultant for numerous multilateral and government agencies. A former Canadian diplomat, he wrote *Climate and Human Migration: Past Experiences, Future Challenges* (Cambridge University Press, 2014).

"*Migration and Displacement in a Changing Climate* draws on evidence from the social and natural sciences, and from examples from across the globe, to provide an authoritative, balanced, and comprehensive guide that cuts through the hyperbole and points to constructive ways to respond to this powerful emerging risk to social order."

Professor Jon Barnett, University of Melbourne

"*Migration and Displacement in a Changing Climate* provides an excellent resource for those new to the topic of climate-related mobility as well as those with years of experience. The authors skilfully build a foundation through definitions and a review of research on migration drivers, and then build on these foundations with careful review of research findings in both social and natural sciences. Compelling case studies illuminate the lessons learned. Especially important and useful in today's conflict-ridden world is that the authors never lose sight of the fundamental humanity inherent within population movement, both now and in centuries past."

Professor Lori M. Hunter, University of Colorado, Boulder

"Best, Ober, and McLeman provide a unique reference of case studies on climate-affected migration in the US and elsewhere while positioning them in the context of interdisciplinary theory and policy. I finally have a resource I can use in my own class that includes all of the fundamental material on the subject in one place."

Professor Valerie Mueller, Arizona State University

MIGRATION AND DISPLACEMENT IN A CHANGING CLIMATE

KELSEA BEST

Ohio State University

KAYLY OBER

Georgetown University

ROBERT A. McLEMAN

Wilfrid Laurier University

CAMBRIDGE
UNIVERSITY PRESS

Shaftesbury Road, Cambridge CB2 8EA, United Kingdom

One Liberty Plaza, 20th Floor, New York, NY 10006, USA

477 Williamstown Road, Port Melbourne, VIC 3207, Australia

314–321, 3rd Floor, Plot 3, Splendor Forum, Jasola District Centre, New Delhi – 110025, India

103 Penang Road, #05–06/07, Visioncrest Commercial, Singapore 238467

Cambridge University Press is part of Cambridge University Press & Assessment,
a department of the University of Cambridge.

We share the University's mission to contribute to society through the pursuit of
education, learning and research at the highest international levels of excellence.

www.cambridge.org
Information on this title: www.cambridge.org/9781009449595

DOI: 10.1017/9781009449625

When citing this work, please include a reference to the DOI 10.1017/9781009449625

First published 2025

A catalogue record for this publication is available from the British Library

A Cataloging-in-Publication data record for this book is available from the Library of Congress

ISBN 978-1-009-44959-5 Hardback
ISBN 978-1-009-44960-1 Paperback

Additional resources for this publication at www.cambridge.org/best

Contents

Preface

The Summary for Policymakers of the Intergovernmental Panel on Climate Change (IPCC 2023) Synthesis Report begins with this statement:

Human activities, principally through emissions of greenhouse gases, have unequivocally caused global warming, with global surface temperature reaching 1.1°C above 1850–1900 in 2011–2020. Global greenhouse gas emissions have continued to increase, with unequal historical and ongoing contributions arising from unsustainable energy use, land use and land-use change, lifestyles and patterns of consumption and production across regions, between and within countries, and among individuals.

The report's next main statement is:

Widespread and rapid changes in the atmosphere, ocean, cryosphere and biosphere have occurred. Human caused climate change is already affecting many weather and climate extremes in every region across the globe. This has led to widespread adverse impacts and related losses and damages to nature and people. Vulnerable communities who have historically contributed the least to current climate change are disproportionately affected.

Among the widespread impacts on people referred to by the IPCC is a growing number of people on the move. Rarely a day goes by without news of people somewhere being forced to flee their homes in the face of a storm, flood, or wildfire, or of drought conditions causing hunger or worse in a low-income country. These and other extreme weather events occur on every inhabited continent, and in many regions their severity, frequency, and/or geographical range are increasing. On average, more than 20 million people each year worldwide are displaced by weather- and climate-related disasters, and that average is trending upward. No country is exempt; the worst affected countries include high-, medium-, and low-income countries (e.g. USA, China, and Bangladesh). Some Pacific island nations have already begun relocating coastal communities and the government of the Maldives is constructing artificial islands that will house tens of thousands of citizens in apartments built high above rising sea levels. The number of people on the move for climate-related reasons is almost certain to continue growing for three key reasons:

- our collective unwillingness to reduce greenhouse gas emissions;
- the growing number of people living in areas exposed to climate risks; and
- the slow progress in building sustainable prosperity in low-income countries and communities.

If these three trends persist, the World Bank (Clement et al. 2021) has projected that 200 million people or more may need to move from their homes by the impacts of climate change by the 2050s – that is, within this lifetime of most readers of this book.

The good news is that this does not need to happen, and hopefully won't. The technology needed to transition away from the fossil fuel economy is now widely available and rapidly falling in price; all that is now lacking is the political will to accelerate the adoption of these new technologies. Companies and countries that make a lot of money from the fossil fuel industry are resisting this transition, but alternatives to coal and oil are not only cleaner but also cheaper and more efficient, and market realities will overcome their resistance. As many have correctly noted, we didn't switch from driving horse-drawn wagons to automobiles because we ran out of food for the horses; we switched because automobiles are a superior technology. Coal-fired power plants and gasoline-powered vehicles are entering the horse-and-buggy stage of obsolescence.

There are over eight billion people on the planet. That's a lot for the planet to support, all of them needing food, water, sanitation, shelter, employment, health care, education, and the other basic services most people living in high- and middle-income countries take for granted. The UN Department of Economic and Social Affairs projects that we will add another billion people before the human population starts to plateau, again in the 2050s, and most of those extra people will be living in low-income countries where the necessities for a happy and healthy life cannot always be taken for granted. Further, many of the areas of the world experiencing rapid population growth are locations that are highly exposed to extreme weather events and climate hazards. If we can accelerate progress toward meeting the UN Sustainable Development Goals, we can ensure that people living in countries that are poor today are less poor tomorrow and that we can meet the basic needs of everyone who is already here on the planet and all those extra people who will join us in coming decades.

If we reduce greenhouse gas emissions and build sustainable prosperity, not only can we avoid nightmarish scenarios of hundreds of millions of people having to flee their homes, we can actually lower the current average of 20 million people displaced by climate-related disasters each year. We can likely never eliminate entirely the risk of floods or fires forcing people to move at least temporarily from their homes when they occur – these events are an inherent part of nature – but we can certainly reduce the ensuing hardship and suffering and help people recover quickly, if we so choose.

Over 10,000 studies and papers have been published on topics related to climate and migration, according to the CliMig database maintained at the University of Neuchatel. There are a dozen or more edited scholarly collections of chapters contributed by researchers working on the subject. But there are very few monograph-style books that provide one-stop, in-depth coverage of the essential information needed by students, aspiring researchers, and people working for governments, multilaterals, and nongovernmental organizations. That's what this book offers. This is not the first such book; that was *Climate and Human Migration: Past Experiences, Future Challenges*, written by one of the three authors of the present book and published by Cambridge University Press a decade ago. It continues to be widely used as a course textbook, despite the fact that a lot has changed in both the science of understanding climate-related migration and events occurring around the world. It's time for an update.

Nine of the ten hottest years recorded have occurred since that book was published in 2014. Most of the present book was written in 2023, which was the hottest year on record. That year there were devastating floods in Libya, a quarter million Canadians had to evacuate from wildfires twice the size of any previously known, the strongest and longest lasting cyclone ever recorded struck southeast Africa, and Lahaina – the former capital of Hawaii's kings and of tremendous importance to Hawaiian people still – was destroyed by a wildfire. The economic impacts of climate disasters are spiraling upward, with the human impacts being increasingly and disproportionately felt by disadvantaged and marginalized people around the world. The impacts of the COVID-19 pandemic continue to ripple through societies. Political turmoil, civil conflicts, and wars have emerged in Africa, the Middle East, and the former Soviet Union, and authoritarian parties and rulers are experiencing newfound popularity in high-income countries. Wealthy countries are building border walls and using military technologies to prevent asylum seekers from reaching their borders, and migrants are often vilified.

Despite these dark developments, there have also been many notable achievements, and progress is being made. The 2015 Paris Agreement outlines a pathway forward for countries to reduce their greenhouse gas emissions and help vulnerable countries build capacity to adapt to the impacts of a changing climate, and – notwithstanding what the skeptics say – a great many countries are working actively toward achieving its goals. In 2023, an agreement was reached to channel funds from wealthy nations with historically high greenhouse gas emissions to poorer nations, to help them recover from losses and damages suffered from the impacts of climate change, including the displacement of people by weather-related disasters. Global "Compacts" were reached through the UN process that provide guidelines to countries on how to prepare for and manage in a humane way involuntary displacements caused by climate-related disasters. Countries in Africa and Latin America have built regional agreements to help one another deal with climate-related displacements. Powerful new computing technologies, methods, and data allow researchers to better identify how climate events affect migration and displacement and allow for increasingly more sophisticated projections and future scenarios to assist policymakers and decision-makers plan for people on the move in a rapidly changing climate. International human rights decisions are improving protections for people unable to return to areas they fled because of the impacts of climate change.

This book takes a deep dive into the connections between the earth's climate, the changes being forced upon it by human activity, and the implications these changes have for voluntary migration, involuntary displacement, and immobility (people being unable to move or not wanting to move). Using the latest findings from natural and social science, we describe the causal interactions that shape current climate-related migration, displacement, and immobility; project future changes in those patterns and how we might model them; summarize key policy and governance instruments available to us to manage the movements of people in a changing climate; and conclude with some thoughts on future research needs and opportunities.

A key departure point for this book is that migration is neither inherently good nor inherently bad. It is simply something that people have always done and will continue to do. It is not something we should fear; it is not something we should attempt to control at all costs. Migrants are people who are simply doing what they believe is best or necessary to maintain and (hopefully) improve their own well-being and the well-being of their family members and loved ones. There is nothing inherently problematic in any of this. The circumstances under which people migrate, under which migration occurs – *that* is what is important. Migration generates benefits for sending communities, receiving communities, and migrants themselves when people are able to make informed decisions, are able to move through legal channels, are able to enter the regular workforce, and enjoy the same social protections as preestablished residents of the places they move to. When people are forced from their homes without a choice, when the only options for moving are to sneak across borders at night or hire organized criminals to help them do so, when they are denied legal rights and social protections upon arrival, when the only jobs open to them are in the informal economy, and when they are separated from families, this benefits no one. And yet this is how a growing number of countries treat migrants. This must stop.

This book describes migrants and migration processes as being potential (but not necessary) outcomes of individual and household decisions of how to respond to climate risks within the context of the much wider social, cultural, economic, and political processes that shape everyday life in a given place at a given moment in time. Although we use the terms migration, displacement, and immobility throughout the book, they are not separate, distinct outcomes; decisions to migrate (or not) fall along a continuum of agency, from being purely voluntary to completely involuntary. In the face of a particular climate event, some people may move away temporarily and then return, others may leave and not return, others may not leave at all, and still others may move into the affected area from elsewhere. Some of these outcomes may be voluntary, others less so. The factors that influence migration are complex and context specific, and so are the outcomes.

The book is organized as follows. Chapter 1 starts with a brief overview of the longer history of how climate has affected human population distributions and migration, followed by a crash course on the science of climate change and the risks that emerge. It then reviews key concepts and theories that underpin how we understand adaptation to climate risks and how (and why) people move. The goal is to provide the reader with all the background concepts and terminology to begin deep dives into migration and displacement related to specific types of climate hazards, including extreme weather, cyclones (hurricanes, typhoons), and floods (Chapter 2); extreme heat, droughts, and wildfires (Chapter 3); and coastal hazards associated with rising sea levels (Chapter 4). In Chapter 5, we review techniques researchers use to model and predict climate–migration interactions, and in Chapter 6, we review policy options directly related to managing climate-related migration at global, regional, and national scales. We conclude the book with thoughts on key areas for future research (Chapter 7). We chose to end with a discussion of future directions in climate-related migration as a reminder that this field is continuously evolving. As researchers use new methodologies, collect new data, and elevate the voices of people who have previously been excluded, our understanding of the complexities of climate-related migration will improve. Some of these important future directions include consideration of the gendered dynamics of climate-related migration, inclusion of Indigenous knowledge and experience (and a related consideration of how climate-related migration interacts with cultural heritage), planning for and identifying critical thresholds in climate-related migration dynamics, and reflection on unforeseen outcomes and deep uncertainty about the future of climate–migration interactions and outcomes. We suspect this book will need to be updated 10 years from now to describe new advancements in our knowledge, and what we have yet to learn.

We conclude this preface with a short description of our approach to writing this book and our own backgrounds as researchers and scholars. This book was a collaborative effort by three authors who never all met in person at the same time. It was a product of many video calls, shared documents, virtual brainstorming sessions, and online "writing hacks" – a writing approach many scholars learned to use during pandemic lockdowns.

Kelsea Best is a professor of civil, environmental, and geodetic engineering and city and regional planning at the Ohio State University in Columbus, Ohio, United States. She has worked with interdisciplinary teams to research climate-related migration in coastal communities in Bangladesh using a wide range of methods, including machine learning, agent-based modeling, and mixed-methods approaches. Her research broadly focuses on understanding how climate change interacts with human societies and infrastructure, how people may adapt to climate change effects, and how climate adaptation measures can be designed and implemented in a just and equitable way. Her work is highly interdisciplinary and strives to connect methods, disciplines, and researchers from across geographies and fields.

Kayly Ober is a second-generation migrant based out of Washington, DC who has worked at the intersection of climate, migration, adaptation, and conflict issues for the last 15 years. She has worked in academia, advocacy, and multilateral organizations, which have analyzed and shaped the climate–migration–adaptation policy landscape. She has previously served as the Senior Program Officer for the Climate, Environment, and Conflict program at the United States Institute of Peace; Senior Advocate and Program Manager of the Climate Displacement Program at Refugees International; and a member of the Task Force on Displacement established by the Executive Committee of the Warsaw International Mechanism for Loss and Damage. She has also worked at the Asian Development Bank, the ODI Global, the Woodrow Wilson Center, and the World Bank, where she coauthored the flagship publication "Groundswell: Preparing for Internal Climate Migration." She is currently a senior advisor at the US Department of State and an adjunct professor at Georgetown University.

Robert McLeman is a professor of geography and environmental studies at Wilfrid Laurier University in Waterloo, Canada. He has been doing research on climate change and migration for two decades, using a variety of qualitative, quantitative, and geospatial methods, with a heavy emphasis on dryland and drought-prone regions, especially the North American Great Plains. He served as a coordinating lead author for the IPCC's Sixth Assessment Report, for the chapter that assessed the impacts of climate change on human health, well-being, migration, displacement, and conflict. Academia is his second career; from 1990 to 2002, he worked as a Canadian foreign service officer specializing in migration and visa work at consulates and embassies in Belgrade, New Delhi, Hong Kong, Seattle, and Vienna.

We hope that you, the reader, find our work helpful and accessible, and that if it doesn't inspire you to pursue further research and scholarship in this field, that it at the very least helps you see climate and migration (and the future of it) in a different light.

Companion materials including resources for instructors teaching this book are available at www.MigrationInChangingClimate.com.

The views and opinions expressed in this book are those of the authors and not necessarily the views and opinions of the United States Government.

Acknowledgments

The authors would like to thank Cambridge University Press for the opportunity to create this book, and especially Matt Lloyd and Maya Zakrzewska-Pim for their support and assistance. We also thank the five anonymous reviewers who shared their time and thoughts with us, and which made the final product much stronger as a result. Trina King and Alana Peroff of Wilfrid Laurier University are thanked for their assistance with cartography and editorial management, respectively. Charity Parr-Vasquez and the Office of Research Services at Laurier provided financial support to assist us with producing this book.

Kelsea Best would like to thank the many mentors and colleagues who have collaborated with her and supported her in her research career. She would especially like to thank Dr. Deb Niemeier for being an unwavering advocate and for pushing her to be a better scholar. Kelsea would also like to thank her mom, Kim Best, who always provides wisdom and support. Most importantly, Kelsea would like to thank her number one cheerleader and other half, Ash Gillis. This book would not have been possible without their endless encouragement and love. Finally, Kelsea would like to thank her dog, Henry, and her cat, Dany. Pets make everything better!

Kayly Ober would like to thank first and foremost the rock in her life that made the extra effort of this book possible: Ashley Nash-Hahn. Her mom and dad, Dora and Joe Ober, also provided much needed encouragement. She would also like to thank those that helped to mentor and support her work on the issue of climate-related migration through the years, including especially Dr. Geoff Dabelko, Dr. Patrick Sakdapolrak, Dr. Harald Sterly, Beth Ferris, Dr. Kanta Kumari Rigaud, Margaret Arnold, Smita Nakhooda, Hardin Lang, Eric Schwartz, Dr. Yael Schacher, and countless others that have been members of the Climate, Migration, and Displacement Platform. Evalyn, Christian, Andrew, Helena, Nina, and Salote – here's to you!

Robert McLeman would like to thank the many scholars and students who have collaborated with him on climate-migration research over the years – far too many to name here. He would also like to thank his fellow coauthors on the Intergovernmental Panel on Climate Change AR6 Working Group II team, particularly those who contributed to Chapter 7 and the cross-chapter box on migration. His support team of Coleen, Anna, and Sophie McLeman can't be thanked enough. Finally, sincere thanks to Barry Glendenning; hearing his voice first thing Monday mornings is a grand way to start the work week.

1

People on the Move in a Changing Climate

1.1 Introduction

People are regularly on the move; it is a defining feature of our species. We move on a daily basis within our homes, neighborhoods, and communities: doing chores, going to work or school, meeting friends for coffee, checking in on loved ones, or dropping off or picking up children at activities or childcare. We move to meet basic needs, which might involve going to a grocery store to purchase food or going out into a field or barn where we produce our own food. We also, though less frequently, move across longer distances and for extended periods of time. We move to live with other people, to seek out new opportunities, to vacation, to pursue higher education or skills training, or to retire somewhere pleasant (if we can afford to do so). Young adults are often tempted to go out and "see the world," for the sheer adventure of travel and being somewhere new. Unfortunately, we also sometimes move because we have no choice, as factors beyond our control force us from our homes, sometimes never to return. Some people are continuously on the move because of a lack of permanent shelter, no place to call home.

This book is about particular types of movement – migration (a voluntary move resulting in a change of residence for an extended period of time) and displacement (being forced to move) – and how they are shaped by climate and changes in the climate. There are more migrants and displaced people today than at any previous time in human history, and their numbers are growing. As of the year 2020, over 280 million people were living in countries other than the one in which they were born – that is, they are international migrants – and hundreds of millions of other people moved within the geographical confines of their home countries (International Organization for Migration 2024). In the 1970s, the total number of international migrants was less than one-third of the current number. In addition to people who moved voluntarily, at the end of the year 2023 an estimated 75.9 million people globally were involuntarily displaced within their home countries, primarily for reasons related to conflicts, violence, and environmental hazards, and over 40 million people had fled from their home countries as refugees, seeking asylum in another country (Internal Displacement Monitoring Centre 2024, International Organization for Migration 2024).

Most people on the move today, whether voluntarily or involuntarily, are moving or have moved for reasons not directly linked to climate. But some have moved for climate-related reasons, and their numbers are growing. In the year 2023, approximately 20.3 million people around the world were displaced by floods, storms, droughts, wildfires, and other climate-related weather hazards (Internal Displacement Monitoring Centre 2024) – an average year based on statistics that have been kept since 2010, with the previous year (2022) having set the record for weather and climate-related displacements at 32 million people globally. It is not so easy to estimate the number of people who move voluntarily because of climate (we explain the reasons why in subsequent chapters), but it is safe to say that they number in the tens of millions at very least.

It is not surprising that large numbers of people are on the move for reasons related to climate, and their numbers are set to grow for two key reasons. First, there are simply more people on the planet with each passing year – over twice as many today as in 1970 – and more people than ever are living in locations that are highly exposed to climatic hazards. Second, we are rapidly changing the climate itself by pumping ever-increasing amounts of carbon dioxide, methane, and other heat-trapping greenhouse gases (GHGs) into the atmosphere, thereby increasing the frequency and severity of floods, storms, droughts, wildfires, and climate hazards. In addition, sea levels are starting to rise because of the growing amount of heat accumulating in the surface layer of ocean water. The mean rate of sea level rise was initially slow – an increase of roughly 1.8 mm per year in the first part of the twentieth century – but seas are now rising at 3.6 mm per year, and the pace is accelerating (Oppenheimer et al. 2019). The combination of more people living in high-risk locations and more climate hazards taking place will inevitably lead to more people being on the move because of climate change. The World Bank has warned that over 200 million people in low- and middle-income countries could move due to climatic hazards by the year 2050 if we do not act collectively to control GHG emissions and make significant advancements in sustainable economic development in those countries (Clement et al. 2021).

The actual number of people who end up moving in the future for reasons directly or indirectly related to climate change will depend heavily on three key factors:

- the extent to which we act to control and reduce GHG emissions (which will in turn determine the frequency and severity of future climate hazards),

- the ability of people and communities to adapt to climate hazards through means other than moving
- how governments approach migration and displacement policy and management

This book takes a deep dive into how each of these three factors will influence the movement of people in a changing climate and in doing so details the available data, the key natural and social science concepts on which current knowledge is based, and the policy options available to us.

There are many books on migration and displacement, and many about climate change and its impacts, but there are relatively few that focus directly and exclusively on climate-related migration and displacement. There are two important reasons why books like this one are needed. First, climate change is not simply just another variable to be added to long lists of other things that influence migration decisions and outcomes, such as household incomes, social networks, political processes, and cultural norms. Climate change not only interacts in a compounding way with other variables to influence migration decisions, but its impacts also have the ability to fundamentally alter the other variables – by undermining livelihoods, depriving people of the basic necessities of life, generating political and economic stability, rupturing social networks, and making locations where people currently live unlivable. Unchecked, climate change will make the physical environment unlike anything people have experienced before, and we will consequently see migration and displacement patterns unlike any we have experienced before.

Second, a common vocabulary and clear understanding of the connections between climate and migration is needed to support national and international actions to address anthropogenic (i.e. human-caused) climate change. The key international agreement for doing so is the UN Framework Convention on Climate Change (UNFCCC), and late each calendar year its signatories meet at an annual Conference of the Parties (COP) to negotiate next steps in reducing GHG emissions and helping vulnerable people and countries adapt to the impacts. Climate-related migration and displacement are increasingly featuring in these negotiations, and a Loss and Damage Fund agreed to at the 2023 COP will almost certainly, once fully established, become a vehicle through which low-income countries will seek financial assistance to help them cope with climate-related displacements. As more government agencies and multilateral organizations that have not historically been involved in migration management and policymaking become concerned with it because of climate change – or indeed, as more organizations that have historically been concerned with migration and displacement become engaged with climate policy – there is a need to ensure everyone is speaking a common language and has a common, fact-based understanding of the fundamentals of climate processes, migration/displacement processes, and their interactions. This book aims to provide such an understanding.

The remainder of this chapter provides:

- an introduction to climate-related migration and displacement in the distant and more recent past
- an overview of the basic natural science processes behind anthropogenic climate change for readers that require one (feel free to skip past if you do not)
- a review of how the impacts of climate change in a general sense present risks to individuals, households, and communities, and how vulnerability and adaptation shape these risks
- a summary of the social science on how migration decisions are made and the general types of patterns and outcomes that emerge
- a consolidated picture of how climate hazards interact with nonclimatic processes to shape migration and displacement

A series of text boxes are included to help readers understand concepts and terminology with which they may not already be familiar (Boxes 1.1 and 1.2). By the end of this chapter, the reader will have the background information and key concepts necessary to take a deep dive in subsequent chapters into how particular types of climate hazards affect migration and displacement, modeling tools being used by researchers to expand our understanding of climate-related migration, and the options available to policymakers hoping to respond to it. We wrap up the book with a look to the future, identifying important areas for future research.

Box 1.1

Key Terms: Mobility, Migration, Displacement, Immobility, and Climate-Related Migration

Humans are a mobile species. We move for a variety of reasons and purposes, across distances short and long. *Mobility* is a generic, umbrella term used to describe movement of any type, distance, and duration. It includes daily movements of people as they travel from their homes to places of school or work; short-duration trips taken for vacations, visiting friends and relatives, and similar purposes; and other journeys that do not involve changing one's ordinary place of residence. Although these types of movements can be influenced by climate, they are not a subject of interest for this book. Instead, we focus on two other broad categories of mobility and, perhaps counterintuitively, situations where people might move but do not. The first and broadest category of movement addressed in this book is *migration*, which refers to circumstances when people change their place of residence for a permanent or semipermanent period of time (this definition is taken from a widely cited paper by Lee [1966]), and in doing so, they travel some distance longer than simply moving to a different home in the same town, city, or immediate vicinity. Also important is that the term migration implies a voluntary decision to move. People who move voluntarily are referred to in this book as *migrants*.

A second type of movement discussed frequently in this book is *displacement*, which refers to situations when people have no choice but to move in the face of immediate threats to their lives, livelihoods, and/or well-being. The two most common reasons why people become displaced are (1) acts of violence or conflict and (2) environmental hazards. In this book, we are specifically interested in displacement that is caused directly or indirectly by a specific subset of environmental hazards, namely climate-related hazards. Other types of environmental hazards, such as geotechnical hazards (e.g. tsunamis and earthquakes) and environmental toxins, may also lead to displacement but are not assessed in any detail in this book. People who experience climate-related hazards might be displaced for short periods of time, after which they return immediately to their place of residence, or indefinite periods, sometimes never being able to return to their former homes. The distances they move during their period of displacement can vary considerably. People who are displaced are generally referred to in this book as *displacees*, and where appropriate we use the term *evacuees* to describe people who are displaced for very short periods of time and expect to return to their homes quickly after a hazardous event, and the terms *involuntary migration* and *involuntary migrants* in reference to displacees who must relocate permanently elsewhere.

A third aspect of mobility we discuss often in this book is *immobility* – a condition where people might move or might be expected to move under particular circumstances but do not. Immobility can be *voluntary* or *involuntary*. As with migration and displacement, in this book, we consider immobility only in the context of climate-related events and conditions. For example, we describe cases in Chapter 4 of people who live on small islands threatened by rising sea levels, some of whom want to move to safer islands but lack the means to do so, and others who plan to remain where they are, even in the face of severe risks to their lives and livelihoods.

Many terms have been used over the years by scholars, governments, the media, and the wider public to describe people who move for reasons directly or indirectly related to climate, including *environmental refugees* and *climate refugees*. We deliberately avoid using the word "refugee" in this context, as there is a clear, internationally recognized definition of a refugee under the 1951 *United Nations Convention Relating to the Status of Refugees*, and this definition does *not* apply to people who are forced to move for reasons related to weather, climate, or the environment more generally. It is possible that individual governments or the international community might choose to recognize people who move for environmental reasons as "refugees" under national or international law (see Chapter 6), but for the purposes of this book – and indeed, in any general discussion of the connections between climate and the movement of people – it is important that the term "refugee" be reserved for people fleeing across international borders for fears of violence and persecution.

In this book, we use the adjective "climate-related" to describe migration, displacement, and immobility that are directly or indirectly influenced by weather events, climatic conditions, and/or longer term changes in the climate. *Climate-related migration*, climate-related displacement, and climate-related immobility are not legal terms, but they are clear and consistent with the broad base of scholarly research that has emerged in recent decades. They are also the same terms used in the Intergovernmental Panel on Climate Change (IPCC) 2022 Working Group II Assessment Report (Cissé et al. 2022), which is important, for it is the key reference document used in international negotiations of the UNFCCC, which is in turn the main international agreement through which global action is coordinated to tackle climate change and respond to its impacts, including migration, displacement, and immobility (see Box 1.4 for more information on the IPCC). Too often, academic researchers unnecessarily use language and terminology that are unfamiliar to decision-makers and the wider public, causing their work to be poorly understood or ignored. We choose not to make that mistake here.

We use the term climate-related migration and not the shorter "climate migration" to reflect how there is rarely a single reason why people move (or don't move). Mobility decisions are often multicausal. Even when confronted with an imminent, potentially life-threatening climate hazard, people's short- and longer-term decisions about whether to stay or leave, when to leave, where to go, and whether they go back to the places where they once lived are influenced by a wide range of economic, social, cultural, political, and other nonclimatic factors (Black et al. 2011). People who work in the field often summarize the reasons for migration as being a combination of *push factors and pull factors*: things that make people leave one place and things that make another one more attractive. Climate can work as both a push factor and a pull factor. The strength of its push or pull influence varies from one individual, household, or community to another and from one climate event to another. For some people or in some circumstances, climate may be the primary factor influencing their mobility decisions. For others, it may be a secondary reason, or just one reason among many. And in many cases, climate has no particular influence at all. Using the term "climate-related" provides a clear, consistent, generic way of capturing those circumstances under which climate (or climate change) has an influence on migration, displacement, and immobility from those when it does not.

1.1.1 *Migration in the Context of Human Adaptation to a Naturally Changing Climate*

Apart from Antarctica and the hottest, driest parts of a small number of deserts, there are few terrestrial spots on this planet where people have not lived or attempted to live on a permanent basis. The history of our species since it emerged approximately 200,000 years ago is of people on the move, and for most of that history, climate played an important role. Radiating out from our origins in East Africa, *Homo sapiens* have over the millennia moved and adapted biologically, behaviorally, and technologically to a wide range of environments. Hot or cold, wet or dry, rugged or flat, continental or island, if people could get there, they did, and attempted to establish homes and livelihoods. Sometimes "home" was not a single location, but a wider territory over which people would travel as they hunted, fished, and gathered, often in synchronicity with the seasons. In other locations where resources and climate permitted, people established fixed settlements, some of which over time prospered and grew into towns and cities, while others dwindled and were abandoned (McLeman 2011). The act of moving around in search of favorable locations was an important component of larger processes of behavioral adaptation that made our species so successful, if we measure "success" in ecological terms such as the number of individuals in a species (eight billion and counting in our case) and the wide variety of habitats we occupy.

There are, however, limits to physical and behavioral adaptation. Humans are biologically incapable of withstanding exposure to very hot or very cold temperatures for extended periods of time, and we must have a supply of water to drink daily. In this sense, climate places an important check on the types of places to which people are able to move; it defines our geographical range, to import another term from ecology. Within areas that are biologically and climatically viable for us, the distribution and density of human settlements is neither uniform nor random; instead, settlement patterns historically reflected the availability, quality, and distribution of resources critical for our survival, with climate again playing a determining factor. Its influence varies across scales from the global to the local. Resources potentially available for human use vary considerably among Arctic, temperate, and tropical environments due to climate; they also vary between the north side and the south side of a hill, and windward and leeward sides of a coastal mountain. Large- and small-scale variations in climatic conditions interact with other ecological processes to render certain locations more desirable than others for human settlements.

The geographical expansion of the human population over the millennia has been further shaped by continuous and ongoing changes in the Earth's climate over long and short periods of time due to natural processes. Human populations have learned how to adapt to the seasonality of the Earth's climate and the inevitable and generally predictable year-to-year variations in temperatures, precipitation, growing season length, and other weather conditions that affect livelihoods (see Box 1.2). People have also had to adapt to changes in climate that unfold over longer periods of times, along with unexpected short-term fluctuations triggered by natural processes that occur within and beyond the Earth's atmosphere. One of the more obvious long-term climatic processes that has shaped human population

movements and patterns has been the slow climatic cycle of ice age to warm interglacial period and back again to ice age. These glacial/interglacial periods play out
over thousands of years, driven principally by Milankovitch cycles, named after
the physicist who first documented them (NASA 2020). These are small variations
in the orbit of the Earth around the Sun, in the tilt of the Earth's axis, and in the
steadiness of the Earth's spin on its axis. Milankovitch cycles change the distance
of the Earth from the Sun (the closer we are, the more solar energy we receive, and
average global temperatures warm up) and the orientation of the Earth's surface
toward the Sun (locations that are more perpendicular to the Sun receive more
radiation and become warmer than others).

Box 1.2

Distinction between "Weather" and "Climate"

Tropical cyclones, thunderstorms, tornadoes, and blizzards are among many
phenomena that are alternatively described as "extreme weather events" and "climate-
related hazards." This begs the question, what is the distinction between *weather*
and *climate*? The official – and somewhat vague – definition of *weather* given by
the World Meteorological Organization (WMO) is, "the state of the atmosphere at a
particular time, as defined by the various meteorological elements." "Meteorological
elements" can describe a wide range of phenomena, with the most commonly
measured ones being temperature, precipitation, wind, humidity, and air pressure.
The key thing is that weather refers to conditions experienced at a specific place and
time. Weather is temporary in nature, described and recorded in short increments of
time: hours, days, weeks, months, and seasons. The term *climate* refers to average
weather conditions as measured over an extended period of time, with 30 years being
the shortest measuring period typically used by scientists. The term *climate change*
therefore refers to changes in long-term average weather patterns and conditions as
observed over multiple decades, centuries, or longer. Scientists further distinguish
between natural and *anthropogenic climate change*, the latter referring to change
that is attributable to human activities such as the burning of fossil fuels and removal
of forest cover. When referring to the range of potential fluctuations in climatic
conditions over a given period of time (as opposed to changes in average conditions),
scientists typically use the term "climate variability." For example, average daily July
temperatures in Ottawa, Canada, are approximately 21°C and in January are –10°C.
However, overnight, July temperatures can easily dip below 10°C (an especially
mild January day might also see temperatures reach 10°C). A hot summer July
day in Ottawa might see temperatures soar into the mid-30s – temperatures more
associated with the American capital of Washington DC than Canada's capital city.
All of these are examples of the inherent variability of the Ottawa climate, and none
are examples of the impacts of climate change – unless they happen so frequently
year after year that they cause longer term average temperatures to shift.

Global climate patterns can fluctuate irregularly due to natural processes such as volcanic activity and fluctuations in solar radiation. Recent centuries have seen only occasional large volcanic eruptions and a somewhat steady occurrence of small ones, but in periods when there has been a lot of volcanic activity fluctuations in global climate conditions have occurred. The specific effects of volcanic eruptions on climate are complicated to disentangle (Chim et al. 2023). Ash, particulate matter, and sulfur gasses emitted into the air during eruptions exert a cooling effect on the climate, but carbon dioxide and water vapor that are also emitted into the air trap heat, and the combined effects may be to temporarily cool the climate and then subsequently warm it (but not always). Temporary variations in the amount of energy received from the Sun can also stimulate changes in the climate. On average, the amount of solar energy received from the Sun equals 341 watts per square meter, but there can be variations over relatively short periods of time. The amount of energy emitted by the Sun varies very slightly over the course of a 11-year cycle, and on occasions there are storms on the surface of the Sun that cause flares or bursts of additional energy to be emitted (often referred to as "sunspots," for that is how they appear to us from the Earth) (NASA 2024). Should a solar flare happen to point toward the Earth when it occurs, our planet can receive a slight increase in the amount of energy received, in turn leading to a slight temporary warming. In addition to variations in climate stimulated by orbital variations, volcanic activity, and solar flares, there are multiple naturally occurring, cyclical oscillations in the Earth's climate driven by interactions between oceans and atmosphere, the best known of which is the El Niño Southern Oscillation (ENSO) (Box 1.3).

Throughout the longer course of human history, there have been many times and places when naturally occurring short- and long-term changes in climatic conditions have led to large-scale movements of people into and out of particular areas. In the past thousand years, two periods stand out as being particularly notable for the extent to which climatic changes affected the movements and distribution of people. The first is the Medieval Warm Period (MWP) that ran from the tenth to the early fourteenth centuries CE, a time when average temperatures across much of Europe, Central Asia, Africa, and the Americas were as warm or warmer than they are today (Mann et al. 2008). Excluding present-day climate, average temperatures during the twelfth-century peak of the MWP were likely the warmest experienced since the last ice age (i.e. the last ten thousand years). The impacts of the MWP were favorable for people in Europe, where the climate of the preceding thousand years had been highly variable, characterized by long periods of harsh winters, droughts, extreme storm events, and notable sea level change along the coasts (Lamb 1995, Grove 2002). The MWP created a benign climate that

---

Box 1.3
El Niño Southern Oscillation (ENSO)

The ENSO is a natural climate event that occurs on an irregular basis roughly once or twice per decade and lasts on average from one to two years. The exact circumstances that trigger an ENSO event are not entirely known. ENSO events pass through three phases, first the El Niño phase, then a neutral phase, and finally a La Niña phase before returning to neutral (National Ocean Service 2024). In its first phase, surface winds over the central Pacific Ocean which normally blow from east to west weaken or reverse direction. This allows warm surface waters from shallow parts of the western Pacific to migrate eastward and spread north and south along the west coast of the Americas. South American fisherman gave the phenomenon the name El Niño to reflect how they would notice these ocean warming events in December, near Christmastime (El Niño being the Spanish word for a boy child, the baby Jesus). The El Niño phase has distinct impacts on weather conditions worldwide and leads to more frequent extreme events in many places (World Meteorological Organization 2024). Warmer and drier than average conditions tend to occur over Canada and the northern US, wetter conditions in the southeastern US, droughts in Central America and northern South America, and wet conditions in southern South America. In Africa, southern and western regions often experience severe droughts, while other regions that are typically dry may receive unusually high amounts of rain. Southeast Asia, an area associated with warm, wet weather, is usually drier than usual during this first phase of the cycle, and wildfire risks may increase. Average global temperatures tend to rise during this phase.

After an intermediate phase in which atmospheric conditions revert to neutral, the third La Niña phase sees a re-intensification of east-to-west winds over the central Pacific and a cooling of sea surface temperatures there. Regions of the world that are typically wet become wetter than usual, dry regions may become drier, cold regions become colder and warm regions warmer than usual – in other words, the weather conditions we would ordinarily expect in a given area do not immediately return to normal, but go past this to exhibit an amplification of expected conditions for several months before returning to normal.

---

led to increased European agricultural productivity and expansion northward of warm weather crop production, even leading to the establishment of vineyards in southern England (Grove 2002). During the MWP, Scandinavian Norse (popularly known as the Vikings) began migrating to and establishing permanent settlements on the Faroes, Iceland, and Greenland (the coastal areas of which were indeed green during the summers of the MWP), and the building of smaller, ephemeral hunting and fishing settlements on Newfoundland (Dugmore et al. 2012). Inuit settlements across the North American Arctic and Greenland also expanded during this period of relatively mild conditions (Friesen et al. 2020). Elsewhere, MWP

Figure 1.1   Remains of Anasazi cliff dwellings at Mesa Verde, present-day Colorado. These apartment-style settlements carved out of sandstone cliff walls are found at multiple locations in the US southwest and supported thousands of people until their sudden abandonment in the twelfth century CE. Photo by R. McLeman.

climate conditions were not so benign. A series of severe, extended droughts coincided with the collapse and depopulation of the great cities of the Mayan empire in Central America in the tenth-century and the twelfth-century abandonments of large pueblos (hilltop settlements) and Anasazi cliff dwellings in what is today the southwestern US (Haug et al. 2003, Wahl et al. 2007, Lekson and Cameron 1995) (Figure 1.1). Inter-city conflict, deforestation, and an inflexible political hierarchy likely amplified the effects of MWP droughts on Mayan cities (Shaw 2003, Orlove 2005); it is less clear whether droughts alone were responsible for settlement abandonments in the Pueblo and Anasazi territories or if other factors also came into play. Across Asia, climatic conditions were highly variable during the MWP (this was actually a cold period in East Asia), and some scholars have suggested the expansion of the Mongol emperor of Genghis Khan and his successors into China and Eastern Europe may have been necessitated by persistent dry conditions in Central Asia in the twelfth century (Fagan 2004).

The MWP was followed by a period of falling average temperatures and greater climate instability in the northern hemisphere that reached its nadir between the sixteenth and nineteenth centuries in what has been described as the "Little Ice Age" (Grove 1988). During this period, European winters became longer and colder, ice skating was widely enjoyed in the Netherlands, and England's

River Thames froze over on several occasions (Huntley 1957, Robinson 2005). Agricultural settlements in marginal areas of Europe were abandoned, as were Norse settlements on Greenland (Lamb 1995). Tremendous storms struck the northern European coast on several occasions, causing erosion and the abandonment of many coastal settlements (Clarke et al. 2002). In the eastern North American Arctic, colder temperatures coincided with a southward movement of Inuit settlements (Friesen et al. 2020). Indigenous settlements known as Greater Cahokia in mid-continental North America were abandoned during serious droughts in the years 1350–1450 (Pompeani et al. 2021). In West Africa, the great empire of Mali collapsed in the late sixteenth century during a period of repeated catastrophic flooding in the Niger River valley followed by severe droughts upstream at Timbuktu (Makaske et al. 2007). In India, the recently built capital city of Fatehpur Sikri had to be abandoned due to drought and lack of water (Hillel 1991).

The examples given earlier are just some of many past instances when large-scale population displacements and migrations coincided with notable climatic events and conditions (see McLeman 2011 for further examples). The simultaneous occurrence of a major climate event and a significant event in human history does not prove that the former caused the latter. With migration events of long ago our access to details about human systems and local climate conditions is limited, so it is important to be cautious about assuming a migration event was caused by climate and to not ignore other concurrent events of a political, social, or economic nature that may have had a causal influence. Nonetheless, it is reasonable to view the MWP and the Little Ice Age as being noteworthy periods of human history when migration and displacement patterns were heavily influenced by climate change.

This raises the question: Have we entered a new period when climate change – now being driven by human activity as opposed to natural processes – is once again becoming a dominant influence on migration and displacement? It is too soon to say for certain, but the warning signs are clearly there. The climate is clearly changing rapidly – much more rapidly than at any time humans have been on the planet – and this will test our ability to adapt unlike ever before. Fortunately, because human activity is the cause of the changes now occurring to the climate, it is within our ability to prevent further changes. Whether we will collectively decide to do so is also too soon to say. As of the time of writing this book in 2024, GHG emissions are rising steadily with only modest signs of slowing despite many promises made by politicians from around the world. For at least the next several decades, the climate will continue to warm, generating more frequent and severe hazards that will almost certainly lead to higher rates of migration and displacement in the most highly exposed areas.

1.2 Climate Change: Physical Science, Impacts, and Risks

In this section, we provide a summary of how the Earth's climate functions, how anthropogenic emissions of carbon dioxide, and other GHGs are altering it, the impacts and future risks that result, and how these risks are a product of human exposure to particular climate hazards, our vulnerability to them, and how vulnerability and risks can be moderated through adaptation. Readers seeking a more detailed examination of the physical science of climate change, its implications for human and natural systems, and the options for reducing human impacts on the climate are encouraged to consult the most recent assessment reports of the IPCC (visit www.ipcc.ch) and the many national and sectoral assessments that have been published in recent years.

1.2.1 Basic Functioning of the Earth's Climate and the Effect of Greenhouse Emissions

The Earth's climate is regulated by interactions between solar radiation arriving on a continuous basis from the Sun and the gaseous composition of the Earth's atmosphere. The amount of energy entering the Earth's atmosphere via solar radiation is balanced over the long term by an equal amount of energy being radiated back into space. If this were not the case, there would be no long-term stability of the climate. For example, if the amount of incoming solar radiation exceeded the amount of energy leaving the atmosphere, temperatures on the Earth's surface would rise rapidly. Conversely, if the amount of energy escaping the atmosphere exceeded the amount coming in, the Earth would cool rapidly. On average, the Earth receives just over 340 Watts per meter squared of energy in the form of solar radiation on an ongoing basis. Radiation is produced in different wavelengths depending on the temperature of the emitting source. These wavelengths range from exceedingly tiny gamma rays – roughly the length of the nucleus of a single atom – to radio waves that may be hundreds of kilometers long. The shorter the wavelength, the more concentrated the energy that is carried. Solar radiation enters the Earth's atmosphere in a range of wavelengths that include the near-infrared part of the electromagnetic spectrum, visible light, and ultraviolet light (Figure 1.2). Most of the ultraviolet radiation is prevented from entering lower portions of the atmosphere and reaching the Earth's surface by ozone molecules in the stratosphere – a function that makes life on the Earth possible, since ultraviolet radiation damages the cells of living organisms.

To maintain the energy balance, the incoming radiation from the Sun must equal the amount of energy leaving the Earth's atmosphere. Slightly less than 30% of the incoming solar radiation is immediately reflected back into space from shiny

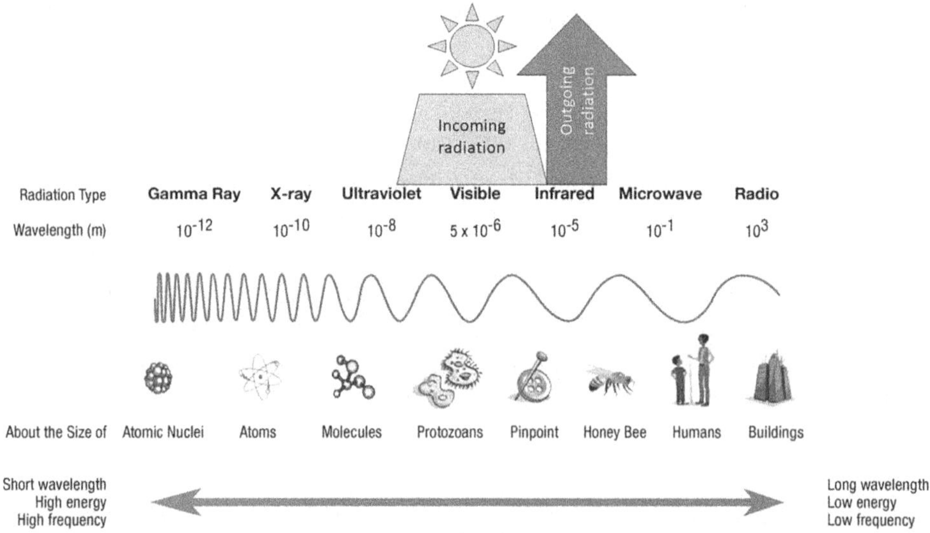

Figure 1.2   The electromagnetic spectrum. Solar radiation arriving at the Earth's atmosphere is primarily in the near-infrared, visible light, and ultraviolet portions of the spectrum, with ultraviolet light being intercepted by ozone molecules in the stratosphere before it can reach the Earth's surface. Radiation leaving the Earth's atmosphere is in wavelengths in the far infrared part of the spectrum. Adapted from MyNASAData.

surfaces, such as the tops of clouds and snow cover on the ground, and therefore has no impact on the Earth's climate (Figure 1.3). Just under half of the incoming solar radiation reaches the Earth's surface and is absorbed, with the remaining 23% of the incoming radiation being directly absorbed by gasses in the atmosphere before it reaches the surface. Energy that is absorbed by the surface and in the atmosphere gets re-emitted back into the air; however, it is re-emitted at a longer wavelength than incoming radiation, in the far part of the infrared spectrum, because the emitting source is not as hot as the Sun. In the absence of an atmosphere (such as is the case on the Moon), this re-emitted radiation would escape directly to outer space. However, only about 12% immediately escapes to space, because our atmosphere contains small concentrations of gasses that absorb infrared radiation and re-radiate it in all directions, including back toward the Earth's surface. This causes air temperatures at the Earth's surface to warm up. These same gasses – carbon dioxide (CO_2), methane (CH_4), and nitrous oxide (N_2O) – do not interfere with the short-wave radiation arriving from the Sun, but only with the infrared radiation trying to escape the atmosphere. In this way, they perform a function similar to glass in a greenhouse, which allows sunlight to enter and warm the interior surfaces but slows heat from escaping, creating temperatures inside the greenhouse that are warmer than the outside air. For this reason, carbon dioxide, methane, and nitrous oxide

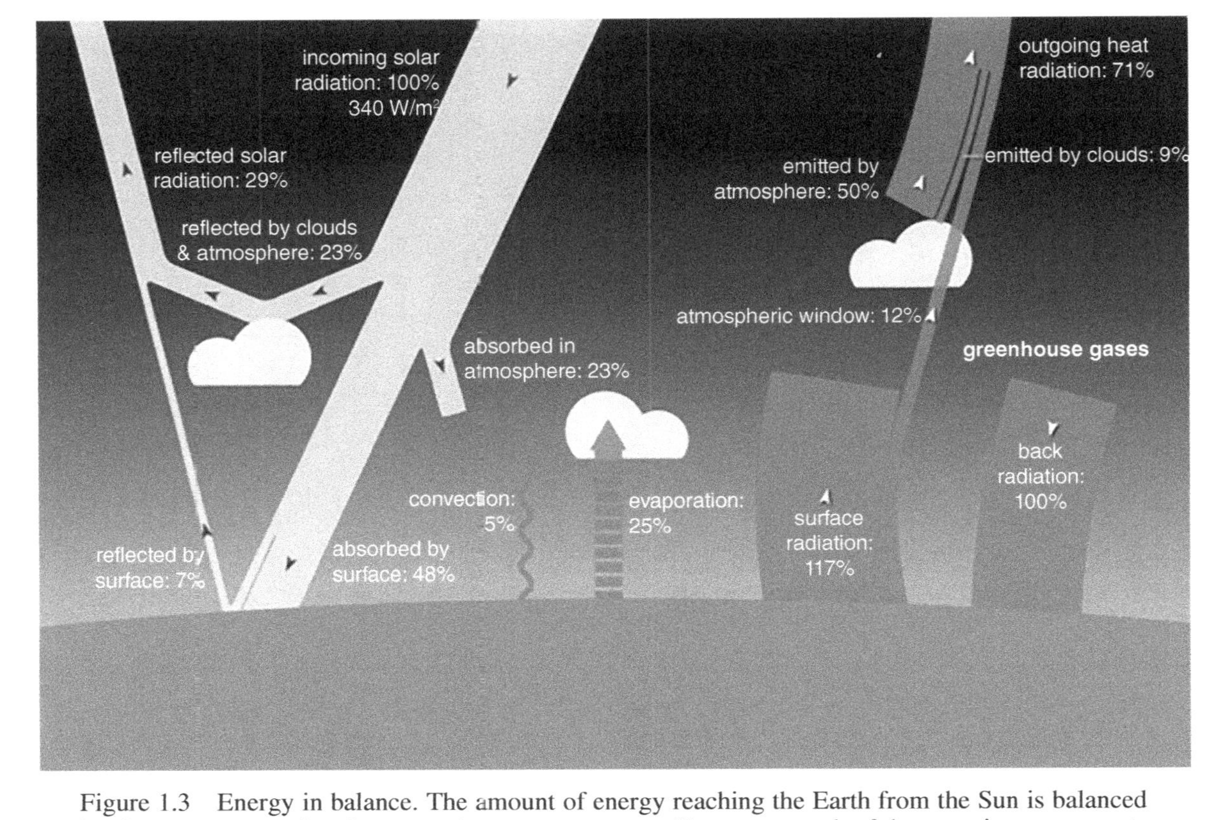

Figure 1.3 Energy in balance. The amount of energy reaching the Earth from the Sun is balanced by the amount escaping the atmosphere to outer space. However, much of the outgoing energy gets intercepted by carbon dioxide and other trace gasses in the atmosphere and re-radiated multiple times before it escapes, causing heat to build up at the Earth's surface. This phenomenon is referred to as the greenhouse effect. For a detailed explanation, visit https://earthobservatory.nasa.gov/features/EnergyBalance).

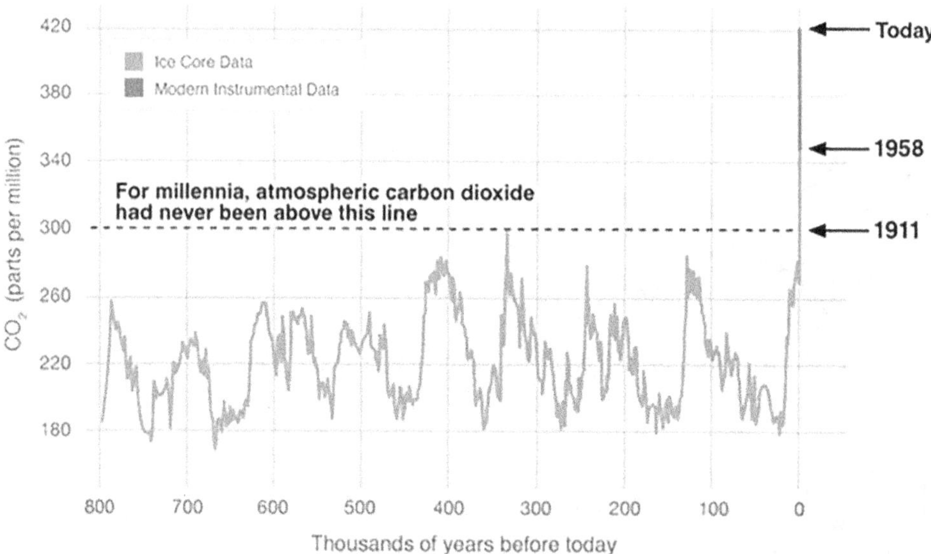

Figure 1.4   Atmospheric concentrations of carbon dioxide over the past 800,000 years. Since 1958, concentrations have been measured directly on a continuous basis. Concentrations for previous millennia are measured from air bubbles trapped within cores of ice taken from Greenland and Antarctica. As snow and ice is deposited over time, air becomes trapped. The lower down the core, the older the air. Source: NASA 2024.

are referred to as GHGs because of their heat trapping *greenhouse effect*. If there were no GHGs in the atmosphere, the Earth's average surface temperature would be approximately −15°C, almost 30°C colder than it actually is.

GHGs are not naturally abundant; concentrations of carbon dioxide have historically fluctuated within a range of 180–300 molecules per million molecules of air, and methane and nitrous oxides are even less common, measured in parts per billion molecules of air. Water vapor is also a GHG; its concentration in the air varies significantly from one location to another, being high in moist tropical environments and near zero in desert environments. Globally, water vapor averages roughly 1% of the air in lower parts of the atmosphere. Starting in the mid-1800s, atmospheric concentrations of GHGs, particularly carbon dioxide, began rising rapidly (Figure 1.4). This increase is not caused by any natural processes described in Section 1.1.1, but by human activity: the consumption of fossil fuels; the production of cement and steel; and, rapid reduction of the Earth's forest cover. Fossil fuels include coal, oil, and natural gas that are extracted from underground and burned, with the combustion process releasing carbon dioxide gas into the air. Burning 1 kg of coal (which is nearly pure carbon) releases over 3 kg of carbon dioxide gas into the atmosphere (the mass of the carbon atom plus the mass of two oxygen atoms joined to it). The industrial processes used to make cement and steel

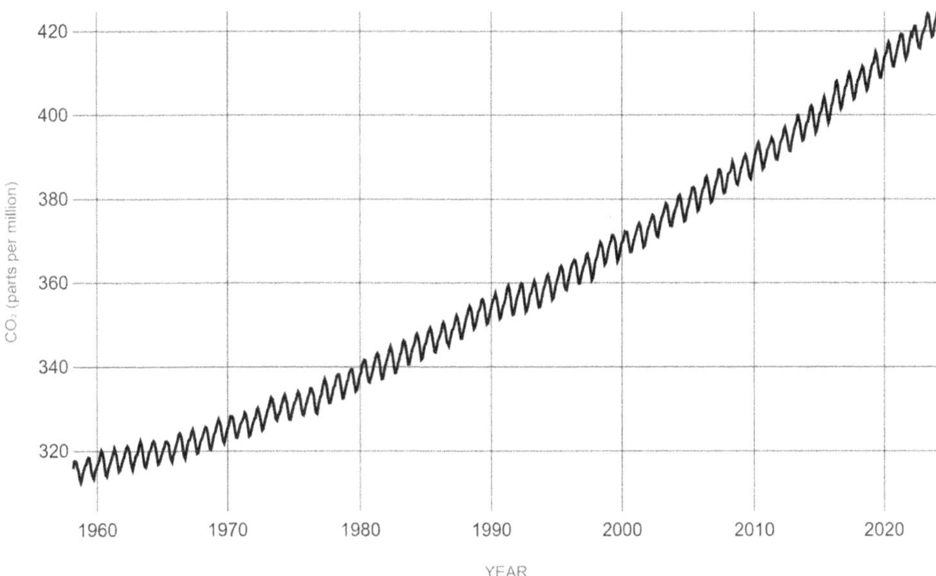

Figure 1.5　Daily measurements of atmospheric carbon dioxide made at Mauna Loa, Hawaii. The overall trend since the late 1950s shows a steady increase; the continuous up-and-down fluctuation reflects the seasonality of photosynthesis levels of terrestrial vegetation. This graph is popularly known as the "Keeling curve" after the scientist who first began taking such measurements. Source: NASA 2024.

also emit carbon dioxide into the air and collectively comprise up to roughly 5% of total anthropogenic GHG emissions. Plants – and forests in particular – remove large amounts of carbon dioxide from the atmosphere and store it in their tissues via photosynthesis. When it is summer in the northern hemisphere – which has a much larger land area than the southern hemisphere and therefore the largest mass of terrestrial vegetation – the amount of photosynthesis occurring is so great it reduces the amount of carbon dioxide in the air temporarily. When the northern winter arrives much of the vegetation goes dormant, and carbon dioxide levels rise again. This seasonal effect of vegetation on atmospheric carbon dioxide is seen in the annual up-and-down fluctuation in measurements of carbon dioxide recorded since the 1950s at the weather observatory at Mauna Loa, Hawaii (Figure 1.5). Deforestation reduces the amount of photosynthesis that occurs, thereby weakening the effectiveness of a natural process that removes carbon from the atmosphere. This process is known as a *carbon sink*, two other key carbon sinks being photosynthesizing phytoplankton that float on ocean surfaces and direct absorption of carbon dioxide from the air by the ocean surface to form carbonic acid, making seawater gradually more acidic. The collective capacity of the Earth's three main natural carbon sinks is unable to keep pace with human emissions of carbon dioxide, hence its rapid accumulation in the atmosphere.

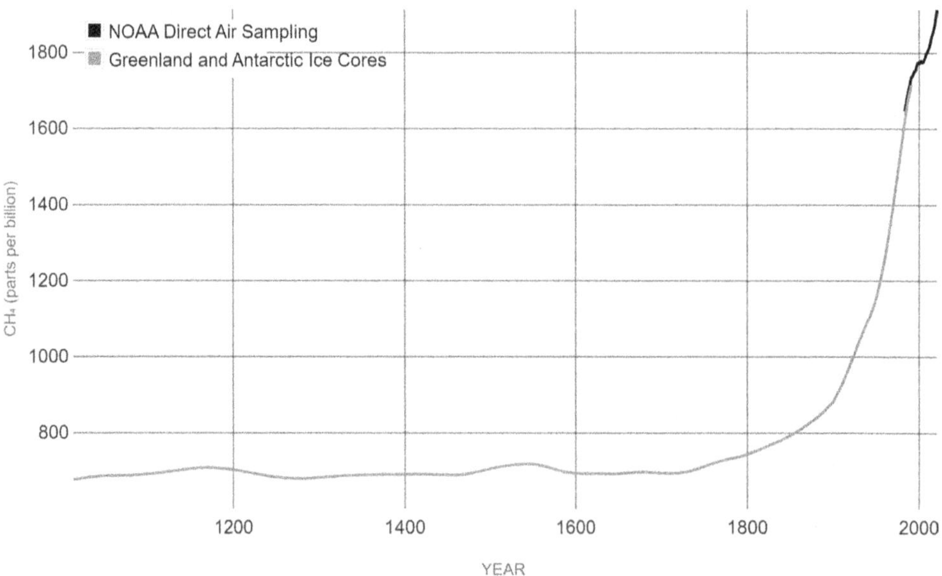

Figure 1.6   Atmospheric concentrations of methane over the past thousand years.
Source: NASA 2024.

Although carbon dioxide is the most abundant GHG (apart from water vapor), atmospheric concentrations of methane have also been rising rapidly over the past 150 years (Figure 1.6). Individual methane molecules trap far more heat in the atmosphere than individual carbon dioxide molecules, but methane molecules remain in the atmosphere for much shorter periods of time than carbon dioxide molecules (7–12 years vs hundreds of years). Methane forms whenever the decomposition of organic material occurs in oxygen-deprived environments, such as buried trash, the stomachs of animals, and wetlands. The fossil fuel known as "natural gas" is methane. Key anthropogenic sources of methane emissions are agriculture, landfills, and oil and gas production. Atmospheric concentrations of nitrous oxide, a GHG that is released as a by-product of combustion and the manufacture of synthetic fertilizers, have risen by approximately 18% over the past century (NASA 2024).

A consequence of rising atmospheric concentrations of GHGs has been a rapid increase in average temperatures at the Earth's surface. Widespread global coverage of thermometer-recorded daily temperatures dates back to the mid-1800s, and so scientists use average global temperatures for the period 1850–1900 as a baseline or reference period for comparison, referring to it as *pre-industrial temperatures*. Any significant deviations in average global temperatures from this reference period are described as *temperature anomalies*. In 2017, average global temperature for the first time rose to more than 1°C warmer than the pre-industrial reference period, with most of this increase occurring since the year 1950 (IPCC

2018). Scientists expect temperatures will continue to rise so long as anthropogenic GHG emissions continue to rise, with average global temperatures reaching 1.5°C warmer than pre-industrial by the late 2030s or mid-2040s; the significance of this will be explained shortly. Figure 1.7 charts average annual global temperatures from 1850 to 2024. Each year's temperature is compared against the mean average for the entire period, represented by the horizontal line set at 0°. Years with temperatures colder than the long-term average have bars pointing downward and those with temperatures warmer than average pointing upward. As can be clearly seen, every year since the mid-1970s has been warmer than the average, and although there is year-to-year variability, the trend is toward consistently warmer global average temperatures.

This raises the question of how much warmer average global temperatures will get, and at what rate of change. The simple answer is it will depend upon future rates of GHG emissions. If GHG emissions continue to rise at current rates, average global temperatures by the year 2100 will be approximately 2.8°C warmer than pre-industrial levels (World Meteorological Organization 2024). If GHG emissions rise faster than current rates and no action is taken to attempt controlling them – what scientists refer to as a *high emissions scenario* – average global temperatures at the end of the century could be 4°C or warmer than pre-industrial temperatures (Arias et al. 2021). Conversely, if the international community acts rapidly to reduce GHG emissions such that they plateau over the next decade or so and are reduced to near zero by the year 2050, average global temperature by the end of the century could be between 1.5°C and 2°C warmer than pre-industrial average temperatures. These differences in future warming trajectories are important, because the impacts of future warming – including floods, storms, droughts, wildfires, heat waves, crop failures, and the spread of infectious disease vectors – increase in frequency, severity, and/or geographical scope as temperatures rise (reviewed in detail in Chapters 2 and 3). Sea levels that are currently rising at a rate of approximately 3.4 mm per year will begin to rise faster, threatening ever larger numbers of people living in coastal areas (Chapter 4). Scientists have warned for decades that an increase of 1.5°C over pre-industrial temperatures would have dangerous consequences for human well-being and for ecological systems, and under the Paris Agreement of the UNFCCC, the international community has committed to not allowing temperatures to exceed that threshold (see Chapter 6). An assessment conducted by the IPCC found that even the seemingly small difference between a 1.5°C and 2°C increase in average global temperatures would increase the number of people exposed to severe climate risks by several hundred million, and cause sea levels to rise by an extra 0.1 meters, which would in turn threaten the homes of an additional 10 million people (IPCC 2018). As a result, it is in our collective best interest to avoid every last fraction of a degree of warming possible.

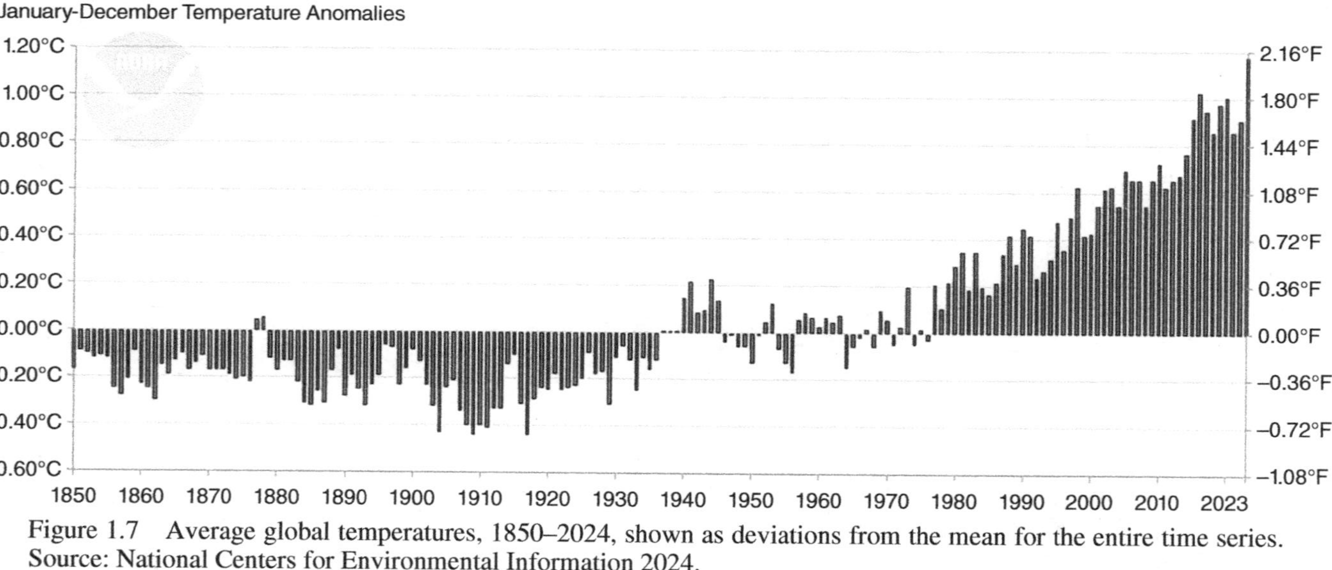

Figure 1.7   Average global temperatures, 1850–2024, shown as deviations from the mean for the entire time series. Source: National Centers for Environmental Information 2024.

1.2.2 Climate Change Risks

When we discuss the potential future impacts of anthropogenic climate change – including the possibility that large numbers of people might be displaced from their homes – we are talking about things that might happen but that could potentially be avoided. There could also be unwelcome things that happen but that we are not able to foresee based on current evidence. The IPCC uses the term *risk* to describe any potentially adverse impacts of climatic variability or change. Risk is in turn a function of four key components:

1. *Hazard*: The potential occurrence of a climate- or weather-related event or condition that may cause loss of life, injury, or other health impacts, as well as damage and loss to property, infrastructure, livelihoods, service provision, ecosystems, and environmental resources.
2. *Exposure:* The presence of people; livelihoods; species or ecosystems; environmental functions, services, and resources; infrastructure; or economic, social, or cultural assets in places and settings that could be adversely affected by a hazard.
3. *Vulnerability*: The propensity or predisposition to be adversely affected by a hazard, which may include a particular sensitivity or susceptibility to harm and/ or lack of capacity to cope and adapt.
4. *Adaptation*: Adjustments and responses to actual or expected climate and its effects, in order to moderate harm or exploit beneficial opportunities.

Figure 1.8 illustrates the interaction of hazard, exposure, vulnerability, and adaptation in the creation of risk in a generic sense. Hazard, exposure, and vulnerability are all positively related to risk; that is, as any of these factors increase, so too does risk (and conversely, if any shrink, so does risk). Adaptation – when it is done well – operates in the opposite way: as adaptation increases, risk decreases (and vice versa). The term "response" is also included in Figure 1.8 because sometimes peoples' responses to climate hazards are simply reactive in nature, and may not address the underlying causes of risk – such as rebuilding a waterside house that has been damaged by a flood exactly as it was before, without making any adjustments to prevent it from being damaged by future floods (this happens more often than one might expect). In such a case, a response has occurred, but it is not a beneficial adaptation.

A simple example that we will return to in later chapters explains how the concept of risk works in a practical sense: In the Pacific and Indian Oceans, there are hundreds of thousands of people living on low-lying coral islands known as *atolls*, where none of the land is more than a meter or two above sea level. Simply by virtue of living on an atoll, one's home, livelihood, community, and infrastructure

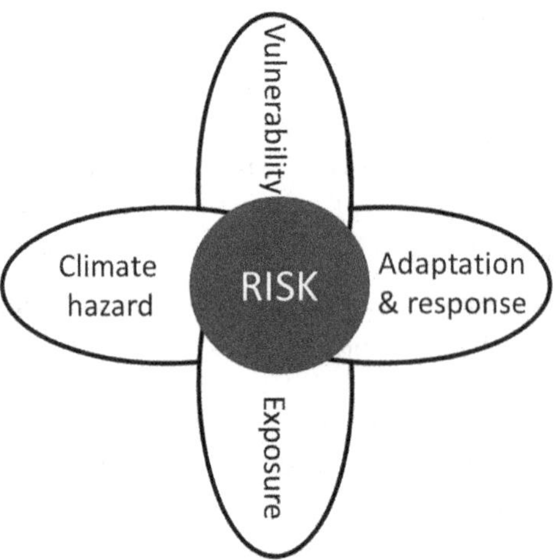

Figure 1.8   IPCC diagram showing how climate-related risk is formed through interactions of hazard, exposure, and vulnerability, with adaptation and other responses having the potential to moderate risk. Diagram and definition of terms in preceding paragraph adapted from IPCC AR6 WGII Report, Chapter 1 (Ara Begum et al. 2022a) and Figure 1.5 Panel (b) in Ara Begum, R., R. Lempert, E. Ali, T.A. Benjaminsen, T. Bernauer, W. Cramer, X. Cui, K. Mach, G. Nagy, N.C. Stenseth, R. Sukumar, and P. Wester, 2022: Point of Departure and Key Concepts. In: *Climate Change 2022: Impacts, Adaptation, and Vulnerability*. Contribution of Working Group II to the Sixth Assessment Report of the Intergovernmental Panel on Climate Change [H.-O. Pörtner, D.C. Roberts, M. Tignor, E.S. Poloczanska, K. Mintenbeck, A. Alegría, M. Craig, S. Langsdorf, S. Löschke, V. Möller, A. Okem, B. Rama (eds.)]. Cambridge University Press, Cambridge, UK and New York, NY, USA, pp. 121-196, doi:10.1017/9781009325844.003. Reprinted with permission.

are *exposed* to a variety of coastal *hazards*, including storm surges that cause damaging floods, penetration of salt water into agricultural soils and underground fresh water supplies, and rising sea levels that may one day lead to complete inundation of the atoll. However, although all atolls are exposed to similar hazards, the *vulnerability* of atoll communities and nations varies from one to another, and as a result, the relative level of risk differs. For example, to reduce its vulnerability and risk, the government of the Maldives – a country situated on a chain of atolls south of India – is building artificial islands with apartment complexes well above the zone of potential flooding or inundation, and these are capable of housing tens of thousands of people (Jayasinghe 2023). To finance this, Maldivians have established high-end tourist resorts that generate significant revenues from international visitors and entered into close political relationships with the government of China, with Chinese companies helping to construct the artificial islands. Other atoll nations, such as Kiribati, Tuvalu, and Vanuatu, are too distant and isolated to attract deep-pocketed international tourists or foreign benefactors. Their populations face a far more imminent risk of being displaced by climate hazards, a situation we explore in greater detail in Chapter 4.

The targeted use of tourist revenues and construction of artificial islands is an example of *adaptation* in action. Adaptation can be initiated at any societal scale, from the international community or nation-state level down to local communities, households, or individuals. Artificial islands, sea walls, and hazard early warning systems are examples of adaptations that typically require institutions and higher levels of government to coordinate, finance, and implement. Examples of individual- or household-led adaptations might include people living in flood-prone areas storing their valuables on the second story of their homes or farmers in drought-prone areas planting crops that require low amounts of soil moisture. These are just a few of the many types of *in situ* adaptations (i.e. adaptations that do not involve moving someplace else) for particular climate hazards described in chapters that follow. Migration is also a possible form of adaptation, which we explain in Section 1.6. It is also important to note that adaptation may be initiated proactively – that is, in anticipation of a potential risk – or after a hazard event has been experienced, so as to reduce the potential for loss or harm from future similar events. The ability (or inability) to adapt successfully is referred to as *adaptive capacity* and is determined by a wide range of non-environmental factors, including political processes, economic wealth, cultural norms, social relations, technological expertise, knowledge, skills, and acumen.

Box 1.4
What Is the IPCC?

The IPCC was created in 1988 by the WMO and the United Nations Environment Programme (UNEP), and its role is to provide governments with scientific information that they can use to develop climate policies and support international climate negotiations (IPCC 2024). The IPCC publishes cyclical reports every 6–7 years that assess the current state of scientific research on climate change, as well as occasional special reports on topics requested by the 195 member governments. The first IPCC assessment report was published in 1990 and consisted of four volumes that assessed the physical science of the causes of anthropogenic climate change; the impacts of climate change and vulnerability and adaptation to them; options for reducing atmospheric accumulation of GHG emissions (also known as *mitigation*); and, a synthesis of the other three reports. This four-volume model has been kept to this day and is reflected in the structure of the Sixth Assessment Report (AR6) that was published in 2021–2022.

IPCC reports are prepared by volunteer scientists and professionals who have been nominated by member countries and vetted by the IPCC secretariat to ensure that author teams for each chapter possess the requisite expertise and reflect a diverse range of countries and backgrounds. Reports go through multiple rounds of external reviews during the drafting process and are presented as objective assessments of the best current knowledge. Reports maintain standardized terminology and language, and key statements about climate change and its current or potential impacts must include guidance with respect to certainty. For example, something that is "very likely" to happen means there is a 90–100% chance of it happening, while "virtually certain" means the probability of it happening is 99–100%. Something "very unlikely" to happen has a probability of between 0% and 10% of occurring, and "exceptionally unlikely" has a probability of 0–1% (other standardized terms are used for probabilities between 10% and 90%). Intergovernmental Panel on Climate Change reports are non-prescriptive, meaning that authors are not permitted to offer opinions as to what governments ought to do, and are required to do no more than outline options for action and the potential costs and benefits associated with them. Each report begins with a summary for policymakers (SPM) that is reviewed by IPCC country representatives and must be agreed to unanimously before a report can be published. A result is that the SPM may not be as emphatic about climate risks and adaptation or mitigation needs as the substantive chapters within the report, and a common critique of the IPCC is that its statements about climate risks may be too conservative.

1.2.3 Hazards: Sudden-Onset, Slow-Onset, Compound, and Cascading

Hazards come in many different types. The ones assessed in this book fall within the larger category of what scientists call *natural hazards*. There are non-natural hazards as well, such as chemical spills, leaky pipelines, fires deliberately set by people, and so forth, but we do not consider them here. While all natural hazards have the potential to generate migration and displacement, not all have links to climate – volcanoes, earthquakes, and tsunamis being common exceptions – and these are also not assessed in this book. The climate hazards we assess in Chapters 2–4 differ in many ways, one important one being the speed at which they occur. Some, such as tropical cyclones, severe floods, and wildfires can materialize within a matter of days – and in the case of tornadoes, within a matter of hours or minutes – giving exposed populations little time to prepare for their onset. Climate scientists often refer to these as *sudden-onset* or *rapid-onset hazards*. By contrast, droughts, sea level rise, and rising average temperatures that may eventually make some areas of the world too hot to live in are examples of hazards that emerge over the course of months, years, or decades, allowing people time to prepare and adapt. We call these *slow-onset hazards*.

Climate hazards – both slow- and rapid-onset – are often perceived by the public and governments as being exceptions from "normal" weather patterns, but in reality, they are inherent features of the Earth's climate system that simply occur less frequently than do benign weather conditions (Hewitt 2019). In the broader scholarly field of natural hazard research, it has long been recognized that infrequent climate events and conditions become hazardous to human well-being due to social, economic, cultural, political, and other non-climatic processes that concentrate vulnerable people in dangerous locations without adequate protection and/or ability to cope (Burton et al. 1978, Blaikie et al. 1994, Hewitt 2013). Consistent with this, a key focus in subsequent chapters of this book is how climate processes and human systems interact to create hazards, exposure, vulnerability, and risk, and the options available to us to reduce such risks.

For analytical purposes, this book focuses on summarizing generally observed migration/displacement patterns and outcomes in response to specific types of sudden- and slow-onset climate hazards. However, reality is often more complex. Climate hazards do not always occur as discrete, unique events; a drought may be followed by flooding (or vice versa), and heavy rains associated with tropical cyclones may generate floods or landslides far from the storm's center. Climate hazards can also occur at the same time a community is dealing with a non-climatic hazard, such as an economic downturn or a health crisis. Scientists refer to these types of cases as *cascading hazards* (i.e. when successive hazards are experienced one after another) or *compound hazards* (when two or more hazards are experienced at the same time) (Zscheischler et al. 2020). As an example of a cascading hazard, the COVID-19 virus arrived in the Bahamas only six months after Hurricane Dorian caused widespread loss of life, injuries, and damage to homes and infrastructure, and recovery efforts were not completed when authorities had to implement responses to the public health risks of the pandemic (UNDP 2022). An example of a compound hazard occurred in the 1930s on the North American Great Plains, when widespread severe droughts struck in the midst of the Great Depression, leading to widespread farm abandonment and migration out of affected areas (McLeman and Smit 2006). Both compound and cascading hazards have the potential to overwhelm the adaptive capacity of exposed populations that might have been able to cope with a single hazard, leading to migration outcomes that would not have otherwise emerged.

1.2.4 The Relationship between Climate-Related Migration and Risk

Breathing is not a risk – it is something people do naturally – but breathing polluted air *is* a risk. Similarly, migration is not in and of itself a *risk,* it is simply something people do. When people move voluntarily to seek opportunities or to help their families adapt to climate hazards, such acts of migration do not cause harm and are therefore by definition *not* a risk (Gilmore et al. 2024). However, when migration occurs involuntarily or if people are undertaking migration that is inherently dangerous, such as paying organized criminals to transport them clandestinely across oceans or deserts, it is a risk. This is not simply an academic distinction. The words we use are important, and by associating migration with the term *risk* without any regard for the implications of doing so can feed into anti-immigrant sentiments and narratives in receiving countries. Migration out of communities exposed to climate hazards can, depending on circumstances, serve an important role in building adaptive capacity in the sending community, thereby reducing vulnerability and risk. An example of this would be out-migration of household members to seek wage labor opportunities with the goal of remitting money home to help the household build more robust housing capable of withstanding severe weather. Conversely, immobile individuals or groups that lack the ability to migrate may miss out on opportunities to gain remittances that facilitate adaptation and consequently see their risk increase over time.

Having completed our summary of climate science and climate risks, we now turn to the general question of why people move. At the conclusion of this chapter, we link climate science and climate risks with migration and adaptation processes, setting the stage for the more detailed analysis of migration and displacement in the context of specific climate hazards in the chapters that follow.

1.3 Migration, Displacement, and Immobility: The Fundamentals

In its narrowest sense, migration (whether voluntary or involuntary) is the phenomenon of people moving from one place to reside in another (recall Box 1.1). But it is in reality much more than this. The act of migration has implications for the individual migrant, their family and friends, the community they leave behind, and for the community they eventually join. Migration changes the future social and economic trajectories of all, because a migrant takes with them knowledge, talents, abilities, and yet-to-be realized potential. The decision to move (or its corollary, the decision to stay and not move) is usually not undertaken lightly, and not for simple reasons.

To understand how climate change is likely to affect migration and displacement in the future, it is important to first review how migration processes function in a general sense (see Box 1.5 for more on the origins of modern migration research). After all, most migration and displacement currently occurring around the world are motivated and shaped by factors other than the climate, such as economic opportunities, social networks, political conflicts, and cultural norms. Even when climate risks come into play, the migration and displacement patterns that result filter through economic, social, political, and cultural processes. In the remainder of this chapter we provide a general review of what these non-climatic factors are, the mechanics of how they shape decisions to move (or not move), and how they interact with climate risks in a general sense.

Box 1.5
The Origins of Modern Migration Research

Research on why people move has evolved over a relatively long period of time. Modern scholarship on the subject traces back to a series of publications in the 1880s by E. G. Ravenstein on what he called "The Laws of Migration." Using census data from the UK – a data source still commonly used in migration scholarship – Ravenstein described how most migration tended to flow from rural to urban areas, that individuals were more likely to move than entire households, that most migration took place over relatively short distances within countries, that migration outcomes differed by gender, and that migration outcomes were influenced by a range of economic, social, and other factors. With some nuance, all of these conclusions are still relevant characteristics of migration patterns that we observe today. Ravenstein also introduced a concept that is still used regularly to describe factors that shape migration decision-making processes: *push and pull*. Ravenstein concluded from the data he examined that migration is shaped by the combined effects of forces operating at a migrant's place of origin and at potential destinations. "Push" forces are those that predispose or incentivize an individual or household to consider migration, while pull forces shape the comparative attractiveness of possible destinations. Examples of push factors might be political instability, high unemployment, or a desire to travel and see new things. Examples of pull factors might be job vacancies, educational institutions, or social opportunities. The push–pull concept is a simple and useful reminder that migration decisions are usually influenced by multiple factors operating in sending and receiving areas, but it is important not to underestimate the complexity of processes that shape migration.

1.3.1 Migration and Displacement: Complex Causes, Diverse Outcomes

Every migrant's experience is unique and personal to that individual. The reasons that lead up to the decision to move, the timing of the decision, the destination (or destinations), the reasons for selecting one destination over other possibilities, whether migration is temporary or becomes indefinite, the costs and benefits that result from migrating – these all differ from one person to another (see Box 1.6 for a discussion of temporal and spatial scales of migration). Yet in academia, in policymaking, and in popular discourse, we tend to describe migrants and migration not as unique individuals on the move, but as flows and stocks of anonymous people, and assign them labels according to generally understood categories of why people move: economic migrants, job seekers, guest workers, international students, nomads, asylum seekers, refugees, trafficking victims, and snowbirds, to name just some of the labels. These labels are easily recognizable, but the reality is that individuals may be on the move for many reasons. Consider someone who applies for a visa to study at a college or university in another country. They may simultaneously have multiple intentions and motivations. While studying, they might work part time and remit money home. They might be planning to remain in the destination country indefinitely and pursue a career there, or they may plan to return home after graduation and hope that having international qualifications improves their future career prospects. Or maybe they're not sure what they will do next. Perhaps during their studies, they will fall in love with a fellow student, and this might alter their previously held plans and aspirations. Perhaps they will become unwell or do poorly at school and be forced to return home without the diploma or degree they hoped for. The key is to think of the act of migration not as simply a movement of person X from point A to point B, but as part of the larger trajectory of an individual as they move through the physical world and through the social world over the course of a lifetime. By doing so, we realize people may move multiple times to multiple places, sometimes returning back to places where they once lived, perhaps never going back. Every migrant's story is unique and continually changing.

Now, having established that we ideally ought to think of every migrant as an individual on a unique journey through physical and social space, we authors contradict ourselves for the remainder of this book by generalizing the most common reasons why people move.

Scholars have many broad theories and explanations of why people migrate. Some theories relate to the decision to leave one place for another, other theories consider what draws people to particular destinations and how they become incorporated (or not) into the receiving communities, and still others consider the factors that mediate the migration process as a whole. What they all have in common is an understanding that migration outcomes are diverse, and are the product of complex interactions of cultural, economic, environmental, political, and social processes operating at multiple scales, from the local to the global. Some of the factors that influence a given individual's decision to move (or not) are particular to that individual's household or family, such as relationships with parents and siblings, the health of family members, the household's livelihood(s), and financial savings, among many others. Other factors that influence migration decisions are beyond the influence or control of the household, driven by what we may refer to as macro-level processes, and include a wide range of possibilities such as interest rates, the costs of travel, cultural norms, and the presence or absence of political instability in a given country. These and other factors that shape migration decisions operate at multiple scales and are continuously interacting with one another, and so the conditions under which migration occurs are highly dynamic and always changing. Figure 1.9 provides a simplified illustration of how these complex interactions between macro-level processes and household characteristics affect the decision of an individual to migrate (or not). Each of the factors shown in the figure is now summarized in turn.

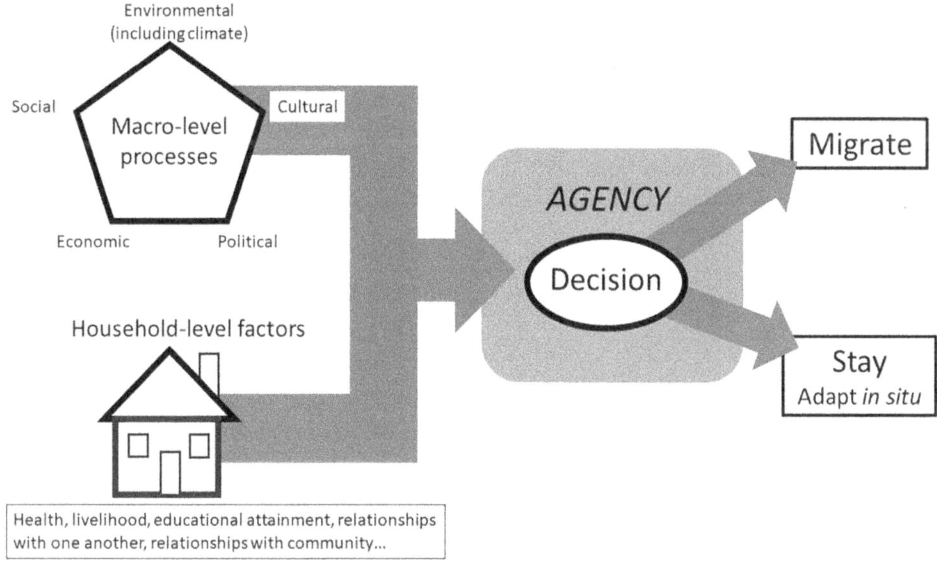

Figure 1.9 Simplified representation of how macro-level processes and household characteristics shape migration decision-making. Adapted from Foresight 2011, McLeman 2014.

Box 1.6
Temporal and Spatial Characteristics of Migration

Migration can take on any number of characteristics in terms of the distance traveled, the frequency with which it occurs and its duration. A simple typology from Gonzalez (1961) categorizes the various temporal characteristics of migration, and we follow this terminology throughout this book:

Temporal category	Example
Seasonal	Migrant harvest workers
Temporary non-seasonal	Migration to seek temporary wage labor when/as needed
Recurrent	Transhumance pastoralists (moving back and forth between the same locations each year)
Continuous	Itinerant workers; nomadic pastoralists
Indefinite	Migrants seeking permanent residence in another country; asylum seekers and refugees

In this book, we use the following standardized terms to describe the geographical and spatial dimensions of migration:

Term	Description
Internal migration	Migration that takes place within the political boundary of a nation or state
International migration	Migration across state boundaries; may be further described as *intra-regional migration* (movement between contiguous countries within the same geographical region) or *inter-regional migration* (movement to more distant countries)
Rural–urban migration	Movement of people from rural areas to cities, towns, and suburbs
Rural–rural migration	Movement of people from one rural area to another
Urban–rural migration	Movement of people from cities, suburbs, and towns to rural areas

1.3.2 Economic Processes That Influence Migration

Economic conditions are an important influence on migration. Places that experience economic growth and a corresponding increase in employment opportunities often attract migrants from other places with slower growth and/or lower wages (Todaro 1969, Karras and Chiswick 1999). The relatively large number and range of potential income opportunities in cities is an important driver of rural–urban migration, especially for young adults of working age (Todaro 1969, Sassen 2001). Migration can stimulate additional economic growth in the receiving area, as new arrivals bring skills that can enhance the productivity of businesses and increase demand for goods and services, assuming that newcomers are able to integrate and participate fully in the economy (Engler et al. 2020). Urban centers that develop large pools of economic and human capital may hold a stronger attraction to potential migrants than other places (Sassen 2001). In some economic sectors, such as agriculture, mining, forestry, and construction, large numbers of workers are required at specific locations and/or at certain times of the year, and temporary migrant workers may be recruited from elsewhere to fill such jobs as needed. In other economic sectors, such as light manufacturing and the garment industry, production facilities can be located anywhere that has reasonably good facilities for shipping goods to market and access to electrical power; in these industries, companies may deliberately seek out suppliers in countries where large numbers of low-wage workers are available. This became an important influence on migration patterns within many countries starting in the 1970s and 1980s, when global production of textiles, clothing, and footwear shifted dramatically from high-income countries to low-income countries in Asia and, to a lesser extent, Latin America (International Labour Organization 1996). Governments in low-income countries have often attempted to attract such facilities by establishing special zones where offshore companies can operate with low or no taxes, so long as production is for export. These concentrated areas of low-wage jobs draw laborers from across the host country and/or adjacent ones, generating new migration flows in directions that did not previously exist (Chang 2015). Oil-rich, low-population countries in the Middle East region have in recent decades become a key destination for wage-seeking migrants, and now host over 24 million temporary migrant workers, 83% of them male, drawn from countries across Asia and Africa (International Labour Organization 2024).

Adverse economic conditions, as experienced through such phenomena as high rates of unemployment, commodity price swings, changing interest rates, price inflation, and changing currency valuations, can also have a strong influence on migration. For example, during the Great Depression of the 1930s, large numbers of unemployed North Americans migrated from city to city in search of work, some crisscrossing the continent on railway cars or on foot in search of work (Mitchell 1996). During the "Great Recession" of 2007–2008, immigration to the US from other countries slowed temporarily because economic growth slowed and unemployment rose sharply (Singer and Wilson 2010).

The specific ways in which economic considerations influence migration decision-making are context specific. Economists have historically assumed that migrants (or people thinking about possibly migrating) are primarily opportunity-seekers that weigh the monetary costs and benefits of whether or not to migrate and assess potential destination options based on the information available to them (Todaro 1969). Labor markets and wage differentials combine to create potential migration pathways for people with particular types of job skills, training, aptitudes, and knowledge (Sjaastad 1962, Todaro 1969, Borjas 2001). The influence of wage differentials on migration is most evident in movements of people within the borders of countries (i.e. internal migration), where there are typically fewer restrictions on movement as compared with international migration, where high-wage/high-income countries often restrict entry of low-skilled labor-seeking migrants but deliberately favor and facilitate the entry of higher skilled workers from other countries. For example, under the North American Free Trade Agreement, Mexican citizens that hold university degrees in particular professional programs are eligible to be sponsored by prospective US employers for temporary work permits, but other Mexican workers are not, and may not have legal pathways to working in the US. The European Union takes a different approach to labor migration under the Schengen agreement, allowing citizens of member countries freedom to move about and seek employment. Because borders are less open to them, lower-skilled workers seeking to move internationally to higher-wage countries may get pushed into clandestine, riskier forms of migration that raise their potential to become victims of exploitation by untrustworthy employers and criminal organizations (Salt and Stein 1997).

A different economic interpretation of migration behavior is known as the *new economics of labor migration* (or NELM), first described in the 1980s (so not especially "new" anymore). Instead of focusing on the cost-benefit analysis made by individuals, NELM considers the household to be the key decision-making unit, and suggests that migration is one of a variety of strategies that households may use to diversify income sources and reduce exposure to hardship and risk (Stark and Bloom 1985). An important point here is that the desire to avoid losses or harm can be as powerful a motivation for migration as is the desire to acquire wealth. So it is, for example, that in many cultures, young adults will leave the family home in search of job opportunities in distant locations and remit money back (Adger et al. 2002, Connell and Conway 2000). Massey et al. (1993) suggest that this type of migration strategy is more likely to be prevalent in areas where local income-making opportunities are scarce and other options for managing risk are unavailable, such as insurance, lending institutions, and financial and commodity markets. As an example, Rain (1999) describes in rich detail how rural households in West Africa send young adults to cities each dry season to reduce the number of mouths to feed while at the same time hoping they will find wage labor and send home cash remittances (Box 1.7). In that example, the migration is cyclical in nature; when the rainy season returns, so too do the migrants, as their labor is once again needed at home. In other examples in other places, household members may migrate away indefinitely, on the understanding that the migrant will repay the costs of their migration and absence from the household workforce by remitting money home (Erdal 2022).

Box 1.7

The Importance of Remittances

The term *remittances* refers to money, goods, or other resources that migrants send to family members remaining in the place of origin. The World Bank has estimated that over US$860 billion is remitted globally each year, primarily from higher income countries to lower income countries (Ratha et al. 2023). Remittances have been shown to be especially important in helping households and communities respond to climate hazards and in supplementing household incomes in highly seasonal environments (Banerjee et al. 2011, Le De et al. 2013, Scheffran et al. 2012). Households may use remittance money in a variety of ways, including improving their homes, businesses or farms, purchasing additional land or consumer items, and reinforcing food security (Davis and Lopez-Carr 2014, Mabrouk and Mekni 2018, Moniruzzaman 2022). In the wake of extreme weather events, affected households often find they are able to receive financial assistance through remittances much more rapidly than from governments and humanitarian agencies, helping them recover more quickly (Le De et al. 2015, Bragg et al. 2018). However, pressure to provide remittances can have adverse impacts on the migrant's own economic security and well-being, particularly when the need is great in the home community (Obokata and Veronis 2018).

1.3.3 Social Processes and Structures That
Influence Migration

Some types of migration are inherently driven by social motivations, such as the desire to live near friends, to be reunited with family members, to get married, or to move into a retirement community of people of similar age and need. Social and economic considerations often go hand in hand when making a migration decision, so they are not entirely discrete factors, but from an analytical point of view, it is useful to distinguish social factors because of their potentially larger and longer-term effects in initiating and perpetuating migration movements. Even over short distances, migration has direct financial costs for the migrant and their household, such as transportation, visa fees (if applicable), tuition (of an international student), and the need to find accommodation at the destination. Intermediaries such as labor recruitment agencies, matchmaking services, and smuggling organizations (in instances where legal migration options do not exist) may also require up-front payments. There are also the indirect costs of migration to consider, such as the household's loss of the migrant's labor, the migrant's absence from participation in family life, and the opportunity costs associated with giving up a known economic and social system for a less well-known one. Social networks can help migrants and those considering migration overcome some of the direct and indirect financial costs, particularly once an initial group of migrants has successfully established itself in a new destination and is able to facilitate the subsequent migration and incorporation of family, friends, and acquaintances from the sending community (Massey and Espinosa 1997, Nee and Sanders 2001).

Examples of the benefits social networks provide include detailed information about how to cross borders, assistance in securing transportation over long distances, introductions to potential employers in the settlement area, and assistance in finding accommodation upon arrival (Massey and Espinosa 1997, Palloni et al. 2001). International migrant networks lead to the formation of what are known as *transnational communities* through which people, money, resources, and ideas flow between origin and destination countries, examples being the transnational Kurdish community connecting western Europe to Turkey (Faist 2010) and the transnational community connecting the Dominican Republic to New York City (Georges 1992). Although not a truly transnational community – for Puerto Ricans are US citizens – pre-existing migrant networks linked to cities in New York and Florida were critical in helping people in Puerto Rico recover from widespread destruction caused by Hurricane Maria in 2017 (see case study in Chapter 3). Not all migration movements automatically give rise to strongly interconnected transnational communities (Riccio 2001). The types and nature of social connections between sending and receiving areas can vary significantly from one case to another depending on the cultural group and the degree of assimilation or incorporation into the destination culture. Deeply ingrained social, religious, and class structures can have a significant influence on where migrants and would-be migrants perceive they can and cannot move and the extent to which receiving communities are willing to accept newcomers and facilitate their incorporation (Box 1.8) into the community. For example, in India large numbers of low caste workers, many heavily in debt to landlords, migrate each year to perform difficult and dangerous work in cities where they know they may be exploited and subject to abuse (Acharya 2021). They do so because the caste structure allows them few alternatives to earn a livelihood.

Box 1.8
What Do We Mean by "Incorporation"?

When people move to a new place, they may experience a wide range of short- and long-term experiences. Ideally, newcomers will be welcomed and accepted, but they might also experience rejection, marginalization, and hostility. The process by which migrants and people living in destination communities learn to live with one another is referred to as *incorporation*. In some situations, people living in the receiving community may expect that newcomers discard their own traditions, cultural practices, language, and ways of doing things and adopt those of the host population (also known as *assimilation*). In other instances, the host population and newcomers may informally end up adopting some of one another's practices and evolve toward a *multicultural* community. In still other cases, newcomers and host communities may have limited social interactions with one another, and newcomers may be virtually excluded from social and economic life.

Incorporation is often a slow process, and it may take several generations before newcomers no longer feel "new" to the community; even then, the descendants of migrants may still feel like outsiders. Challenges of acceptance and integration may not be as great when people move within their home countries as compared with international moves between countries with different cultures and/or languages, but even internal migrants can sometimes be met with suspicion or hostility. For example, in the 1930s, large numbers of Americans moved to California from the Great Plains during a period of drought and economic recession. The numbers became so great, the Los Angeles sheriff's department attempted to blockade the state's border for several months in 1936 to prevent would-be migrants from entering (Giczy 2009). Many migrants from the Great Plains spent extended periods living in government-operated tent camps in California, in the open along roadsides and drainage ditches, and in segregated "Okie-towns" of low-quality housing ("Okie" being a derogatory term used for the migrants). The migrants were openly discriminated against in some California towns and cities, being barred from movie theaters and restaurants, denied access to public medical clinics, and their children forced to sit on the floor at school while other children were seated at desks (McLeman 2006). Many towns in California's Central Valley still bear the cultural characteristics of 1930s Great Plains migrants, whose descendants have risen to positions of power and influence with the passage of time. In Chapter 4, we describe another example of internal climate-related migrants being received with some hostility – that being the case of Carteret islanders seeking relocation to Bougainville island.

As many cities and countries become increasingly diverse through migration of people from other cultures, the dynamics of integrating new arrivals from new groups can take on a wide range of forms (Grzymala-Kazlowska and Phillimore 2018). The experience of newcomers has a significant impact on the social, economic, and cultural trajectories of destination and sending areas. When migrants receive similar legal rights and protections as do resident populations, and are able to rapidly integrate and enjoy economic and social success in their new communities, it may serve as an incentive for others to migrate as well, creating flows of people moving back and forth between sending and destination areas. However, the lack of an open and welcoming reception at the destination is not necessarily a deterrent to further migration, especially where the economic rewards of moving are great and/or where socio-economic or political conditions in the sending area are poor. For example, laws in France that ban Muslim women and girls from wearing headscarves in schools have hurt the educational achievements of those affected (Abdelgadir and Fouka 2020) but have not deterred migration of Muslim people to that country, just as migrant detentions and aggressive border patrol tactics have not discouraged people seeking entry and asylum at the US border with Mexico (Romero 2023).

1.3.4 Cultural Influences on Migration

Cultural norms and understandings about migrants and migration can be influential at all stages of the migration process, from decision to migrate to the incorporation of migrants into the destination population. The relationship between culture and migration varies considerably across cultural groups. Relevant factors arise from the migrant's own culture as well as from the culture of the dominant population at the destination, assuming they differ. For some groups and peoples, migration is a fundamental element of their culture. This is particularly the case for many of the estimated 200 million people worldwide who practice pastoralist livelihoods centered on moving herds of livestock on a seasonal or annual basis across diverse geographical settings, from Central Asia's steppes to Sudano-Sahelian Africa to Scandinavia's Lapland region (Kaufmann et al. 2019). The cultural practices, values, and beliefs of pastoralists are often markedly different from those of other, more politically dominant groups in the same geographical regions, and the mobility of the pastoralist lifestyle does not mesh well with private property regimes favored by the political majority in many countries. This has sometimes led to political instability and occasionally violent conflicts between pastoralists and their non-pastoralist neighbors, and between pastoralists and governments that have sought to restrict their movements or force them to move into fixed settlements (Fernandez-Gimenez and Le Febre 2006, Penu and Paalo 2021).

In other cultures and communities, undertaking migration at a particular life stage, typically young adulthood, is an encouraged or at least widely accepted practice (Gabriel 2006, Geisen 2012) (Box 1.9). The underlying reasons may be primarily economic, such as seeking out higher-paying employment elsewhere and remitting money home. In some cultures and communities, households that receive remittances acquire not only greater wealth but greater social status within their communities, and this dynamic can place pressure on other young people to migrate as well (Sana 2005, Suksomboon 2008). In other cases, the motives for youth migration may be primarily lifestyle related, with young people seeking out adventures and new experiences. As just one example, the ski resort industry depends heavily on large numbers of young workers originating from a wide swathe of countries, whose primary goal is to have ongoing access to skiing and snowboarding opportunities (Thorpe 2017). Young people who migrate with economic motives may be eager to integrate into the destination culture, thereby enhancing their employment prospects, but those motivated by lifestyle considerations may be less concerned about integration and more interested in the experience of learning about new places and people (Samuel Hall 2023). Youth culture around the world is changing rapidly in the era of social media, and so, too, are youth attitudes and experiences with respect to migration in terms of undertaking it themselves and in their views of migrants arriving in their own communities from elsewhere.

Marriage-related migration is an important feature of many cultures. The dynamics of how marriages are arranged or facilitated, the expectations of the families involved, and the desired social and economic outcomes vary considerably from one cultural group to another (Yeung and Mu 2020), and so it is important to be careful about making generalized statements about it from the perspective of a culture where it is not widely practiced. Marriage-related migration can lead people to move within countries and internationally, and in most cultures where it is widely practiced the intention is for young people to enter into a heterosexual marriage with a partner from a family with a similar cultural background, religious faith, and/or economic status (Ullah and Chattoraj 2023). Although not universal, in most cultures, it is the woman who is expected to move to join the husband,

---

Box 1.9
Migration Over The Life Course

The likelihood of any individual to consider migrating somewhere else voluntarily changes over the course of one's life. Certain types of migration behavior or events are more likely to occur among certain age cohorts, reflecting how the social and economic advantages and disadvantages of migration vary over the course of one's lifetime (Plane et al. 2005). Until they are teenagers, children have little choice but to live where their parents live, and parents with children tend to be less likely to migrate as compared with adults older or younger than them. By contrast, young adults are generally much more mobile than other age cohorts. Their migration decisions and patterns are often strongly influenced by opportunities in the labor market and in higher education. Rural-to-urban migration in many countries is heavily dominated by young adults seeking work. As people age and settle into more established employment, relationships, and communities, non-economic factors become increasingly influential to their decisions of where to live. As people reach the end of their working lives, quality of life issues, desire to seek out environmental amenities like a warm climate, and concerns like health services and proximity to loved ones or caregivers become important considerations about where to live. During the post-retirement period, people will often make a final move to the place where they intend to spend their remaining years.

---

although arrangements may differ if the woman is working abroad in a higher income location and is able to bring her new spouse to join her. Higher levels of vulnerability in terms of personal safety for women and girls in societies can result from marriage-related migration, especially those moving from low-income households where the marriage is seen as an opportunity to raise the family out of poverty. It is estimated that 22 million people worldwide, two-thirds of them female, have been coercively forced into marriages they did not want, most often by their parents (International Labour Organization, International Organization for Migration and Walk Free 2022).

Marriage migration is just one aspect of the wider gender dimensions of migration of all types and motivations (see further discussion in Chapter 7). Although too numerous to summarize here, there are trends distinctive to female migration that warrant mention. Roughly half of all international migrants are female, and although this percentage has not changed much in recent decades, the proportion of women international migrants who move in search of employment has grown significantly (Global Migration Data Portal 2024). There are, however, notable variations between countries and regions. In many high-income countries of Europe and North America, the migrant labor workforce is disproportionately female, with migrant women making up growing numbers of workers in sectors such as domestic work and health care. By contrast, labor migration to oil-rich states in the Middle East is disproportionately male, the key employment sectors being industry and construction. The expansion of light manufacturing and garment, textile, and footwear production to low-income countries has in many instances led to increased female migration to fill such jobs (Jones 2020). This has in many cases run contrary to traditional migration practices in such countries, where it was once primarily men that moved and stimulated changes in the perceived role of women in society.

Education shapes migration behavior and patterns in a variety of ways. The greater availability of educational opportunities can be an incentive for rural people to migrate to cities (van Maarseveen 2021). In some countries, rural people with higher levels of education are more likely to move to cities than are other people, but this is not the case across all countries (Ginsburg et al. 2016). Educational attainment is often closely linked to international labor migration flows, with higher skilled workers often migrating in search of better wage opportunities and government regulations tending to make their movement across borders easier than for lower skilled workers. However, migrants' education and skills credentials may be seen by employers in high-income countries as being less valuable, leading to newcomers to work in jobs below their skill levels and earning lower average wages than established residents as a result (OECD 2022). Many western countries encourage international student migration not so much from a sense of enlightenment but as a way of generating additional revenues for schools, colleges, and universities (international students contribute an estimated US$40 billion to the US economy each year (Greenfield 2023)).

1.3.5 Political Influences on Migration

Three important ways by which political factors and processes influence migration behavior are through:

- the policies and regulations implemented by states and other political actors to shape migration and citizenship;
- general governance structures and processes that make a given jurisdiction more or less politically stable and economically prosperous relative to others;
- violence carried out by the state against members of its own population or those or its neighbors.

Government regulations on mobility and migration are an obvious constraint on the agency of migrants and potential migrants. States that allow their citizens a high degree of freedom with respect to mobility and travel tend to have higher numbers of people migrating internationally over long distances as compared with states where mobility and travel rights are restricted (Karamera et al. 2000). Under international law, states have the right to enforce entry across their borders and to make determinations with respect to legal right of residency within their borders and citizenship. Most states maintain entry controls at land border crossings, ports, and airports; some also maintain exit controls in order to have greater influence over anyone who might wish to leave. The passport and the visa are basic tools used by states to regulate the movement of people across borders. A *passport* is a government-issued identity document for the purpose of international travel that follows standards set by the International Civil Aviation Organization for its appearance and required contents. The passport allows the issuing state to exert some control over its nationals' ability to travel abroad, and allows receiving states to establish the identity and citizenship of those seeking entry. A *visa* is a document issued by a state that allows non-nationals the right to appear at its border control points and seek admission. Its effect is to enable the issuing country to pre-screen those who would seek entry, keeping them at a physical distance until the traveler has demonstrated at a consulate, embassy, or authorized agent that they meet admissibility requirements. Such controls do not necessarily prevent unauthorized migration (see Box 1.10 for more on why this terminology is preferred); millions of people each year enter other countries using what are known as "irregular" channels (i.e. entering a country without authorization and without appearing at an official point of entry). The actual criteria for gaining admission and the length of time for which non-nationals are permitted to stay vary from one state to another.

Box 1.10

Avoiding the Term "Illegal Migrants"

It is common among politicians and the media to use the term "illegal migrants" to refer to people who enter a country without official documents or who overstay their authorized period of stay. It is inappropriate to do so. In most countries, not having the proper migration documents is not a criminal offense. Under international law, everyone has the right to leave their country, to seek permission to enter another country, and to have their request for entry considered fairly. Further, people fleeing persecution or violence in their home country have the right under international law to seek protection in a safe country – even if they do not possess the necessary documents for entry – and to not be deported back to the country that has persecuted them. As a result, the United Nations High Commissioner for Refugees requests that governments and the general public instead use the terms *undocumented migrants* and *irregular migration*, and we do the same in this book. For more discussion, see Crépeau and Vezmar (2021).

States also determine questions of citizenship which typically includes an automatic right of indefinite residence within a state's borders. Other rights and benefits accrue with citizenship, which vary according to the state. Common ones include the protection of the state, access to the labor market, and to state-provided services such as health and education and, in democracies, the right to participate in the selection of political leadership. In return, citizens defer to the state's authority and undertake certain obligations, such as the requirement to pay taxes. Some countries may also require that citizens perform a period of military duty or other form of national service. Citizenship can be obtained through three general mechanisms, the specific details of which vary from state to state. One is consanguinity, whereby a child acquires the citizenship of its parent. Variations of consanguinity exist, with some states having broad rules that allow people with familial ties more distant than simply parent and child to claim citizenship on the basis of their ethnicity. A second way of acquiring citizenship is through place of birth, whereby countries automatically confer citizenship upon anyone born within their territory, regardless of the citizenship or residency status of the child's parents. Not all countries do so, however, and a child born in a country that does not confer automatic citizenship can become stateless if the parents' home country does not automatically confer consanguine citizenship (see Box 1.11). The third possibility is naturalization, whereby people who have migrated from another state may acquire citizenship according to some prescribed administrative process. Each state maintains its own rules and regulations regarding naturalization. States that allow little or no possibility of naturalization, such as many oil-rich Middle Eastern states, are committed to a policy of ethnic citizenship only, having no interest in integrating foreign workers.

Box 1.11
Statelessness

There are tens of millions of people globally who do not have any official proof of identity or citizenship and are thus considered to be stateless (Strode and Khanna 2021), a situation that drastically curtails their mobility. Statelessness can arise for a variety of reasons. It is a phenomenon often associated with discrimination against ethnic minorities or arises as a by-product of conflict or political instability, and its effects can span generations; millions of Palestinian people and their descendants were rendered stateless by the 1948 creation of the state of Israel (UNHCR 2024). Statelessness can also result from restrictive laws on the transmission of citizenship to children, such as in the case where the identity of a child's parent is not properly documented or where the parents are not citizens or legal residents of the jurisdiction where a child is born, rendering the child stateless (Manly 2007). Some states reserve the right to revoke citizenship of individuals who violate the laws of the state, reside abroad, or marry a foreign national, thereby rendering such individuals stateless after the fact.

Many states impose strict controls on the entry of foreign nationals, and will actively detain and deport those who violate immigration rules. This does not necessarily prevent people from attempting to enter; in a single month (December 2023) patrols encountered nearly 250,000 people trying to enter the US from Mexico at locations other than official border crossings, most of whom were immediately returned to Mexico (Gramlich 2024). Governments in many western countries are increasingly using remote sensing devices, drones, and other advanced technologies to prevent undocumented entry at their land and sea borders (McLeman 2019). As a consequence, migrant smuggling organizations have become more active in organizing migration into the European Union from Africa, the Middle East, Turkey, and Asia (İçduygu 2021, Frowd et al. 2023, Triandafyllidou 2022) and into the US via Central America and Mexico (Kim and Tajima 2022). Migrant smugglers are often embedded within larger organized crime groups whose methods are violent and exploitative, and prey upon the vulnerability of people on the move (Cleaveland and Kirsch 2020). This has in turn led to the occasional occurrence of migrant "caravans" – large groups of overland migrants who travel together toward the US border through Central America and Mexico in order to protect themselves from criminals, corrupt officials, and others that might exploit them (Frank-Vitale 2023).

Some governments impose controls on population movements within their borders to achieve particular political or economic ends (Coutin 2010). For example, the People's Republic of China, has since the 1950s operated a system of household registration known as *hukou* (Cheng and Duan 2021). Under this system, the personal details of each household member are registered with the local government registration office, along with any changes due to births or deaths. Those registered in the hukou are entitled to basic social benefits such as education and primary health care, and are legally eligible to work in that jurisdiction, but they are not officially allowed to change their place of residence without permission. The massive industrial expansion in China that began in the 1990s created employment opportunities far better than those on offer in rural areas, leading hundreds of millions to move without official permission to large cities and industrial centers. They are sometimes referred to as China's "floating population."

There are many other ways by which the policies and practices of governments have a direct influence on migration. The legacies of colonialism and past conflicts continue to have impacts on migration and displacement at local, regional, and global scales, and amplify the vulnerability of previously displaced peoples to climate change and other environmental hazards (Sadiq and Tsourapas 2021, Farrell et al. 2021). In some parts of the world, colonial-style removals of traditional and Indigenous peoples from their lands continue to occur, to make way

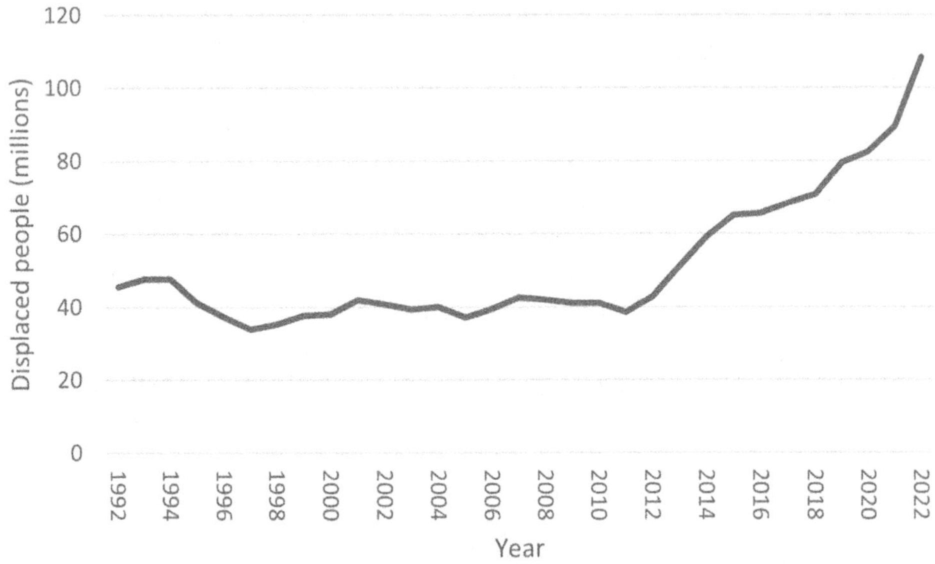

Figure 1.10   Estimated global number of displaced people, 1992–2022. Statistics include asylum seekers, refugees, and estimates of all internally displaced people, regardless of reason. Data source: UNHCR 2024.

for such activities as forestry and export-oriented industrial agriculture, in a phenomenon known as land-grabbing (Rude et al. 2021, Vigil 2022). In politically unstable states that lack responsible government, such as Syria, Somalia, Sudan, and Afghanistan, a lack of personal security and periodic episodes of violence can result in large-scale internal population displacements and displacements across borders. The number of displaced people globally has risen significantly since 2010, exceeding 100 million people by the end of 2022 (Figure 1.10).

1.3.6 *Environmental Influences on Migration (Non-climatic)*

Weather and climate are only one subset of environmental factors that have the potential to influence migration decisions. Some environmental conditions are beneficial for human livelihoods and well-being, and places that feature these can become attractive destinations for migrants and experience population growth. Valleys of wide, slow-flowing rivers, coastal estuaries, and deltas generally make attractive locations for human settlements given the natural availability of wild foods, fertile soils, potable water for human use, and flat surface waters that allow for easy transport of people and goods. It is not a coincidence that the most densely populated regions on the Earth are located within such geographical features. Over 400 million people live in the Ganges River floodplain (GRID

2015), and of the world's ten largest cities, only Mexico City is not located on a major river or in a coastal area. Other types of environments can also be seen as attractive for certain groups of people and/or for people at certain stages of their lives. For example, wealthy retirees will often move for part or all of the year to locations with pleasant temperatures, especially those with mild winters (earning such migrants the moniker *snowbirds*). Other people may be attracted to locations of scenic beauty with natural features such as mountains and lakes, a phenomenon known as *environmental amenity migration* (Abrams et al. 2012). The term *digital nomads* has been coined to describe people who earn their living through online work that does not require them to live in any fixed location, and so they move to places that have natural or cultural attributes that appeal to them (Chevtaeva & Deniczi-Guillet 2021). The emergence of the online economy is just one example of how technological innovations have affected the relative attractiveness of particular locations and environments over time. For example, in the early decades of European settlement and colonization of North America the southern Great Plains were labeled on maps as the Great American Desert (Dillon 1967), mountains were seen as barriers to be overcome and refuges for only the poorest settlers (Case et al. 2008), and the state of Florida – now home to over 20 million people – was lightly settled, seen as being a sweltering, malarial wetland (Faust 1951). Technological innovations such as mechanized pumps to power irrigation and drainage systems, air conditioning (and cheap electricity to power it), insecticides, railways, interstate highways, and affordable air travel combined to transform places that were once to be avoided into places now seen as having favorable amenities (Gutmann & Field 2010), and led to the late twentieth-century growth of large cities in otherwise unlikely locations such as Phoenix, Las Vegas, Denver, and Orlando.

Environmental conditions – or changes in them – can also have adverse impacts on livelihoods and well-being, and thereby act as stimuli for people to leave, or in the worst cases, flee particular locales. Sometimes such changes occur for natural reasons, other times from human carelessness. Potable water is a basic requirement for human settlements, and loss or contamination of it by pollution quickly makes a location unlivable and triggers outmigration and eventual abandonment (McLeman 2011). Fertile soils on which food systems and rural economies depend can be easily degraded by poor land management practices. Land degradation exacerbates the impacts of droughts and can potentially increase the potential for out-migration from affected areas (Hermans & McLeman 2021); this dynamic contributed to the large outflow of migrants from the southern US Great Plains during the 1930s Dust Bowl era (McLeman et al. 2014). Although the present book focuses on climatic hazards, there are other types of naturally occurring environmental hazards as well, such as earthquakes,

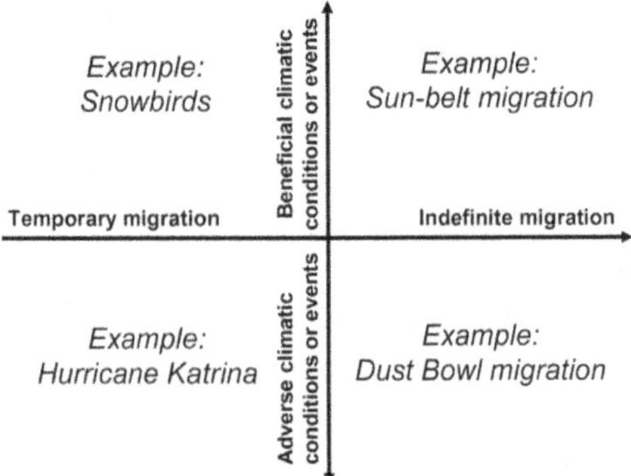

Figure 1.11 Environmental influences on migration, with examples. From McLeman (2014), reprinted with permission from Cambridge University Press.

volcanoes, and tsunamis, and these cause destruction of homes and infrastructure that can lead to temporary and potentially permanent out-migration from affected areas, depending on the circumstances (Barclay et al. 2019, Basile et al. 2024, Rigg et al. 2008)

A simple way of illustrating the range of potential migration responses to environmental phenomena is shown in Figure 1.11, in which duration of the migration event is represented on the horizontal axis and the nature of the environmental stimulus (i.e. hazards vs amenities) is on the vertical axis, with known examples of climate-related migration from the US given for each of the four quadrants. The two quadrants above the horizontal axis reflect the potential migration outcomes where environmental conditions are beneficial or attractive. The upper right quadrant describes migration undertaken on a permanent basis to take advantage of favorable environmental conditions, an example being high rates of population growth in the Sun Belt states in the era of air conditioning, much of it being fueled by north-to-south migration (Gutmann & Field 2010). The upper left quadrant reflects temporary or seasonal migration to access favorable environmental conditions, the previously described snowbirds being an easy example. The lower half of Figure 1.11 describes situations where changes in environmental conditions have adverse impacts on livelihoods and well-being, the lower right quadrant describing long-term or indefinite migration out of affected areas such as the 1930s Dust Bowl migration, and the lower left quadrant describing short-term or temporary migration in response to adverse environmental conditions, which often occurs following sudden-onset hazard events such as floods and extreme storms. Hurricane Katrina is one such event we assess in greater detail in the next chapter, and the multiple types of migration responses to it would sprawl across both lower quadrants of Figure 1.11.

1.4 Household-Level Factors That Influence Migration Outcomes

Sections 1.3.1–1.3.6 provided a brief summary of the wide range of macro-level processes outside the direct influence of households that (1) affect their health, wealth, and well-being, and that (2) give rise to situations where a household or its members may need or want to make a decision about migration. The context within which migration decisions are made is influenced by the combination of those macro-level factors and the specific characteristics of individual households, which vary within and across communities. Entire books and scores of journal articles have been written about household-level considerations that enter into migration decision-making; we intend to be much briefer here and tease these out in subsequent chapters with respect to migration decisions made in the face of specific climate hazards. For now, we can summarize household characteristics that are known from the wider body of migration scholarship to shape migration decisions as including, but not limited to:

- the age of household members and where they are in their respective life courses,
- the health of household members,
- the gender composition of the household,
- the livelihood(s) pursued by household members,
- the wealth of the household and the nature of its assets,
- the geography of the household's social networks,
- whether members of the household have migrated in the past,
- the household's past experience with events that might prompt migration or displacement,
- the perception of risks and opportunities.

These characteristics are not all independent of one another; age and health often go hand-in-hand, and livelihood options are often shaped by wealth (and vice versa). Elements of many of these have been mentioned in previous sections and others will be discussed throughout the following chapters, but a few additional comments are warranted here. Households and individuals experiencing poverty, poor health, and/or having few family connections or weak social networks are less potentially mobile than people enjoying good health, wealth, and strong social networks. Certain types of livelihoods, from pastoral nomads to digital nomads, are inherently more mobile than crop farming or working in a place-specific occupation. Economic wealth in general opens up options and opportunities for households that possess it, but certain types of wealth and assets are less portable than others, and thus more likely to discourage migration. People who own farmland, homes, or other fixed capital resources (e.g. small businesses based on shops or manufacturing) may be much more hesitant to move elsewhere, even in the face of considerable climatic or other risks. Conversely, people who rent their homes, landless rural workers, and people who work in easily learned or easily transferable high-demand occupations may be much more likely to move. The geographical characteristics of a household's social network shapes whether members of a household are more likely to migrate or not, and where they might move if they do. People with strong local social connections are less likely to contemplate moving to a distant location as compared with people who have geographically extensive social networks. The chances of migration are also increased if a household or any of its members have previously migrated, because past experience provides (1) insights into the costs and benefits of potential future moves and (2) social connections to other places. Past experience with not just migration but with potential drivers of migration, such as extreme weather events or economic crises, can also shape a household's perceptions of emerging or future risks and whether migration would be an appropriate response (Koubi et al. 2016).

1.5 Agency and the Decision to Migrate or Stay

The term *agency* refers to the context within which individual migration decisions are made, and as was shown in Figure 1.9, agency is formed through interactions between macro-level processes and the characteristics of individual households. The decision to migrate or stay and the outcome of that decision are shaped by the agency of the individual or household making the decision. People with high agency have a wide range of options to them, and are relatively free to decide whether or not to move elsewhere should the need or opportunity to make such a decision arise. They have the latitude to make their decision with a broad base of knowledge about the potential costs and benefits, and may have many possible options of where and when to move. By contrast, people with low agency may have few choices available to them in terms of potential destinations, the timing of their decision, or alternative options – or perhaps they may have no options at all. Migration that occurs under low-agency conditions is often involuntary and/or may result in outcomes that leave the individual or household in a worse position than before the decision had to be made. Agency is a critical feature of many of the examples of climate-related migration that appear in the following chapters.

The alternative to moving (i.e. mobility) is staying in place (i.e. immobility). Although the title and focus of this book is people on the move in a changing climate, it is important to recognize that not all people who experience changing climatic conditions will want to move, and not all people who want to move will be able to do so. The act of staying in a place that is exposed to a climatic hazard is considered to be *voluntary immobility* if the people in question choose

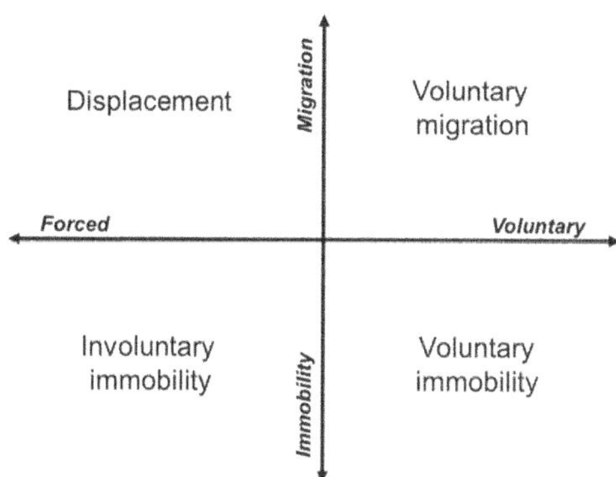

Figure 1.12   Four potential outcomes of the migration decision-making process.

Box 1.12
Trapped Populations

In the early days of climate-related migration research, the concept of "trapped" populations emerged as an important counterpoint to the prevailing research focus on how climate change would increase the number of people on the move (Black and Collyer 2014, Foresight 2011). The idea was that, for some populations, the effects of climate change and climate-related hazards may inhibit mobility rather than magnify it. "Trapped" populations were described as those that lack the capacity to migrate away from places impacted by climate change, and were likely to include the poorest and most vulnerable members of their respective communities. For them, the decision to migrate to someplace safer is not an option. At the same time, they may lack the necessary resources to respond, recover, and adapt in place, meaning that their exposure to risk and adverse impacts will grow over time. Since its initial introduction to climate-related migration discourse, the concept of "trapped" populations has evolved (Ayeb-Karlsson et al. 2018). It is useful to highlight inequities and differences in access to resources that make migration an adaptation option, it also obfuscates individual agency in the decision not to move: Some people don't move because they choose not to, and are therefore not "trapped" at all. Just as decisions to migrate are complex, decisions to stay in a place affected by climate change are also complex and multifaceted.

not to move and *involuntary immobility* if they would prefer to move but cannot (Schewel 2020). Groups of involuntarily immobile people are sometimes alternatively described in research literature as *trapped populations* (Box 1.12). The outcomes of the migration decision process described in Figure 1.9 end up in one of the four quadrants shown in Figure 1.12. Examples of voluntary migration, displacement, voluntary and involuntary immobility in the face of specific climate hazards are found throughout Chapters 2–4.

1.6 Climate Risk, Adaptation, and Migration: Bringing It All Together

It is now time to revisit the concepts of climate hazards and risks described in Section 1.2 and link these explicitly to migration processes and outcomes described in Sections 1.3–1.5. Recall that the risks to human lives, livelihoods, and well-being are a function of:

- the nature of the climate *hazard*,
- the extent to which people are *exposed* to the climate hazard,
- the *vulnerability* of those exposed,
- the *adaptation* measures that have been put into place and the *responses* people take after a hazard has occurred.

Let's focus particularly on the question of adaptation and response. When confronted by a climate hazard, people may adapt in a wide variety of ways. People living in a flood-prone coastal community might decide to raise funds and build a protective sea wall or construct their homes on stilt-like supports so that the living space is situated above floodwaters. If in a tropical area, they may attempt to vegetate the shoreline with mangroves, which act as a natural barrier to storm surges and help stabilize the shoreline. These are all examples of adaptations discussed in greater detail in later chapters. None of them require residents of the community to move, but are designed to help them continue living in the same place. These are known as *in situ adaptations*. But in some cases, the scale of the potential future flood risk may be so great that residents decide it is prudent to move somewhere else. Individual households might arrive at this decision independently, in which they autonomously decide to relocate, or there may be cooperation between residents and higher levels of government to organize a *planned relocation* of some or all members of a community (a process we discuss in detail in Chapter 4). These are not *in situ* adaptations, but responses where migration and relocation are components of the adaptation process, sometimes described simply as "*migration as adaptation*" (McLeman & Smit 2006). In other cases, people may choose to continue living in a location highly exposed to flood hazards because they have no choice (involuntary immobility) or because the social and economic benefits outweigh the risks (voluntary immobility). If and when the hazard event arrives, people may be displaced from their homes temporarily and then quickly return, others will migrate elsewhere, some for a short period of time and others permanently (we provide many specific examples in chapters that follow). Whether these responses can be considered to be "adaptation" depends on whether they reduce the vulnerability and risk of those affected over the long run.

Figure 1.13 illustrates the connected processes of climate risk, adaptation, and migration in their totality. Moving from left to right in the diagram, we start with the emergence of a climate risk, which may be related to slow- or sudden-onset

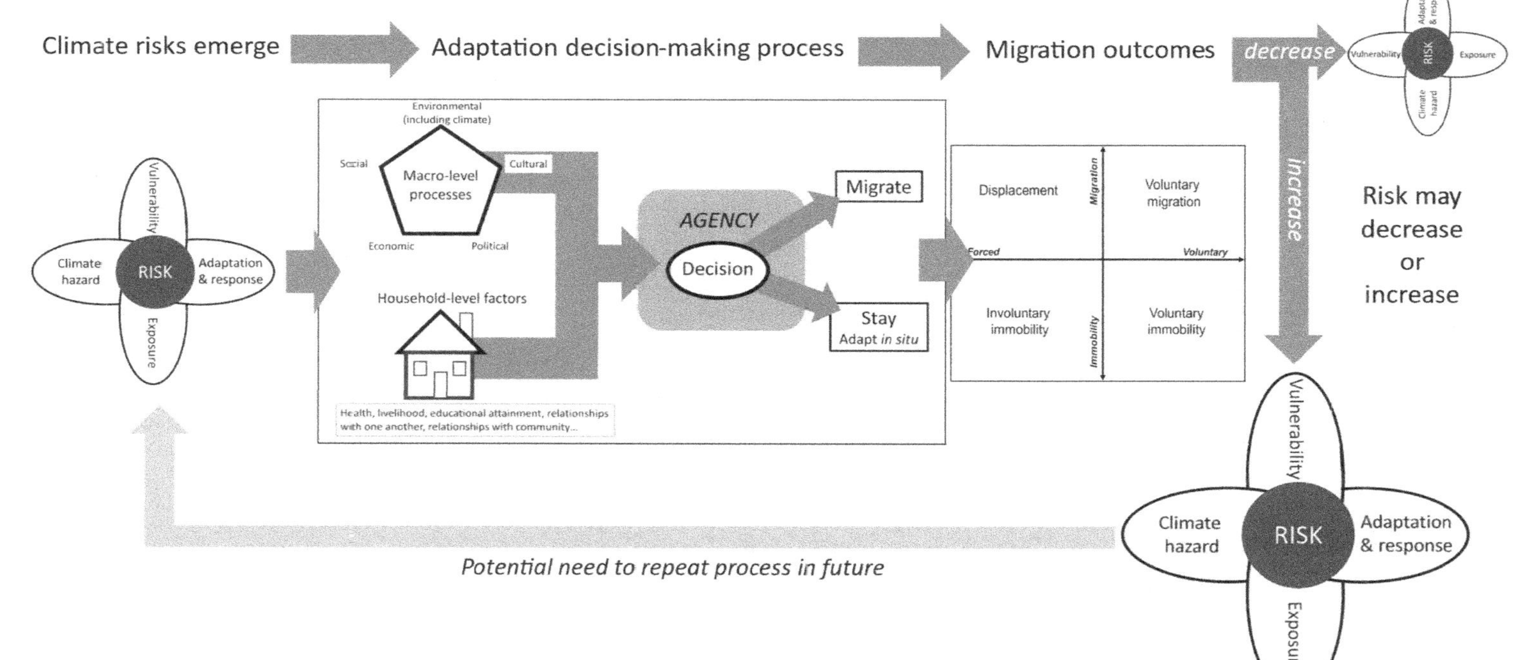

Figure 1.13   Diagram showing progression from left to right of emergence of climate risk, adaptation decision process, migration outcomes, and changes to risk that may ensue. Adaptation and migration decisions may need to be revisited, particularly in cases where risk increases after the initial decision to move or stay has been made.

hazards or some combination of both. The nature and scale of the risk is moderated by the nature of the hazard, the nature of exposure to it, and the vulnerability of the people exposed. Decisions will need to be made about how to adapt and respond, and these decisions may or may not include people moving. These decisions are shaped by macro-scale cultural, economic, environmental, political, and social processes and by the characteristics of the households and individuals that are exposed to the hazard. Depending on their agency, people may choose to move or be forced to move, they may choose to stay and adapt, or they may have no other choice but to stay and adapt as best they can. The outcomes of these adaptation and migration responses affect the nature of risk moving forward. If they reduce people's vulnerability and/or exposure to climate hazards, their future risk shrinks. If they leave people still exposed to climate hazards and/or at higher levels of vulnerability than before, their future risk grows – in which case, they may end up having to repeat the process and make difficult decisions about adaptation and migration once again.

This brings us to the conclusion of this Chapter 1 that summarizes the concepts necessary to understanding the general connections between climate hazards, risk, adaptation, and migration, and allows us to begin a deeper dive into how climate change is likely to affect future migration and displacement. Chapter 2 investigates migration associated with sudden onset hazards including extreme storms and floods, Chapter 3 investigates migration in response to droughts and heat-related hazards, and Chapter 4 investigates risks to coastal communities exposed to rising sea levels. The concepts and terminology used in Chapter 1 are used consistently and repeatedly in the following chapters. Readers are encouraged to refer back to this chapter and in particular to Figure 1.13 whenever clarification or a reminder about key terms or concepts is needed.

2

Migration and Displacement Associated with Extreme Weather Events

Tropical Cyclones, Severe Storms, Heavy Rainfall Events, and Flooding

2.1 Introduction

Extreme weather events such as tropical cyclones (also known as hurricanes and typhoons), severe convective storms (including those that generate tornadoes), heavy rainfall events, and floods collectively displace more people around the world each year than all other climate-related events and conditions combined. The combination of large volumes of rain and high winds generated by extreme weather creates risks of storm surges and mudslides/landslides in addition to floods, raising the potential for extensive damage to housing, farms, businesses, and critical infrastructure, as well as loss of life. Although the specific weather event may be sudden in its onset and short-lived, the physical and socio-economic impacts may endure for weeks, months, or years afterward. There is strong evidence that extreme precipitation exacerbates long-term income inequality in affected areas, with low-income and rural populations being especially hard hit (Palagi et al. 2022). Although anyone living in an area that is hit by an extreme weather event may experience loss, harm, and/or temporary displacement, socio-economically marginalized people are often most vulnerable and most at risk of being indefinitely displaced.

This chapter reviews how climate change is projected to affect the frequency, severity, and/or spatial distribution of tropical cyclones, severe storms that generate tornadoes, and floods; the factors that influence people's exposure and vulnerability to such events; adaptation options for reducing displacement risks; and, common characteristics of migration and displacement across all categories of extreme weather events. We then focus on specific types of extreme weather and provide more detailed analyses and case studies of migration and displacement events associated with tropical cyclones, tornadoes, and floods. Throughout this chapter and subsequent ones, we use standard terminology developed by the Intergovernmental Panel on Climate

Change (IPCC) to describe interactions between climate and human systems (i.e. the combined interactions of cultural, demographic, economic, political, and social processes) that give rise to particular risks, impacts, and outcomes for individuals, households, and communities. There is an inherent practical value in doing so: Language used by the IPCC is based upon and feeds back into terminology used in the United Nations Framework Convention on Climate Change (UNFCCC) negotiations process, the key international policy tool for responding to the impacts of climate change (described in greater detail in Chapter 6).

2.2 Projected Future Changes in Severity, Frequency, and/or Spatial Patterns of Specific Sudden-Onset Weather Hazards under Climate Change

In the Technical Summary of its 2021 Working Group I Report (Arias et al. 2021), the IPCC summarized a long list of projected changes in the frequency, severity, and/or spatial extent of sudden-onset events by geographical region that are projected to occur in coming decades as average global temperatures rise. The specific nature and extent of future changes in the occurrence of extreme weather events is dependent upon changes in future temperatures, which is in turn dependent upon future greenhouse gas (GHG) emissions. The IPCC uses four standardized emission scenarios for future GHG emissions, ranging from a low-emissions scenario (RCP 2.6) that could occur if an immediate cap was placed on global GHG emissions followed by rapid action to reduce these to zero by mid-century (unlikely given current global emissions trends) to a high-end scenario (RCP 8.5) where emissions accelerate through the remainder of this century (see Box 2.1). Under RCP 2.6, average global temperatures by 2100 would be approximately 1.5°C warmer than pre-industrial levels (at time of publication of this book, global average temperatures are approximately 1.2°C warmer than pre-industrial temperatures), and there would be correspondingly modest changes in the frequency, severity and spatial distribution of extreme weather events from what we currently experience. Under RCP 8.5, average global temperatures would soar to more than 4°C warmer than pre-industrial levels, leading to potentially catastrophic changes in the frequency and severity of extreme weather events.

Given current emissions trends, and assuming that countries follow through on commitments made to date under the UNFCCC to reduce GHG emissions, we are currently on a pathway to an average warming of between 2.5°C and 3°C over pre-industrial levels by the end of this century (Liu et al. 2021, UNEP 2023)

Box 2.1
Representative Concentration Pathways (RCPs)

In the absence of a crystal ball, it is not possible to know with any great certainty whether GHG emissions in future decades will be higher, lower, or similar to current levels – but estimates are necessary in order to make predictions about future levels of global warming and the consequent impacts on climate. To address this, scientists established four standardized scenarios to be used in climate modeling exercises known as "representative concentration pathways" (RCPs) to identify the extent of warming that would result should countries act aggressively to curtail emissions (RCP 2.6) versus warming in a world where fossil fuel use grows steadily throughout the remainder of this century (RCP 8.5), as well as warming that would result under two intermediate scenarios (RCP 4.5 and RCP 6.0). The numbers in the RCP titles refer to the amount of radiative forcing – that is the additional energy retained in the Earth's atmosphere, expressed in watts per square meter – that would result by the end of the century under each scenario. The scenarios were created by taking into consideration plausible combinations of population growth, economic activity, and energy use. The estimated amount of additional warming by 2100 under each RCP is tabulated in Table 2.1. To achieve the low-emissions scenario of RCP 2.6, the international community would need to reduce global carbon dioxide emissions to net zero by the 2050s.

Table 2.1   *Additional warming by 2100 under RCPs 2.6, 4.5, 6.0, and 8.5.*

Representative concentration pathway	Additional warming over pre-industrial temperatures (95% estimation range shown in parentheses) (°C)
RCP 2.6	1.6 (0.9–2.3)
RCP 4.5	2.4 (1.7–3.2)
RCP 6.0	2.8 (2.0–3.7)
RCP 8.5	4.3 (3.2–5.4)

Data source: Arias et al. 2021.

(Box 2.2). Should warming exceed 2°C, the frequency and/or severity of extreme, sudden-onset weather events is expected to change across many regions and sub-regions, as summarized in Table 2.2.

Given the financial costs, damage to housing, loss of livelihoods, injuries, and deaths associated with extreme weather events, any increase in their frequency and/or severity would be catastrophic at global, regional, and local scales. Any such increase would also amplify the number of people involuntarily displaced

Table 2.2 *Changes in floods, tropical cyclones, and extreme precipitation events, by region/sub-region. Summarized from Arias et al. 2021.*

Region	*Sub-regions expected to experience changes in frequency and/or severity of selected sudden-onset event categories*		
	River floods	Extreme precipitation events	Tropical cyclones
Africa	Increase in Central Africa	Increase in all sub-regions EXCEPT Mediterranean countries and southern West Africa	No increase expected; possible decrease in frequency and/or severity in areas of Africa currently exposed to cyclones (i.e. East and Southeast Africa, Madagascar)
Asia	Increase in East, Southeast, and South Asia	Increase in all sub-regions	Increase in East and Southeast Asia
Australasia	Increase in all sub-regions	Increase in northern and central Australia	Potential decrease in number of tropical cyclones making landfall in eastern Australia; uncertain for other parts of the region
Central and South America	Southeastern South America	Northern, northeastern, southern, and southeastern South America	Increase in number in northern Central America; decrease in number but increase in severity of cyclones in southern Central America and northern South America
Europe	Increase in western and central Europe; possibility of decrease in northern, eastern and Mediterranean Europe	Increase across most of Europe; possibility of decrease in Mediterranean areas	Uncertain future trends in severe windstorms of various categories[a]
North America	Increase expected in all sub-regions	Increase in all sub-regions	Increase in central and eastern North America

[a] Tropical cyclones do not form in waters adjacent to Europe, but hurricanes that migrate out of the tropical Atlantic can bring high winds and heavy rainfall to Western Europe.

each year by extreme weather, which currently ranges between 20 and 30 million people (Internal Displacement Monitoring Centre 2024). The remainder of this chapter focuses on the observed impacts of common types of extreme weather on migration and displacement.

Box 2.2
How Do We Measure "Global Warming"?

When scientists use the term *global warming*, they are referring to changes in average global temperatures as measured at the Earth's surface. But change compared to what? There are two general ways temperature change is discussed. The first is simply to describe how current temperatures deviate from a longer-term average, also referred to as the *temperature anomaly*. So, for example, if one were to plot average annual global temperatures from 1880 until the time at which this book was written, one would find that every year since the late 1970s has been warmer than average, and that since the year 2010 annual temperatures have been nearly a full degree Celsius warmer than the average. The IPCC typically refers to changes in average global temperatures as compared with the "pre-industrial period," which it in turn defines as the global mean average of temperatures measured between 1850 and 1900 (approximately 13.6°C). The reason for using this particular time period is that it is the earliest time for which directly measured daily temperature data are available for large areas of the Earth. The mercury-based thermometer was invented in Europe in the first half of the 1700s, but it was not until the mid-1800s that it was being used to systematically collect temperatures at a wide selection of locations on other continents. The geographical distribution of historical temperature records is uneven, with large areas of Africa and South America lacking observational records until well into the twentieth century; there are also large areas of ocean for which temperature records do not exist until the era of weather satellites.

2.3 Current and Future Exposure to Displacement-Inducing Extreme Weather Events

Since 1998, a Geneva-based organization first established by the Norwegian Refugee Council, known as the Internal Displacement Monitoring Centre (IDMC) has been gathering global statistics on people who have been involuntarily displaced from their homes by conflicts and natural hazards. Using a standardized methodology, IDMC has produced reasonably reliable annual reports that estimate the number of people who are newly displaced at national levels each year, along with the number of people who remain displaced at the end of each calendar year. IDMC data for displacements attributable to hazard events are organized according to sub-categories that include non-climate-related hazards (e.g. earthquakes, tsunamis, and volcanic eruptions) and climate-related ones (e.g. floods, storms, and droughts). The 2022 IPCC Working Group II Assessment Report used IDMC data for the period 2010–2020 to provide an overview of the key climatic drivers of displacement across regions. As can be seen in Figure 2.1, floods and storms (including tropical cyclones) together represent far and away the largest drivers of climate-related displacement in every region. Increases in their frequency and/

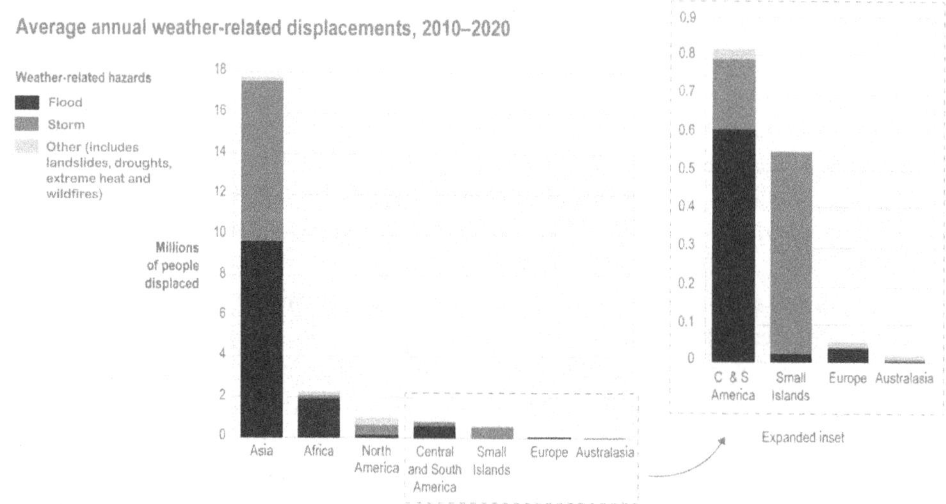

Figure 2.1   Average annual displacements by climate-related events by region, 2010–2020. Source: Figure 7.7 in Cissé, G., R. McLeman, H. Adams, P. Aldunce, K. Bowen, D. Campbell-Lendrum, S. Clayton, K.L. Ebi, J. Hess, C. Huang, Q. Liu, G. McGregor, J. Semenza, and M.C. Tirado, 2022: Health, Wellbeing, and the Changing Structure of Communities. In: *Climate Change 2022: Impacts, Adaptation and Vulnerability*. Contribution of Working Group II to the Sixth Assessment Report of the Intergovernmental Panel on Climate Change [H.-O. Pörtner, D.C. Roberts, M. Tignor, E.S. Poloczanska, K. Mintenbeck, A. Alegría, M. Craig, S. Langsdorf, S. Löschke, V. Möller, A. Okem, B. Rama (eds.)]. Cambridge University Press, Cambridge, UK and New York, NY, USA, pp. 1041–1170, doi:10.1017/9781009325844.009. Reprinted with permission.

or severity will only serve to amplify the potential for larger numbers of people to be displaced in the coming decades in the absence of greater adaptive capacity in exposed populations.

Not only are extreme weather events becoming more frequent and severe in most regions of the world, many regions that already experience such events on a regular basis also have high rates of population increase. This means that the number of people exposed to climate-related hazards is also rising, resulting in a global amplification of risk of more frequent, larger-scale displacements in coming decades. Figure 2.2 maps national rates of natural population increase and identifies those regions expected to experience an increase in the frequency of river floods, heavy precipitation events, and tropical cyclones identified in Table 2.1. As can readily be seen, large portions of every inhabited continent are expected to see rising frequency of one or more types of extreme weather events, and within these areas are many countries experiencing high rates of population growth. It can therefore be stated with confidence that exposure

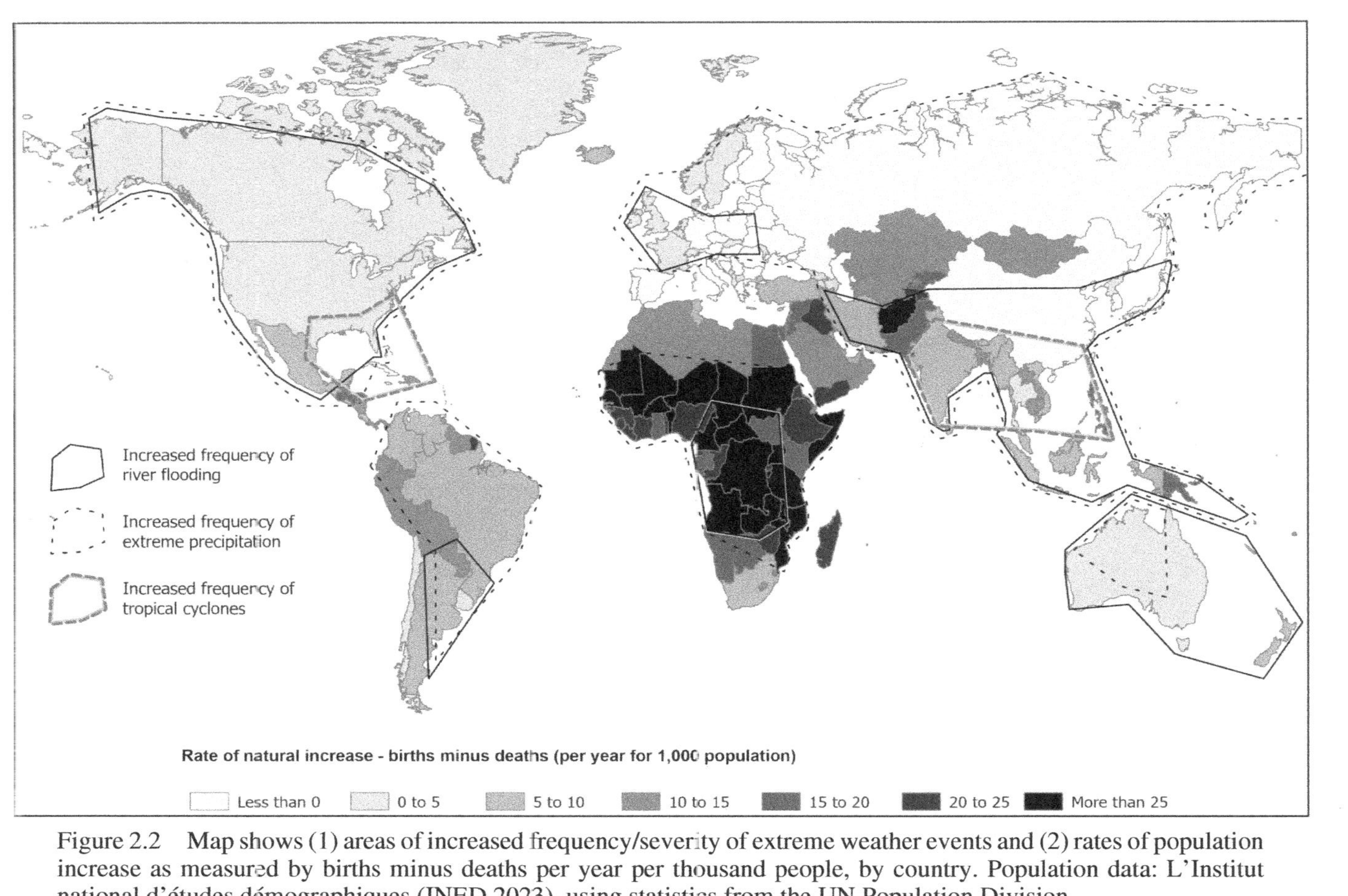

Figure 2.2   Map shows (1) areas of increased frequency/severity of extreme weather events and (2) rates of population increase as measured by births minus deaths per year per thousand people, by country. Population data: L'Institut national d'études démographiques (INED 2023), using statistics from the UN Population Division.

to extreme weather hazards is growing globally from the combined effects of population growth and more frequent and/or severe weather events in highly exposed areas. The final diagnostic steps in assessing the global risk of involuntary displacement due to extreme weather in a changing climate are to (1) identify non-demographic factors that amplify exposure to extreme weather in coming decades, and (2) assess whether there will be any future change in the vulnerability of exposed populations, which may be moderated by changes in the capacity of exposed populations to adapt.

2.4 Non-demographic Factors That Amplify Exposure to Extreme Weather Events

Human settlements and land use are dynamic, and so exposure to extreme weather events is continuously changing as well. Following are examples of well-studied factors known to influence the number of people exposed to weather- and climate-related hazards and the nature of that exposure, thereby affecting the risk of people being involuntarily displaced when such events occur.

2.4.1 Landscape Modification

Human alteration of the landscape can exacerbate the physical impacts of extreme weather events by removing natural features that store precipitation, slow erosion and surface runoff, and blunt the effects of storm surges. The filling of wetlands removes storage for excess precipitation, accelerates precipitation drainage into rivers and other watercourses, and increases flood risks in downstream areas (Opperman et al. 2009). Clearance of forest cover on steep slopes, whether for forestry, agriculture, or settlement, increases the risk of slope failure when soils become saturated by high levels of precipitation, creates the potential for flash flooding, and allows high levels of sediment to erode into watercourses. For example, in 2023, heavy rains from Cyclone Gabrielle triggered massive mudslides on New Zealand's North Island in areas where pines that had been planted decades earlier to prevent erosion were being actively harvested (Bloomberg 2023). Coastal wetlands that absorb storm surges are shrinking in size because of human activity, with climate change-related sea level rise expected to exacerbate wetland losses (Li et al. 2018). Mangroves – a mix of tree and shrub species that thrive in tropical intertidal zones – are especially important in protecting coastal settlements from storm surges associated with tropical cyclones and tsunami. For example, research carried out in coastal Bangladesh following 2007's Cyclone Sidr found that settlements protected by mangroves experienced only half the amount of damage to housing as those that lacked such protection (Akber et al. 2018). Worldwide, mangroves are in rapid decline due to pollution and active clearance by human populations to make way for aquaculture or coastal urban development (Duke et al. 2007). A 2.1% decline in global mangrove cover occurred between 2000 and 2016, with especially high rates of mangrove loss taking place in Southeast Asia Goldberg et al. (2020) – a region with a high level of exposure to tropical cyclones.

Coral reefs and offshore barrier islands perform similar functions of sheltering landward areas from large waves and storm surges, and these, too, are being actively degraded in many settled coastal areas by human activity (Grzegorzewski et al. 2011, Zhang and Leatherman 2011). A study by Carlot et al. (2023) found that when the structural diversity of a barrier coral reef falls by half – as is becoming increasingly common in many coastal areas – the coastal areas sheltered by the reef experience a massive increase in the frequency of extremely high wave events. As wetlands, hillside forests, coastal marshes, reefs, and other natural features that offset the impacts of extreme events disappear from heavily populated areas, the risk of property damage, loss of life, and involuntary displacement from extreme events grows. Costanza et al. (2021) estimate that on an annual basis coastal wetlands save over 4,600 lives and avert nearly US$450 billion in damages to homes and infrastructure.

2.4.2 Residential Location, Housing Quality, and Housing Tenure

Residential location patterns in areas exposed to extreme weather hazards (Box 2.3) can affect the degree of risk of particular groups and communities, and these in turn can be influenced by cultural, social, and economic processes. In some areas, wealthy people deliberately choose to live in highly exposed areas; in others, it is the poor who are relegated to doing so, creating a heterogeneous landscape of risk. In the US, for example, low-income people are more likely to live in high-risk areas of inland floodplains than are other groups, but low-lying coastal areas with high flood risks are more likely to have higher concentrations of older, wealthy people (Qiang 2019). In many countries, coastal waterfront property is considered highly desirable for residential, commercial, and recreational use, notwithstanding the known weather-related risks of occupying such areas (Bin et al. 2008). Coastal property values are pushed upward by trends such as these, and when a tropical cyclone strikes, undamaged housing becomes scarcer and even more valuable (Murphy and Strobl 2009).

In many low- and middle-income countries, poor populations may from lack of choice be obliged to live in self-built housing and informal settlements in locations such as steep hillsides and coastal flats that are highly exposed to floods, storm surges, and other physical impacts of extreme weather events. Refugee settlements and camps for displaced people are also often situated in hazardous locations (Fransen et al. 2024). Although people who live in such locations know they are inherently

Box 2.3
What Makes an Extreme Event "Extreme"?

There is no universal definition of what makes a given weather or climate event "extreme." Most discussions focus on the rarity of the event, the intensity of the event, and/or its destructiveness as compared with some set of relative norms and thresholds (Beniston and Stephenson 2004). What is considered to be "extreme" in one setting may not be in another. For example, studies have shown that in Europe, excess winter death rates due to cold winter temperatures are higher in southern European countries with milder climates than in Scandinavian countries that have much colder winters (Fowler et al. 2015). This paradox reflects the fact that people living in more northerly climates have more experience with cold temperatures and can adapt more easily, and live in homes better suited to the cold. Temperatures that would be considered to be "extreme cold" in Italy or Spain would be an average winter day in Norway or Finland. For example, one study that measured excess winter mortality rates in Portuguese cities (Antunes et al. 2017) used temperatures of 2.5°C and –0.1°C as "extreme cold" temperature thresholds for Lisbon and Oporto; in Helsinki, these same temperatures would be warmer than average from late November through early March. For this reason, to define what constitutes a weather extreme – whether it be temperature, precipitation, or some other variable – scientists typically use an approach that takes the historically experienced range of observations and selects a certain percentile as a threshold (typically the 90th percentile or greater) (Sulikowska and Wypych 2020).

dangerous, they typically have little alternative but to accept such risks, and may not evacuate or abandon such locations even after severe floods, having nowhere else to go (for detailed case studies from Mombasa, Kenya, see Okaka and Odhiambo (2019) and from Bulumbura, Burundi, see Kubwarugira et al. (2019)). When extreme weather events occur, the contrast between impacts in rich and poor neighborhoods can be quite stark. For example, in 2006, heavy rains caused flooding in the El Paso-Ciudad Juarez region along the US–Mexico border. Of the 20,000 left homeless, a disproportionate number were the poorest of the poor, migrants who lived in informal settlements along *arroyos* (intermittent watercourses) on the Mexican side of the border (Collins 2010). Similar dynamics occur in other regions as well. In the fast-growing cities of Manila and Jakarta, both formal and informal housing developments increasingly impinge on flood-prone watercourses, with research by Kefi et al. (2020) projecting that by the 2030s, exposure to flooding will increase by 200% and 80%, respectively, through the combined effects of higher flood frequency and population growth in at-risk locations. These examples are just a small sample of cases that describe a common pattern of socio-economically marginalized people having to live in geographically marginal and unsafe locations in many parts of the Global South.

People whose homes are lost or damaged irreparably by extreme events are at far greater risk of being forced to relocate temporarily or permanently in the wake

Figure 2.3   Metal "hurricane ties" reduce the risk that roof trusses will separate from building walls during high winds. Photo: R. McLeman.

of extreme weather events. Self-built housing in informally built settlements is rarely robust enough to withstand damage from winds or heavy rainfall. One of the barriers in improving housing stock in many low-income countries is that the occupants lack formal ownership of the home or tenure of the land they occupy. Lack of ownership means the occupants have less incentive to improve the quality of the structure and cannot use it as collateral for obtaining a loan to do so (Zazyki et al. 2022). People who live in rented accommodation often have higher rates of exposure to being displaced by extreme weather given that they depend upon the property owner to make investments to improve the quality and structure of the buildings they occupy. The ratio of tenant-occupied to owner-occupied housing may be different in high-risk areas relative to others; for example, research in Houston, Texas, found there were disproportionately high numbers of publicly subsidized homes in areas flooded by Hurricane Harvey in 2017 (Chakraborty et al. 2021). A related challenge is one of high rates of tenant evictions following extreme weather events and/or decisions by landlords to not rebuild damaged housing, a consequence of relief assistance from federal agencies typically being directed to property owners and not tenants (Brennan et al. 2022). Occupants of publicly owned or publicly subsidized housing may find that authorities are slow to rebuild and/or repair properties after extreme weather events, further amplifying the loss of affordable housing (Best et al. 2023, Khajehei and Hamideh 2024).

In high-income countries, prefabricated houses and mobile homes are often most susceptible to storm and flood damage, and their occupants face a much greater risk of harm during tornadoes and tropical cyclones than occupants of more robustly constructed homes (Schmidlin et al. 2009, Rumbach et al. 2020). Tornado-related death rates in the US are especially high when tornadoes occur at night, when mobile homes are most likely to have people inside (Sutter and Simmons 2010). Often, damage to housing stocks from extreme storms is not distributed equally across an affected area, but is concentrated in places where the buildings have similar structural design and the debris from one damaged building is hurled against its neighbors to create a chain reaction of damage (De Silva et al. 2008). Efforts are increasingly being made in areas of high exposure, such as the "Tornado Alley" of the south-central US, to make housing more resilient to extreme weather. The necessary changes are often simple and relatively inexpensive to make, such as using small metal straps known as "hurricane ties" to connect roof trusses to the walls of houses, preventing high winds from tearing the roof off one house and dashing it against its neighbors (Ramseyer et al. 2015) (Figure 2.3).

2.5 How Social Inequality Affects Exposure to Extreme Weather Events

A recurrent theme in the preceding pages and in the case studies that follow is that poverty and social marginalization increase the risk of physical harm and loss of shelter during extreme events. This should not surprise. Families and individuals that are economically secure have greater choice in where they live and work, and have greater ability to avoid locations and activities where they might be exposed to significant environmental hazards. It is not a coincidence that poor and economically marginalized people in every country are more likely to be involuntarily displaced by extreme events given that it is they who live in less sturdy housing, have less access to communications technology and transportation, and lack the financial means to rebuild after an event occurs (Clark et al. 1998, Cutter and Finch 2008). One World Bank study estimates that 1.47 billion people worldwide live in areas exposed to hundred year flood risks (see Box 2.4), of whom roughly one-third have a daily income of less than US$5.5 (Rentschler and Salhab 2020). Exposure to weather hazards often goes hand in hand with housing insecurity (Rao et al. 2023). Groups and individuals

---

Box 2.4
What Is Meant by the Term "Hundred Year Flood"?

A *floodplain* is an area of land adjacent to a watercourse or coastline that might potentially be susceptible to flooding, and a variety of methods are used to delineate it. In general, the farther one moves away from a body of water and the higher the elevation gained, the lower the risk of flooding. Hydrologists use historical records, maps, remote sensing data, and computer models to calculate an estimated probability of a given area being flooded in any given year, and then communicate it to the public in terms of estimated rate of return. So, for example, a location where the risk of being flooded in any given year is estimated at 1% is said to be situated within the "100-year floodplain" – that is that in purely statistical terms, we might expect the location in question to be flooded approximately once in any given hundred year period. Locations with a 2% chance of being flooded in a given year are described as being within the 50-year flood plain and a 5% chance of flooding being within the 20-year flood plain. Governments will often map out such areas for planning purposes, and impose restrictions on certain types of construction and land use within the floodplain boundaries.

There are, however, two key drawbacks to such an approach in terms of communicating flood risks. First, a location that has a 1% chance of being flooded in a given year might be flooded more than once a century, so residents might mistakenly assume that after a flood has occurred they have nothing more to fear during their lifetime. This happened to residents of Peterborough, Ontario, Canada, where a "hundred year flood" occurred in 2002 and again in 2004 (Oulahen and Doberstein 2012). A second challenge is that flood frequency and severity changes over time, with research showing that the frequency of 20- and 50-year floods has increased since the 1970s in many temperate regions of the world (Slater et al. 2021). Finally, land use changes within a watershed or within the coastal area can influence the movement of water and hence affect the susceptibility of particular areas to flooding (Wheater and Evans 2009). For example, forested areas and wetlands tend to absorb precipitation and snowmelt and release it gradually into streams and rivers (thereby moderating flood risks downstream), but urban areas with their hard surfaces tend to accelerate drainage into watercourses and increase the risk of floods downstream. Conversion of forests to farmland and farmland to suburban or urban landscapes has amplified flood risks in many watersheds around the world (Bradshaw et al. 2007). As a result, floodplain maps need ongoing updates, and flood risks expressed in return rates need to be used cautiously.

disproportionately represented among those who lack secure housing include ethnic minorities, children, the elderly, and those in poor health or with physical disabilities, as well as people generally living in poverty (Pham et al. 2020, Tate et al. 2021). In the US, areas of predominantly Black residents face a growing risk of flooding in a changing climate (Wing et al. 2022). Women, especially those who head households, tend to experience higher rates of poverty than men, and so they, too, may have elevated levels of exposure to natural hazards (Mondal et al. 2020, Pathak et al. 2020). Recent research has suggested that transgender and gender non-conforming individuals may also experience higher levels of housing insecurity (Glick et al. 2019). Livelihoods also shape the level of exposure of households and individuals to weather-related hazards. For example, those who perform outdoor work are highly exposed to health risks and injuries associated with a range of extreme weather and heat events (Cissé et al. 2022), while rural households may be doubly exposed to harm associated with extreme weather events due to damage to housing and to livelihood assets (e.g. livestock, fields, and aquaculture ponds) (Fahad et al. 2023, Tate et al. 2021),

2.6 Factors That Amplify Vulnerability to Extreme Weather

In addition to the previously described factors that can influence exposure to extreme weather events, there are a range of additional factors that can increase the *vulnerability* of specific individuals, households, and groups to extreme weather. Research from multiple regions has shown that, regardless of country, elderly people, women, and young children experience disproportionately high rates of injury and death during extreme weather events (García-Hernández 2022, Rhoads et al. 2018, Thomas et al. 2019). In the case of the very young and the very old, heightened vulnerability arises from a combination of physiological characteristics and age-related health needs, and also because they are typically dependent upon other people for their care and well-being. People with pre-existing health conditions are also generally more vulnerable to extreme weather, as are pregnant people, fetuses, and the newly born (Roos et al. 2021). In addition to these individual characteristics, being a member of a racial minority, a traditional community or an Indigenous community often amplifies one's vulnerability to extreme weather, these being groups whose members may face discrimination, prejudice, hostility, and/or inadequate access to resources for adaptation and post-disaster recovery (Ford et al. 2020, Levy and Patz 2015). That said, care needs to be taken to avoid making blanket statements about the vulnerability of any societal group, for it is often the case that particular social groups in a given locale may have strong social networks, traditional knowledge, and/or other skills and attributes that reduce their vulnerability. A well-documented example of this is the minority ethnic Vietnamese community in New Orleans, many members of which lived and worked in areas severely damaged by Hurricane Katrina. Despite having generally lower household incomes than other residents of New Orleans and limited representation in formal governance bodies, Vietnamese institutions, churches, and social networks across the US mobilized resources to help the community achieve a more rapid post-hurricane recovery relative to other groups within the city (Leong et al. 2007, VanLandingham 2017).

2.7 The Importance of Institutional Adaptive Capacity

The human impacts of extreme weather events vary considerably from one location to another, even in cases where the physical and climatological conditions of the event may be similar. A longstanding example from natural hazards research is the contrast in impacts of tropical cyclones in the US as compared with impacts of similar storms in lower-income countries. In the US, the financial cost of storm damage to property has climbed steadily in recent decades (even once adjusted for inflation), but the number of people killed or injured tends to be relatively low. Between the years 2000 and 2020, the Atlantic and Caribbean coasts of the US experienced 130 hurricanes, but only one – Hurricane Katrina, in 2005 – caused large-scale loss of life. In that event, approximately 1,400 people died, most in and around the New Orleans area. In all other calendar years, the total number of people killed in hurricanes never exceeded 150 people, and in most years, the total number of deaths has been much smaller (Insurance Information Institute 2022). Hurricane Katrina was also the costliest US hurricane since 2000, with losses exceeding US$89 billion (in 2021 inflation-adjusted dollars), but five other hurricanes that occurred over that same period caused damages of greater than US$30 billion, even though they cumulatively resulted in a fraction of the loss of life (Insurance Information Institute 2022).

By contrast, the financial costs of cyclone damage in low-income countries, though often rising, is in absolute terms typically much lower than in the US, but the loss of life is often much higher. For example, a single tropical cyclone in 2007 – Cyclone Sidr – killed 3,400 people in Bangladesh (Paul 2009), more than the entire number of people killed in the US by all tropical cyclones between 2000 and 2020. The estimated dollar value of damages attributable to Sidr in Bangladesh was US$1.7 billion, less than 1/30th of the dollar value costs of Hurricane Katrina.

This marked contrast between high- and low-income countries in the impacts of extreme weather events is in part due to factors that influence exposure and vulnerability as described earlier in this chapter – the quality of housing, locations where poor people live, social inequality, and so forth – but it is also shaped by significant differences in adaptive capacity at institutional levels. The US has for many decades operated highly sophisticated meteorological systems that monitor and provide early warning of extreme weather. Such systems are becoming increasingly available in low- and middle-income countries, and the World Meteorological Organization (WMO) is actively working to improve access to such systems globally, for there is clear evidence that they are effective (World WMO 2022). Most Americans have access to broadband internet and/or cellphones, meaning that storm warnings and instructions can be disseminated much more quickly and widely than countries where access to such technologies lags. US states and local governments have emergency management plans that include establishing evacuation routes and emergency shelters, allowing most people the opportunity to get out of harm's way, supported by a federal government that is able to provide large amounts of disaster relief funds to assist storm-affected areas through the Federal Emergency Management Agency (although this institutional capacity does not extend to US-dependent territories, as shown below in the case of Puerto Rico).

Low-income countries often lack the financial wherewithal to dedicate such large financial resources to emergency preparedness and response, but even so, progress on proactive adaptation is being made. For example, in Bangladesh as recently as the 1990s, it was common for cyclones to kill tens or even hundreds of thousands of people per event, but improvements in communications and preparedness have helped reduce exponentially the average number of lives lost in recent decades (Paul 2009).

In short, successful adaptation to extreme events from an institutional perspective means having the economic, technological, and organizational capacity to reduce the number of people injured or killed, to repair or rebuild damaged homes and infrastructure as quickly as possible, and to restore to residents their means of household income and livelihoods. This is not easily done, and requires planning, coordination and financial resources. After an extreme event there is a range of sheltering needs, including emergency shelters, intermediate temporary housing, and permanent replacement housing (Nigg et al. 2006). Meeting these needs means mobilizing a broad range of government and humanitarian resources at multiple scales to supplement the savings and resources individuals and households draw upon. The greater the resources an affected population can mobilize, the lesser the chance members of that population will be obliged to relocate elsewhere for lack of other options. Even in a poor nation with weak formal government institutions, such as Bangladesh, prompt responses and assistance from other sectors of society, such as NGOs and social networks, enable communities to recover quickly from extreme events without large-scale displacement or permanent out-migration (Paul 1998, 2009).

2.8 Household *In Situ* Adaptation When Institutional Adaptation Fails

The term *in situ* is used here to refer to adaptations that do not require migration or mobility on the part of households and individuals. Adaptation to extreme weather events – or to any type of climatic risk, for that matter – can be initiated at any scale of social organization, from an individual or household all the way up to global-scale institutions and multilateral agencies, and can be undertaken proactively (i.e. in anticipation of potential hazards to reduce risk) or reactively (i.e. after an extreme event has occurred). A reality is that governments and other large organizations are more likely to engage in reactive adaptation than proactive adaptation. In many instances, not only do governments fail to plan ahead for the possibility of extreme weather events, they do things that compound the risks and expose more people to potential harm. A good example of this is the failure of governments to plan urban development in floodplains to account for flood risks. In the US, an estimated 1.276 million people live within a river's hundred year floodplain, and another 2.1 million live in coastal areas less than 1 m above sea level, and thus highly exposed to coastal flood hazards (Titus 2023). One would expect that local, state, and federal governments in the US would be working in active, coordinated fashion to prevent additional homes from being built in flood-hazard areas, and providing incentives to reduce the number of people living in such areas. There are a great many options available to decision-makers (Tyler et al. 2019). But this is not happening, and the number of people living in flood hazard areas is actually growing (Titus 2023). The primary adaptive mechanism has been the creation of a national flood insurance program that facilitates the prolonged occupation of hazardous areas and provides partial compensation to people after a flood occurs. The process of establishing floodplains maps has become highly politicized (Pralle 2019), with property owners and would-be property developers often pressuring governments to designate as little land as possible in floodplains and coastal areas as being hazardous, so as to increase the dollar value of existing properties and create new opportunities for building homes and commercial properties that would not require flood insurance as a contingency for mortgages and financing.

These dynamics were revealed in the wake of Hurricane Harvey, a Category 4 hurricane that made landfall in coastal Texas in August 2017, and brought over 1,000 mm of rain to the area over a 6-day period (Jonkman et al. 2018). Catastrophic flooding ensued, especially in the heavily urbanized Houston area, leading to an estimated 70 deaths. Given that millions of people lived in the exposed region, officials issued only targeted evacuation orders to specific neighborhoods and advised other people to shelter in place, leaving 120,000 people to be rescued from flooded areas by authorities after the fact (Jonkman et al. 2018). Research conducted afterward revealed that numerous new housing developments in the rapidly growing city had been knowingly approved in floodplain areas, and that locally built flood control structures that ended up failing had provided residents with a false sense of security (Malecha et al. 2021). The 100-year and 500-year flood maps for the Houston area did not accurately capture the actual flooding that occurred during the storm, and a disproportionate number of homes damaged in that storm were owned by members of minority groups and low-income residents (Collins et al. 2019). There is also evidence that a disproportionate amount of subsidized housing built for low-income people was sited in flooded areas (Chakraborty et al. 2021). In short, instead of helping residents adapt to the inherent risks of living in an area known to be highly exposed to hurricanes, planning practices of local and federal officials made matters worse.

Because governments often cannot be relied upon to undertake proactive adaptation for extreme events, the most meaningful adaptations typically fall to adult members of individual households. Here, too, decisions must be made about whether to take proactive adaptive measures to reduce the household's exposure and vulnerability to extreme weather – assuming the household has the necessary resources and information to make such decisions. Household members may not actually be aware they are living in a hazardous location until after the hazardous event occurs, or may have a poor perception of precisely how greatly exposed they may be, even if they are aware the risk is present, or may not know what steps might be taken to reduce their risk (Sanders et al. 2020). Socially and economically marginalized households may lack the necessary resources for proactive adaptation, even if they have a sound understanding of the risks they are exposed to. The result is that household adaptation to extreme weather – and in particular decisions whether to stay in place and try to better prepare for future events (i.e. *in situ* adaptation) or to move to a less hazardous location – often tends to occur as a reactive response after an event occurs (de Koning et al. 2019).

2.9 General Characteristics of Migration Outcomes in Response to Extreme Weather, Regardless of Type of Event

Extreme weather events can inflict considerable loss or harm on populations exposed to them. This loss or harm can be experienced in a variety of direct and indirect ways, including (but not limited to) impacts on health and well-being through injury, sickness, or loss of life; damage to critical infrastructure such as housing, roads, electricity, and water supplies; economic losses, including damage to crops, livestock, personal property; loss of present or future income or livelihood opportunities. The degree of harm that an individual household may experience is highly context specific, and – as described in previous sections of this chapter – reflects not only the physical manifestations of the hazard but also the cultural, economic, political, and social conditions in which they live. Similarly, the likelihood that a given household may be involuntarily displaced by a hazard event, or the likelihood that the household or some of its members may decide to migrate afterward, are also highly context specific. Whether migration and/or displacement actually happens depends upon the combined effects of responses on the part of institutional actors and the ability of the household to draw upon sufficient economic, social, and cultural capital to recover and rebuild. In situations where institutions and households have robust capacity to respond and adapt, where people's displacement from their homes is temporary and the resumption of livelihoods is quick, indefinite out-migration is unlikely. Conversely, where institutional responses are feeble and/or households lack adaptive capacity, the potential for permanent displacement and migration elsewhere grows.

To restate and simplify, the potential for displacement and/or migration to result from extreme weather events grows with:

- increasing severity of the physical hazard (or hazards, in the case of compound or cascading events) and
- increasing levels of vulnerability among the exposed population.

Conversely, the potential for displacement and/or migration is reduced when:

- households and individuals have sufficient capacity to adjust and adapt to the hazard and/or
- when institutions have implemented proactive measures to reduce the potential impacts of hazards (such as through flood defenses and early-warning systems) and are able to respond robustly to the impacts after a hazard event has occurred.

It is nonetheless important to remember that changes in the *potential* for migration do not automatically translate into observably higher (or lower) levels of migration. As is described in greater detail in following sections, a single hazard event may result in multiple types of migration responses – some people evacuating and returning as quickly as possible, some migrating away for a longer but still temporary period, some leaving and never returning, and others never evacuating or moving away at all. There are even examples of new migrants moving into a location that has experienced an extreme weather event, such as laborers moving to New Orleans to take up employment in the construction sector as that city rebuilt (Fussell 2018). Notwithstanding the general characteristics summarized in this chapter, not all people behave in the same way when confronted with similar challenges and choices. People are not automatons. Mobility decisions in the wake of or in contemplation of hazards are often shaped by non-socio-economic factors such as past experiences and perceptions of risk based on information available to them (Zander and Garnett 2020).

2.10 Migration Patterns Observed in the Wake of Tropical Cyclones

Tropical cyclones (also known as hurricanes and typhoons – see Box 2.5) displace millions of people around the world each year. Between 80 and 90 cyclones form in the tropics each year, although the number varies considerably from one year to another and from one ocean basin to another (Tory and Frank 2010, Weinkle et al. 2012). Cyclones only form where sea surface temperatures are warmer than 26.5°C and when the sea surface is overlain by low-pressure air that is relatively still and moist (Terry 2007). Under such conditions, parcels of warm, moisture-laden air close to the sea surface can begin to rise rapidly, causing moisture to condense and forming a convective cell of circulating air. If such conditions persist, an air pressure gradient forms that causes air to be drawn rapidly toward the center of the convective cell, generating increasingly faster wind speeds. The rotation of the earth (known as the Coriolis effect) causes the air to begin rotating around what emerges as the eye of the storm, and giving cyclonic storms their unmistakable appearance (Tory and Frank 2010). Cyclones tend not to form immediately at the equator, where the Coriolis effect is relatively weak, but in adjacent areas where sea surface temperatures are sufficiently warm (Terry 2007). Once formed, the cyclonic storm will continue to grow and persist so long as the key inputs of moisture and heat are available, which is why they lose strength as they travel out of lower latitudes to cooler waters or pass over land.

Tropical cyclones can travel across long distances. The rate at which they travel varies, and the longer a cyclone lingers in a given area, the larger the volume of precipitation that will be received, increasing the risk of flooding and landslides. The chances of being in the path of a given cyclone decrease with distance from the location where it is generated. Even within cyclone-prone regions, certain locations are more likely than others to experience one. For example, Pielke et al. (2003) suggest that in the Caribbean region, southernmost Florida has the highest chance of experiencing a hurricane in a given year (15%), while Trinidad & Tobago and the Caribbean coast from Venezuela through to Nicaragua have a much lower annual

Box 2.5
Tropical Cyclones: Nomenclature and Categorization

In its early stages of formation, what eventually becomes a tropical cyclone consists of a low pressure area of cloud systems, thunderstorms, and strengthening winds that take on a characteristic circular pattern; meteorologists classify these early-stage storms as "tropical depressions." Once maximum sustained wind speeds exceed 63 km/h (NOAA 2023), the depression becomes officially classified as a "tropical storm" or "cyclonic storm" and a name may be assigned to it. The storm becomes classified as a "tropical cyclone" once maximum sustained winds reach 119 km/h.

Tropical cyclones go by different names depending on the ocean basin in which they occur. Those that form in the Indian Ocean are called Cyclones, those that occur in the north-western Pacific and South China Sea are known as Typhoons (or Super Typhoons if wind speeds exceed 240 km/h), and those that occur in the North Atlantic Ocean, Caribbean Sea, the Gulf of Mexico, the Northeast Pacific Ocean east of the international dateline, and the South Pacific Ocean east of 160°E are called Hurricanes (NOAA 2023). Starting in the 1950s, tropical cyclones have been given distinctive names to help simplify communications about them. This is particularly useful given that in a busy storm season, there may be multiple cyclones active in the same region at the same time. The WMO oversees the naming protocols, with regional WMO offices selecting the naming practices to be used (World WMO 2023). For the Atlantic/Caribbean region, the standard practice is to give the first cyclone to form each year a distinctive name starting with the letter A, the second cyclone is named with the letter B, and so forth. Countries that border the cyclone-prone areas of the Pacific and Indian oceans have contributed lists of culturally appropriate names that are assigned on a rotating basis. The same name is never used in multiple regions, and names of storms that have severe human impacts are retired immediately afterward.

Named storms and cyclones are sub-classified by their maximum sustained wind speeds. Hurricanes are sub-classified according to the 5-point Saffir–Simpson scale, with a Category 1 storm having maximum sustained wind speeds of 119–153 km/h, a Category 5 storm having wind speeds exceeding 252 km/h, and Categories 2–3 and 3–4 separated at 177 and 208 km/h, respectively. Typhoons and Cyclones are categorized using different scales that vary according to national and regional meteorological agencies. For example, the Australian meteorological agency also uses a 5-point sub-classification scale, but with different break points (e.g. a Category 5 storm has wind speeds greater than 280 km/h). The Japanese, Chinese, and Philippines meteorological agencies use 3-point sub-classification scales for Typhoons that have small variations across countries, while the Indian meteorological agency uses a 7-point scale that ranges from tropical "depressions" to "super cyclonic storms."

Regardless of the classification and naming system used, the risks posed to human well-being by tropical depressions, storms, and cyclones are similar across regions (Figure 2.4). Although the focus in this chapter is on tropical cyclones, tropical storms, and depressions can prove to be just as problematic for coastal populations,

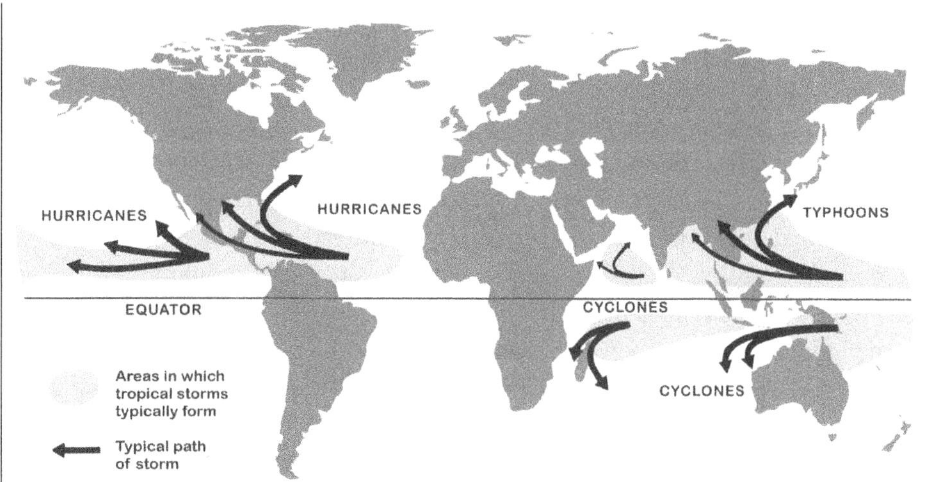

Figure 2.4 Map of regions exposed to tropical cyclones and common storm paths. Image from NASA (2024). https://spaceplace.nasa.gov/hurricanes/en/.

depending on the circumstances. Not all tropical storms or cyclones make landfall before their passage over cold surface water slows their wind speeds, but the remnants can still deliver large quantities of precipitation and gusty winds to coastal areas that cause damage to buildings and infrastructure and can lead to widespread injuries and loss of life. As sustained wind speeds approach 200 km/h, the amount of damage to buildings and infrastructure tends to be catastrophic, and mature trees will be broken or uprooted, regardless of whether it is called a Cyclone, Hurricane, or Typhoon.

risk. Along the US Caribbean–Atlantic coast, the frequency of hurricane strikes varies from one every 2 years along the North Carolina outer banks to one every 3 years along the US Gulf of Mexico coast from Texas to Florida (Keim et al. 2007). Tropical cyclones remain dangerous even as they move inland and/or away from the tropics and weaken, and can still deliver much stronger winds and higher volumes of precipitation than more frequently occurring convective storms.

From the perspective of communities that experience them, tropical cyclones are three extreme events packaged as one:

- high winds that damage the roofs and windows of buildings and knock down trees and power lines;
- large amounts of precipitation that cause wetlands and waterways to flood, and which can trigger mudslides and landslides on steep slopes;
- a "storm surge" of ocean water of several meters or more that builds up in front of the storm and floods inland areas when the cyclone makes landfall. Not only does it force seawater inland, it prevents rain that falls on surrounding inland areas from draining into the ocean until the storm has passed.

The migration and displacement patterns that emerge in the wake of a specific tropical cyclone will vary in size, scale, and duration for reasons related to exposure, vulnerability, institutional response, and household capacity as described in previous sections of this chapter. However, cyclone-related migration and displacement events tend to follow a common, multistage pattern. The first stage occurs in the days immediately preceding a cyclone when authorities may issue evacuation alerts and orders for areas in the path of the storm as it becomes known. Not all people are willing or able to obey evacuation orders; for example, cell phone data suggest that one-third of people living in areas of Texas subject to mandatory evacuation orders issued for Hurricane Harvey did not leave (Marzuoli and Liu 2018). The extent to which people obey evacuation orders depends upon a variety of factors, including the type of homes in which people live, the clarity and timing of instructions given by officials, the age and physical health of individuals, their socio-economic status, access to transportation, shelter options available to them, and their past experience with cyclones and cyclone evacuations (Tanim et al. 2022, Pham et al. 2020, Younes et al. 2021). Generally, low-income people, those who have poor access to transportation, older individuals, and those in poor health may have greater difficulty evacuating. Families with young children are more likely to evacuate than are other groups (Thompson et al. 2017). People who occupy less resilient housing structures, such as mobile homes or self-built housing, tend to travel shorter distances when they evacuate as compared with people that occupy better constructed homes. During the COVID-19 pandemic, the risk of being forced to shelter with large numbers of other people influenced the evacuation decisions of some people (Collins et al. 2021). Pet ownership has also been identified as a significant factor that influences evacuation decisions (Bowser and Cutter 2015).

Evacuees tend to move no farther than is perceived as being necessary, with hotels and homes of friends or relatives living outside the evacuation zone being preferred destinations. Those who do not evacuate prior to the cyclone may need to be evacuated afterward with the assistance of authorities should the area remain flooded for an extended period of time, as was the case of New Orleans following Hurricane Katrina in 2005 (see Box 2.6). Although typically temporary in nature, people who evacuate or are ordered to evacuate prior to a cyclone, as well as people who receive assistance in relocating after the event, are often counted as displaced people in national and international disaster monitoring statistics.

A period of temporary displacement immediately following the event is the second stage of the displacement and migration process associated with tropical cyclones. Its duration depends heavily on the extent of damage to infrastructure, homes, and livelihood assets. This can be as little as a day or two following the passing of the storm, but in the case of Hurricane Katrina, it took nearly 7 weeks to drain the flood waters from the city of New Orleans. As soon as it is safe to do so, emergency crews enter the affected area to assess the damage, identify locations of particular risk (such as fallen live electrical lines and unsafe buildings), and issue updates for those who evacuated. Once evacuation orders have been lifted (and sometimes even before), members of evacuated households will return to learn the extent of their individual losses. Where it is possible to repair (if necessary) and re-occupy their homes, people will typically do so. The timing of return of all members of the household will typically depend upon the extent of the damage to the home and the speed with which it can be repaired, as well as the speed with which authorities are able to repair/restart critical infrastructure that may have been damaged during the storm, including water and sanitation systems, electrical power, and roads. The reopening of critical services such as food stores, schools, and health services is also important. Low-income households may need to return to damaged homes more rapidly than higher income people who can afford the financial costs of remaining outside the affected area for a longer period (Lee et al. 2022). For households that intend to resume living in the affected area, the subsequent time period is referred to as the "recovery" period (Siebeneck et al. 2020), which may last weeks, months, or even years. DeWaard et al. (2019) analyzed consumer credit data for residents of New Orleans, Houston, and Puerto Rico for the years immediately prior and immediately after hurricanes Katrina, Harvey, and Maria, respectively. They found that in the case of New Orleans and Houston, the majority of people who evacuated returned within a year, though significant numbers did not resettle in the same neighborhoods as before. In the case of Puerto Rico, return migration in the year following Hurricane Maria was much lower than the other two locations (see case study that follows).

The third stage of the displacement/migration process emerges during the recovery period, and consists of (1) evacuated households or individuals that do not return to the affected area and (2) households that return but then decide they are no longer able or willing to live in the affected area. A particularly severe cyclone may be followed by a period of weeks or months during which large numbers of people move out of the affected area in search of more permanent shelter and/or employment opportunities. The direction of migrant outflows is typically guided by the geographical distribution of social networks and access to labor market opportunities. For example, in the wake of Hurricane Mitch in 1998, hundreds of thousands of Central Americans in affected areas moved in search of work. These included rural people moving to urban areas within their home countries, people moving from impacted areas to unaffected neighboring countries (e.g. Nicaraguans to Costa Rica) and to the US (particularly Guatemalans and Hondurans) (Kugler and Yuksel 2008, Funkhouser 2009, McLeman 2014). Those who relocate for an indefinite period may eventually return to their homes while others may never return. As is shown in the case study of Hurricane Maria in what follows, it may take years before population movements in and out of the affected region stabilize.

Box 2.6

Lessons from Hurricane Katrina

Much of what we presently know about migration and displacement associated with tropical cyclones comes from research done following Hurricane Katrina, a Category 3 storm that made landfall along the US Gulf of Mexico coast in late August 2005. A massive amount of precipitation and a large storm surge flooded coastal Louisiana and Mississippi and overwhelmed the flood defenses protecting the city of New Orleans. The low-lying city remained flooded for nearly 7 weeks, drainage efforts being delayed by the nearby passage of another hurricane 3 weeks later (Jonkman et al. 2009, Graumann et al. 2006). An estimated 70% of the city's population was able to evacuate prior to the storm, with the remainder needing to be relocated with the assistance of authorities in a chaotic effort, many being transported by buses to shelters in Houston, 550 km distant (Elliott and Pais 2006).

Rates of return to New Orleans peaked approximately 4 months after the storm and then slowed, with researchers noting that the longer evacuees remained away from New Orleans, the less likely they were to return (Fussell et al. 2010). Low-income Black residents were much less likely to return to New Orleans than other evacuees. Within 3 months of Katrina, half of white evacuees had returned, but even after the passage of a full year less than half of Black residents had returned (Fussell et al. 2010). Research showed that these patterns in return rates were strongly associated with damage to housing stocks, home ownership, and the pattern of infrastructure reconstruction. Low-income Black residents were more likely to live in low-lying neighborhoods that were more badly flooded during the storm and more likely to occupy rental housing and/or houses that could not be rebuilt and/or were uninsured (Finch et al. 2010, Groen and Polivka 2010). Predominantly Black neighborhoods were also among the last to see their basic infrastructure repaired, some waiting more than a year to receive municipal water and electricity (Green et al. 2007). The subsequent lack of housing stock resulted in a rise in rents, further preventing the return of low-income residents to the city (Nigg et al. 2006).

Many of those that did not return to the city of New Orleans eventually resettled in surrounding areas of Louisiana and the Gulf coast, an exception being that many of those who were relocated to Houston with government assistance remained there (Curtis et al. 2015, Eyer et al. 2018). In addition to patterns of return and non-return of former residents, researchers have also noted there was a post-Katrina influx of Latino migrants to New Orleans attracted by work in the construction sector, creating a new community of migrants within the city that had not resided there previously (Fussell 2018). As a whole, the case of Katrina emphasizes the role in post-cyclone migration and displacement patterns of previously understudied factors including socio-economic inequality, institutional competence, structural racism, housing stocks, and labor markets.

2.11 Case Study: Migration from Puerto Rico to the Continental US Following Hurricane Maria

2017 was an especially bad year for Puerto Rico. On September 6 of that year, Hurricane Irma passed within 60 miles (100 km) of the Caribbean island, bringing strong winds and heavy rains that knocked down electrical lines and caused flooding and landslides along the north coast. A Category 5 storm on the Saffir–Simpson scale, the storm left an estimated 1.1 million people without power at least temporarily, and another 360,000 were left without water (FEMA 2018). Two weeks later, as the recovery from Irma was still getting underway, the island was hit directly by Hurricane Maria.

Maria was one of the most intense Atlantic hurricanes ever recorded (Box 2.7), with wind speeds that alternated between Categories 4 and 5 on the Saffir–Simpson scale. The storm developed in mid-September, 2017, and its path took it directly across the islands of Dominica, Guadeloupe, the US Virgin Islands, and Puerto Rico, and it passed close enough to Hispaniola and the outer banks of North Carolina to cause extensive damage due to heavy rains. The worst impacts in terms of lost lives and property damage occurred when the storm made landfall in Puerto Rico on September 20. The island's weather radar station, which was designed to withstand winds of up to 133 mph, was destroyed in the storm (Scott 2018). Maximum sustained wind speeds were estimated at 155 mph (249 km/h) (FEMA 2018). A storm surge of between 6 and 10 feet (1.8–2.7 m) was experienced along the east and north-east coasts of the island. Heavy rains of between 380 to 500 mm over a 48-h period led to extensive flooding in most river valleys on the island and triggered widespread slope failures and landslides. Virtually all of the island's electrical grid and cellular telephone network that had not been lost during Hurricane Irma was destroyed by Hurricane Maria, as were most of the island's roads and municipal water systems (FEMA 2018). The official death count published a year after the storm placed the number of deaths at approximately 3,000, although many researchers suspect the actual number to be much higher (Arnold 2019). Over 80% of the value of the island's agricultural production was lost, and roughly half the island's small farms were no longer in operation the following year (Kenner et al. 2023). Hundreds of thousands of homes were destroyed or required extensive repairs, their susceptibility to damage likely amplified by the fact that many were informally constructed and their owners unaware of how to make them resilient to tropical storms (FEMA 2018). Damage to schools, hospitals, and other public buildings was also extensive, and the final financial cost of the storm was estimated to approach US$90 billion (Pokhrel et al. 2021).

Recovery and reconstruction from the combined impacts of the two hurricanes was slow. By the end of October 2017 less than 10% of roads were usable, and it would not be until the end of the year before most schools reopened. The speed and scale of federal emergency management assistance provided to Puerto Rico was quantifiably slower and smaller than assistance provided to the states of Texas and Florida (Willison et al. 2019), which many Puerto Ricans and scholars attribute to indifference and structural racism on the part of the US government (Rodriguez-Díaz and Lewellen-Williams 2020, García-López 2018, Robinson et al. 2023). Because Puerto Rico is a "Commonwealth" (i.e. a colony) of the United States, it has never had full control over its own political destiny or finances as do US states. The storms effectively bankrupted Puerto Rico, whose government was required to apply for reconstruction funding through a US Congressional agency. In October 2017, the federal government promised short-term financial relief of US$4.9 billion, primarily in the form of loans (Brown et al. 2018). The following year an additional US$63 billion was promised, to be delivered in piecemeal fashion through a variety of federal agencies and programs (Marxuach 2021). By mid-2021, less than one-third of the promised funding had been delivered, and in November 2022 – a full 5 years after Maria – the US Government Accountability Office acknowledged that most work completed to date consisted mainly of debris removal, and that much of the damaged infrastructure had yet to be replaced, blaming the delays on "unique challenges" including bureaucratic slowness and a lack of institutional capacity (US GAO 2022).

Two distinct migration patterns emerged in the wake of Hurricane Maria. The first was a large-scale movement of displaced rural workers and small farmers from the countryside to cities seeking work and shelter (Acosta et al. 2020, Kenner et al. 2023). The second was a large-scale movement of people to the continental US. Unlike residents of other Caribbean islands, residents of Puerto Rico and the US Virgin Islands are able to travel to the mainland United States without restriction, and in the 3 months immediately after Maria an estimated 114,000–230,000 people left, most destined for the states of Florida and New York, which hosted large pre-existing Puerto Rican communities (DeWaard et al. 2020). The range of estimates reflects the variety of data sets used by different researchers, including American Community Survey data from the US Census office, US Postal Service change of address applications, FEMA application address data, school enrollment data, social media posts, airline passenger data, and mailing address changes recorded in Consumer Credit Panel data (the system that tracks the credit rating scores of American adults). Even 2 years after the storm, out-migration from Puerto Rico remained considerably higher than pre-storm levels, with return migration being a relatively low 13% (DeWaard et al. 2020). This is almost certainly attributable to the lethargic pace of rebuilding efforts, continued lack of basic infrastructure across much of the island, and depressed economic prospects for residents (Santos-Lozada et al. 2020).

The case of Hurricane Maria is a good example of how migration becomes a household response in the absence of institutional capacity to:

- prepare for extreme weather events in an area known to be at high risk, including ensuring that critical infrastructure is resilient to damage from floods, high winds, and landslides and promoting hurricane resilient buildings in residential construction; and
- lead a rapid post-hurricane recovery and reconstruction effort, and provide adequate assistance to families whose livelihoods and/or housing was damaged or lost.

It further highlights how colonized peoples are made more vulnerable by the political and economic structures imposed by the colonizing power, with migration providing one of the few outlets for people to escape such dynamics. Indeed, the ability to migrate freely to the US represents one of the few beneficial aspects of Puerto Rico's relationship with it in terms of hurricane recovery.

A useful point of comparison is the experience of Dominica, a smaller and less populous independent island state that was also badly hit by Hurricane Maria. There, 95% of housing was destroyed, and 2,000 people from a population of 72,000 were living in shelters a month after the storm. As in Puerto Rico, there were high levels of internal displacement, but migration out of the country flowed primarily to neighboring islands such as Guadeloupe (Cloos et al. 2023). Unlike Puerto Ricans, citizens of Dominica require a visa to travel to the US. Net migration rates (i.e. comparison of migration in and out of the country) showed minimal change in the years following Maria (World Bank 2023). A coalition of multilateral organizations (including the World Bank, the Caribbean Development Bank, and the International Organization for Migration), bilateral development agencies, and humanitarian organizations stepped in to provide relief and reconstruction assistance. In collaborating with these agencies, the Dominican government focused on ensuring that rebuilt homes and infrastructure are more climate resilient than those they replaced. Following the initial rebuilding effort, the island's government launched a 10-year Climate Resilience and Recovery Plan aimed at reducing vulnerability to extreme weather in the future (Robinson and Butchart 2022). Initial implementation of this plan was delayed by the COVID-19 pandemic, and an ongoing challenge is to secure additional funding from international donors. That said, Dominica's progress in recovering from the impacts of Maria appears to be relatively more advanced than it is in Puerto Rico and highlights the fact that household migration is not a substitute for institutional engagement in post-disaster recovery.

Box 2.7

**Is Anthropogenic Warming Causing the Frequency
and/or Intensity of Tropical Cyclones to Change?**

A key question among hurricane researchers is the extent to which the frequency, severity, and/or spatial distribution and tracks of tropical cyclones may be affected by anthropogenic climate change. The general factors associated with the genesis of tropical cyclones are well known, but the specific interactions and circumstances that lead to formation of a cyclone at a specific time and place are complex and difficult to predict. It is known, for example, that sea surface temperatures warmer than 26.5°C are a necessary precondition for the formation of the characteristic circularity of a cyclonic storm. However, a range of other factors including wind shear, the subsurface depth of warm water, and the pre-existence of atmospheric disturbances and thunderstorms are key ingredients in the genesis of cyclones (NOAA 2023). The eventual size, severity, and duration of the cyclone are in turn influenced by the storm's track and the temperature of the water over which it passes, with storms weakening as they pass over cold water and/or land.

Average global sea surface temperatures are estimated to have risen by approximately 0.8°C (1.5°F) since the 1880s, with warming in recent decades appearing to accelerate (US EPA and OAR 2016). As oceans warm due to absorption of additional heat from a warming atmosphere, there is likely to be an expansion of the area of ocean and duration of the annual period in which sea surface temperatures are above the critical 26.5°C temperature threshold (Wu et al. 2022). Over the last 40 years, researchers have observed increased frequency of cyclones in the North Atlantic and central Pacific, which they associate with anthropogenic warming, and project that such trends will continue in the future (Murakami et al. 2020). Tropical cyclone intensity as measured by maximum sustained wind speeds and the amount of associated precipitation is directly linked to sea surface temperatures as well – the warmer the ocean, the higher potential wind speed and precipitation – and climate models suggest these attributes of cyclones will increase in coming decades in a warmer climate (Knutson et al. 2021).

A challenge in understanding current and historical links between anthropogenic warming is that reliable measurements of the frequency and characteristics of tropical cyclones go back only so far as the 1950s. This relatively limited dataset makes it difficult to disentangle the effects of natural variations in sea surface temperatures from the influence of anthropogenic warming (Wu et al. 2022). That said, researchers have in recent years worked to isolate the anthropogenic contribution to features of specific storms, including Hurricanes Irma and Maria. These studies suggest that the extreme levels of precipitation experienced in Puerto Rico during Maria were likely exacerbated by broader trends in atmospheric and sea surface temperatures associated with anthropogenic warming (Patricola and Wehner 2018, Keellings and Ayala 2019)

2.12 Migration Patterns Observed in the Wake of Tornadoes and Derechos

Severe wind storms other than tropical cyclones can occur for a variety of meteorological reasons and can lead to a wide range of impacts on human health, well-being, and, in some instances, lead to temporary or permanent displacement (Goldman et al. 2014). Tornadoes and derechos are examples of especially destructive, sudden onset high wind events (i.e. maximum wind speeds exceeding 100 km/h) that are associated with severe thunderstorms that form through convection, and are often accompanied by heavy rains or hail. Unlike cyclones, they can form in regions outside the tropics. Migration responses to them are much less well studied than those in response to cyclones. The factors that determine displacement, return, and migration following tornadoes share many similarities with those that affect cyclonic storms, but there can be nuances.

Tornadoes can occur on every continent (Figure 2.5), but the vast majority each year (approximately 1,200) form in the midwestern and southern USA, particularly in an area known as "tornado alley" (NOAA 2023). Although tornadoes can form in any month of the year, in the US they are most common between April and June. The countries of Bangladesh and Argentina have the next highest occurrence of tornadoes, although the annual average number in each of these is less than 1/100th of the US. Tornadoes most often form in association with large thunderstorms known as "supercells" that develop a circularly rotating updraft and

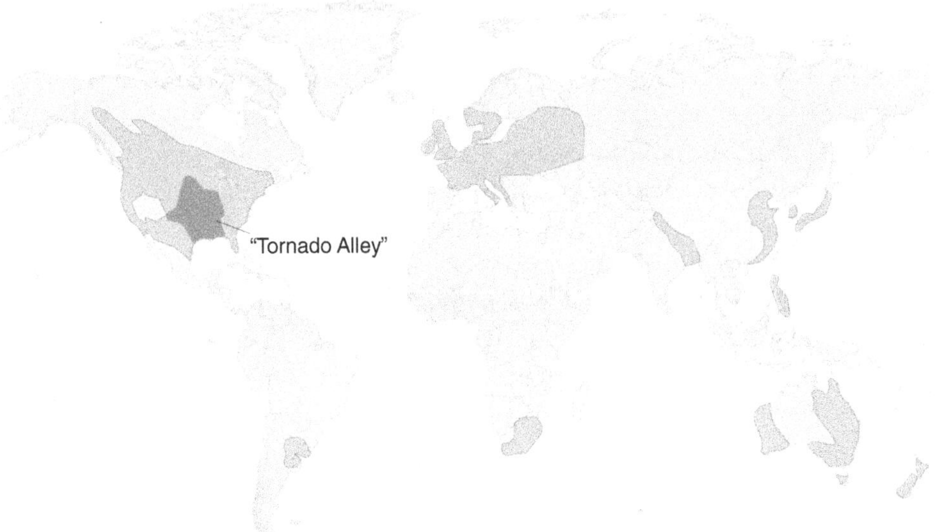

Figure 2.5   Shaded areas indicate those where tornadoes can form, with US "Tornado Alley" shown in darkest shading. Data source: NOAA.

take on the shape of an anvil as the storm grows vertically. The climatic conditions particular to the central US create ideal conditions for supercells to form, as warm, moisture-laden air flowing north eastward from the Gulf of Mexico encounters cool, dry air flowing eastward over the Rocky Mountains or southward from Canada. The mixing of these different types of air creates the potential for steep, vertical wind gradients that can spawn tornadoes. Not all supercells will generate tornadoes, and it is difficult to predict with any certainty whether a particular storm cell will indeed do so. Tornadoes can also form in conjunction with thunderstorms that are not supercells, though they tend to be smaller and shorter lived. Tornadoes can also form in association with the landfall of a tropical cyclone, creating a compound hazard (Nowotarski et al. 2021). The specific duration and path of a given tornado, and the total number of tornadoes that form in association with a particular storm cell, are almost impossible to predict in advance. As a result, unlike tropical cyclones, authorities are typically not able to give people more than a few hours warning of the genesis of tornadoes in a specific area, making proactive evacuation almost impossible. People who live in areas exposed to tornadoes are therefore typically counseled to shelter in place, ideally in a "safe room" where they are protected from flying glass and debris. The need to shelter in place puts people who live in trailer homes and poor quality housing at a higher risk than other people, because such homes contain no safe areas, are often located distantly from public tornado shelters, and are often located far from emergency responders (Strader et al. 2019).

Tornadoes range in size and duration, from narrow ropes of swirling winds no more than a few dozen meters across, to large wedge-shaped storms that create a path of damage several kilometers wide (US National Weather Service 2023). A given tornado may fluctuate in size throughout its lifespan, which can in turn range from a few seconds to a few hours, with the average duration being 5–10 min. The lateral travel speed of tornadoes also varies, with some being virtually stationary while others travel at speeds approaching 100 km/h. The internal wind speeds of the most violent tornadoes can exceed 450 km/h, an intensity that destroys essentially everything in their path. The greatest risks to lives, buildings, and infrastructure associated with tornadoes is flying debris, as broken trees, vehicles, and material from damaged buildings picked up by the winds become missiles that are thrown against other structures and trees. Given the close proximity of buildings, the potential for property damage and loss of life is especially great in urban and suburban areas.

Derechos are strong winds that form as part of a wide, bow-shaped band of thunderstorms. Unlike tornadoes, the winds do not flow in a circular pattern, but instead form a wide, lateral moving current of strong winds. As with tornadoes, derechos tend to occur most commonly in the central and eastern parts of North America. Although the maximum wind speeds of derechos do not approach those of the strongest tornadoes, they are comparable to those of a Category 1 hurricane and have comparable capacity to destroy buildings and topple large trees. A derecho creates a much larger area of damage than does a tornado, with damage paths being over 100 km wide and 450 km in length (Corfidi et al. 2024).

Although there is considerable uncertainty with respect to trends in the frequency and severity of tornadoes, derechos, and other wind storms (Box 2.8), the number of people and housing units in exposed areas of the US, especially in higher risk urban and suburban areas, is rising (Rosencrants and Ashley 2015). It is consequently reasonable to assume that the number of people adversely affected by such events, and therefore potentially at risk of temporary or permanent displacement, is likely to grow in coming decades.

Box 2.8

Are Tornadoes and Other Wind Storms Becoming More Frequent Due to Climate Change?

The effects of climate change on the frequency, severity, and/or geographical distribution of tornadoes, derechos, and other non-cyclonic wind events are not well known. Key limitations are a paucity of data about the historical frequency of such events prior to the 1980s and uncertainty as to the specific atmospheric processes that determine the formation (or non-formation) of such events. Since 1979 there has been no significant change in the frequency of tornadoes in the US, although there appears to be a slightly eastward shift in the areas where the greatest number of annual tornado events occur, as well as a slight shift in the months in which the most storms occur (Gensini and Brooks 2018). Between 2000 and 2019, the geographical area of the US experiencing tornadoes appears to have expanded (Summers et al. 2022). The reasons for such shifts are not known – it may be part of a longer-term, natural variation that cannot be observed given the lack of historical data, or it may be because of atmospheric changes that in turn may be natural or anthropogenic in nature. Another challenge is under-reporting of tornadoes, with those that occur in rural areas being less frequently reported and counted than those that approach more densely populated areas (Potvin et al. 2022).

Tornadoes and derechos are associated with severe thunderstorm events that form through convection – that is the rising of warm moist air from the surface to higher altitudes where the air cools and water vapor condenses. The warmer a parcel of air, the more moisture it is capable of holding. As average surface air temperatures increase due to climate change in tornado-prone areas, it is possible that in the future there will be greater potential for severe thunderstorms to develop in certain regions, such as the southeastern and midwestern US (Haberlie et al. 2022), in turn increasing the risk of tornado and/or derecho formation. However, there is still much research yet to be done to determine precisely how climate change will affect the frequency and severity of such events (Gensini 2021).

Research conducted in the US suggests that severe tornadoes tend to have no long-term effect on the net population of affected areas, but that the demographic characteristics of affected areas may change during the rebuilding and reconstruction process. Low-income residents are at risk of being pushed out of areas experiencing damage – for socio-economic reasons similar to those described in the Hurricane Katrina box above – with those who have the opportunity to rebuild tending to be higher income homeowners, particularly those with private insurance (Raker 2020). In areas that are already socio-economically disadvantaged, tornado damage can exacerbate poverty and inequality should higher- and middle-income households not rebuild but resettle elsewhere (Kashian et al. 2021). People living in trailer homes often lack insurance and lack the financial means to move elsewhere, and the parks in which trailers are sited are often situated in locations that are more hazard-prone than other types of housing (Otsuyama et al. 2021, Rumbach et al. 2020). Immigrants – who are often disproportionately members of low-income populations – may face particular challenges in accessing government disaster recovery resources for a variety of social and informational barriers (Hamideh and Sen 2022). Studies of Moore, Oklahoma – a city of approximately 55,000 people that was hit by tornadoes on three occasions between 1999 and 2013 – found that a range of factors including household level financial impacts, home ownership, the nature of damage to infrastructure, and access to government assistance were important considerations for people deciding whether to rebuild or relocate elsewhere, and that the relative importance of such factors varied by age and family size (Nejat et al. 2018, Mayer et al. 2020). Most residents chose to rebuild and remain in Moore, with place attachment being a key factor in such decisions (Nejat et al. 2018, Greer et al. 2020 [Box 2.9]).

Because tornadoes are often localized in their impacts, media attention tends to wane rapidly, and those displaced are often soon forgotten. When hundreds of thousands of people must be evacuated due to a tropical cyclone, as was the case with Hurricane Katrina, it becomes national news and remains in the headlines for months and years. By contrast, when a small town or city is damaged, such as Moore, Oklahoma, it may be in the news for no more than a day or two. As a consequence, government assistance for recovery efforts following tornadoes may be more limited than for tropical cyclones, floods and other disasters that have much wider spread and better known impacts (Hamideh and Sen 2022). Low-income people, particularly those living in trailer parks, are often perceived as being transient and are consequently stigmatized by local authorities who dedicate fewer resources to helping them recover (Rumbach et al. 2020). An inattention to tornado-related displacements also exists in scholarly research. The relatively limited amount of research on post-tornado displacement and migration decision-making presents an opportunity for future researchers looking to advance our knowledge in this area specifically, and in environmental migration scholarship more generally.

Box 2.9

The Importance of Place Attachment in Post-Disaster Decision-Making

The case of Moore, Oklahoma – a city that has experienced tornadoes on at least 23 separate occasions since 1893 (US National Weather Service 2023), and where 24 people were killed and more than 200 injured in a 2013 tornado – raises the question of why anyone would choose to live there and/or rebuild after the latest tornado, knowing that other tornadoes will inevitably come. Similarly, why do people continue to build, rebuild, and live in coastal Louisiana when that area has experienced dozens of major hurricanes over the last 300 years? Similar questions may be asked of any number of places around the world where people have lived through storms, floods, and other hazards and willingly rebuilt in the same place, knowing that the risk of future displacement is the same as it was before the previous disaster. In climate change and migration research, this unwillingness to move away from hazardous locations is sometimes known as *voluntary immobility* and distinguishes such groups from people that might want to move away from hazardous locations but lack the means to do so (also known as *involuntarily immobile* people (see discussion in Chapter 1)).

An important factor that helps explain the paradox of voluntary immobility is *place attachment*, a natural psychological disposition of people to form strong emotional bonds to the places where they live, and which may be manifest in a variety of symbolic, spiritual, and tangible ways (Swapan and Sadeque 2021). For example, surveys of survivors of Hurricane Katrina found that 84% of residents of the city's Ninth Ward neighborhood – one of the poorer and most badly damaged parts of the city – who returned to live there after the storm stated they did so because there is no other place like New Orleans (Chamlee-Wright and Storr 2009). The specific examples given by respondents included such things as food, music, culture, community spirit, neighborliness, kinship networks, and other non-economic factors specific to the city, and these differentiated the experience of living there from other cities, even those elsewhere in Louisiana.

Measuring place attachment is not an easy task for the researcher, and there are methodological challenges not only in its identification (typically done using tools such as interviews, focus groups, and surveys), but also in attempts to quantify its effects, and to represent its relative presence, absence or abundance in a spatial sense (i.e. what is the level of place attachment in place A versus place B?) (Hernández et al. 2020, Brown et al. 2015). Nonetheless, understanding place attachment is an important consideration when exploring the distinctions between those who migrate and those who do not in a changing climate.

2.13 Migration Patterns Observed in the Wake of Floods

A "flood" can be defined as any event where water flows onto land that is not ordinarily under water (Ward 1978). Floods can occur when watercourses overflow their usual banks and in coastal environments due to storm surges and unusually high tides. Areas adjacent to watercourses and coastal areas that have a high potential for flooding are referred to as *floodplains*. Depending on how they are defined (Box 2.4), floodplains cover somewhere between 800,000 and 2 million square kilometers of the Earth's surface (Tockner and Stanford 2002), and as many as 2 billion people may live within areas exposed to flood risks (Millennium Ecosystem Assessment 2005). A range of weather phenomena can trigger flooding along watercourses, including heavy rains within the catchment area (i.e. the area of land that drains into a given watercourse), landslides that deposit debris into the watercourse and alter its flow, and in more northern latitudes, the formation of ice jams and the rapid melting of snow. In arid and semi-arid regions, watercourses that are dry throughout much of the year (or indeed, across multiple years without precipitation) may experience flash flooding when rare precipitation events occur. This latter type of flooding poses a particular risk for settlements where residents may be unaware of the risk and local authorities do not take measures to warn them or prevent settlement in such locations (Zúñiga 2020). River flooding can also be caused by deliberate human actions, such as decisions to build a dam and flood upstream lands, or unintended consequences of human actions, such as downstream flash flooding caused by the failure of a dam or other water containment structure.

Floods are a widespread and recurrent hazard and account for roughly half of all hazard-related displacements globally each year (Figure 2.2), with China, India, Indonesia, and the US experiencing especially large numbers of floods. Asian countries generally experience over half the global number of flood events each year (Douben 2006). The Global Flood Monitoring Program (www .globalfloodmonitor.org) identified over 10,000 flood events having occurred over a 4-year period between 2014 and 2018 (de Bruijn et al. 2019). Over 80% of floods are attributable to heavy rain events, with snowmelt and rain from tropical cyclones accounting for most of the remainder (Adhikari et al. 2010) (see Box 2.10).

A key factor in exposure to flood risks is population growth and urbanization in river valleys, coastal zones, and other flood-prone locations (Hemmati et al. 2020). The number of people globally living in flood-prone areas grew by an estimated 58–86 million people between 2000 and 2015 (Tellman et al. 2021) with population growth rates since the 1980s being higher in high-flood-risk urban settlements than in less exposed ones (Rentschler et al. 2023). In other words, even in the absence of climate change, the number of people globally living in flood-prone areas – and hence at risk of displacement – is growing (Kam et al. 2021). This conclusion is borne out by research by Kakinuma et al. (2020) that sought to analyze global patterns of flood displacement at

Box 2.10

Is Climate Change Causing a Greater Frequency of Flood Events?

Heavy rainfall events are the most common cause of flooding globally. This makes it difficult to attribute observed changes in flood frequency to anthropogenic climate change because precipitation changes are challenging to measure and model, and the availability of historical data varies considerably across countries. Recent research suggests that flood frequency is increasing in some regions (e.g. central US, northern Europe, and northern Australia) and decreasing in other regions (western North America and southern Australia) (Liu et al. 2022). The IPCC notes that for regions where good observational data exist, there has likely been an increase in heavy precipitation events and that anthropogenic climate change is likely a main driver (Seneviratne et al. 2021). There is evidence that the number of short duration, high volume rainfall events that can lead to unexpected flooding has been increasing across many regions over recent decades (Wasko et al. 2021). This trend is likely attributable to anthropogenic climate change (Paik et al. 2020) and is expected to continue and intensify in coming decades (Tabari 2020). However, changes in precipitation patterns are expected to vary at regional and local scales in coming decades, with some regions (e.g. Central America and the Mediterranean region) expected to see less precipitation while others will likely see more (East Africa and southwest Asia) (IPCC 2021). Overall, the risk of flooding due to extreme precipitation events is expected to grow in most regions in a warming climate (Seneviratne et al. 2021). Kam et al. (2021) have estimated that every half degree Celsius increase in global average temperatures will increase by 50% the number of people globally at risk of displacement due to floods, with population increase in flood-prone areas further adding to the potential risk.

national levels, and found that even countries with relatively infrequent flood events – particularly countries in sub-Saharan Africa – may actually have high risks of flood-related displacements given the number of people living in exposed locations and the limited ability of governments to construct protective infrastructure.

2.14 Variability of Migration and Displacement Patterns Associated with Floods

Migration and displacement patterns related to floods vary from one region and country to another and can vary within a country or watershed as well. To better understand these variations, it is useful to make a distinction between floods that happen on a frequent or regular basis, such as those that occur as a consequence of monsoonal precipitation regimes (Box 2.11), and floods that are triggered by exceptionally high volumes of precipitation in a single, anomalous event. The former type may be referred to as "normal flooding" or "expected flooding," whilst the latter type can be described as "extreme flooding." In low- and middle-income countries where annual flooding is an expected occurrence, rural populations may deliberately incorporate temporary or seasonal labor migration during the flood season as part of a wider set of ongoing adaptations to climatic variability. Extreme floods can generate very large, temporary population displacements, especially in densely populated areas, which in turn raises the potential that those displaced will not return to their communities afterward, for reasons similar to those discussed in the previous parts of this chapter related to tropical cyclones. There is no universal threshold that determines what makes a particular flood event "extreme," but instead there is typically some level of damage to homes, buildings, livelihood assets, and/or infrastructures, which makes relocation and return difficult or impossible. Further, there may be a threshold in terms of the number of flood events that a household or community will endure before deciding to relocate permanently. For example, in rural Pakistan where annual floods are common, households will typically return and rebuild houses and buildings after a flood, but once a home has been lost to floods for a third time the likelihood of resettling permanently elsewhere increases (Ahmad and Afzal 2021).

2.14.1 Rural Migration and Displacement Due to Expected Seasonal Floods

Floodplains are inherently good places for farming. The soils are typically fertile, soil moisture content is high, and there are nearby surface water and/or relatively high underground water tables that can be tapped into for irrigation purposes. By farming in floodplains, residents are knowingly exposing themselves to risks of occasional or even annual flooding – but the benefits may outweigh the potential costs. In regions with monsoonal precipitation regimes – that is, a distinct rainy season or seasons and distinct dry season or seasons – land alongside watercourses is routinely flooded each year during the rainy season. When flood waters recede, the land is recharged with sediments and nutrients that enhance its fertility and agricultural productivity. This annual flood cycle is something to which farming households have over many generations adapted in a variety of ways, which may include temporary out-migration during flood season and subsequent return.

Seasonal migration has historically been an especially important way by which the rural poor across Asia cope with seasonal poverty and lack of food inherent in a region where rice – a crop whose cultivation is tightly linked to the monsoonal precipitation cycle – is a primary source of food, income and labor opportunities (Mobarak and Reimão 2020, Samal et al. 2022). In deltaic areas and river valleys of Bangladesh, India, Nepal, and Pakistan, individuals of working age will often migrate from rural households during floods (as well as in periods during the growing season when the crops require little attention) to nearby urban centers in search of wage labor opportunities, remitting money back to those who remain behind, and returning when their labor is needed at home (Maharjan et al. 2020, Jamshed et al. 2020, Ahmad and Afzal 2021). In most communities, it is young men that are likely to participate in these cyclical migrations, but depending on local socio-cultural norms, working age women may do so as well. The seasonal outflow of working age men creates significant economic, workload, health, and well-being challenges for the women, children, and older adults left behind (Ahmed and Eklund 2021). In some watersheds of South Asia, up to one-third of rural households practice some form of seasonal migration by one or more members on an annual basis, typically those that are more impoverished and that have weak access to government support (Maharjan et al. 2021). Practiced in this way, migration forms part of a household level income diversification and risk-management strategy, described in additional detail in Chapter 1.

However, it is important to note that not all labor migration out of rural Asian floodplain regions is motivated solely by flood adaptation, not all is seasonal in nature, and not all is internal. Even in areas such as rural Bangladesh, where seasonal flooding is an annual phenomenon that has influenced seasonal migration patterns of rural laborers for generations, it is not the only factor that shapes who moves and why (Al-Maruf et al. 2022). Increasingly, a key driver of rural-out migration is poverty itself and the desire to overcome it, with rural migrants traveling longer distances and staying away for longer periods of time in search of wages. In Taiwan, farm workers increasingly come from Indonesia, Vietnam, and other less wealthy Southeast Asian countries, staying for years at a time as either legal guest workers or undocumented workers (Cockel et al. 2023). Jobs in the Middle East and in Malaysia's palm oil plantations attract large numbers of migrants from flood-prone areas of rural Nepal and Indonesia (Sunam et al. 2021). Rainy season migration is not solely an Asian phenomenon; examples are also found in rural areas of Ethiopia (Baylie and Fogarassy 2022) and Ghana (Rademacher-Schulz et al. 2014, Warner and Afifi 2014), among others.

Box 2.11
What Is a Monsoon?

The term *monsoon* refers to seasonal changes in precipitation patterns associated with changes in prevailing wind directions. There are different types of monsoons and driving mechanisms, with the most pronounced monsoons being driven by the annual north–south movement of the intertropical convergence zone (ITCZ) (UCAR 2023). The ITCZ describes an area where the Sun is at its most perpendicular to the Earth, with its location drifting north of the equator during the northern hemisphere's summer and south of the equator during the southern hemisphere's summer. The heat generated at the ITCZ causes warm air to rise and flow away from it. That air will eventually cool and sink to the surface, where it then flows back toward the ITCZ, creating a circular pattern of air flow known as a Hadley cell. In areas where the ITCZ is over the ocean, moisture is drawn up and outward with the rising air, and condenses and falls as rain once it cools. As the ITCZ moves northward in the months of June through August, it delivers large amounts of precipitation to countries in South and Southeast Asia and to countries north of the equator in Central and South America and sub-Saharan Africa. Those areas stop receiving rainfall in subsequent months, and begin experiencing their dry season as the ITCZ returns to the equator and travels farther southward, bringing monsoonal rains to Java, New Guinea, northern Australia, and countries south of the Equator in Africa and South America from December to February. The specific location of the ITCZ varies from one year to another, making some rainy seasons less or more rainy than others. In monsoonal climates, the volume of water in watercourses rises and falls in concert with seasonal precipitation patterns, with flooding being a common phenomenon during the rainy season. In Bangladesh, for example, roughly 20% of the land area exposed during the dry season will be flooded during an average rainy season, with most of the extra water flowing down from the upstream catchment area of the Ganges and Brahmaputra rivers (Monirul Qader Mirza 2002).

2.14.2 Migration and Displacement in Response to Extreme Floods

Even in watersheds where flooding is an annual and expected phenomenon, in some years flooding is so extreme that it extends to locations where flooding is rare, thereby triggering temporary displacement, or causing such severe damage to housing and infrastructure that permanent out-migration ensues (see, for example, Box 2.12). However, it should not be assumed that extreme floods will automatically have a significant effect on patterns of permanent migration. For example, in Bangladesh, extreme floods can have a modest effect in increasing permanent migration of women from affected areas, but the overall effect on migration patterns of the population as a whole is not as great as might be expected – suggesting that other factors have a significant influence (Gray and Mueller 2012, Call et al. 2017). It appears rural small farmers are less likely to migrate after extreme floods than people with higher incomes, greater wealth and/or non-farming labor skills, whose migration decisions appear to be more motivated by economic opportunity-seeking rather than loss or harm experienced from the flood event (Bernzen et al. 2019). Another key factor appears to be whether the flood event (or successive flood events) cause such severe erosion that a given farm is no longer operable, in which case indefinite relocation may become necessary (Pavel et al. 2023). The nature of flood damages sustained by rural households is also important; a loss of only crops or livestock results in temporary migration, but loss of more valuable assets increases the likelihood of permanent relocation (Joarder and Miller 2013). Research done following extreme floods that occurred in 2014 detected a 7% increase in outmigration of poor rural households to other destinations within Bangladesh, and a 3% increase in international migration on the part of higher income households (Giannelli and Canessa 2022), figures consistent with wider findings that even extreme floods have a modest effect on migration patterns in Bangladesh.

The importance of local context and non-environmental factors is seen in flood-related migration responses in other regions as well. Similar to the case of Bangladesh, research in Indonesia has found that heavy rainfall events that cause extreme floods and landslides have less of an influence on migration patterns than might be expected (Bohra-Mishra et al. 2014). In Vietnam and Cambodia, extreme floods that cause severe crop damage and/or housing damage lead to small increases in permanent out-migration from rural areas, typically to other rural areas and urban centers within the same country, particularly among farmers that own only small plots of land (Berlemann and Tran 2020, Nguyen and Sean 2021). In rural Vietnam, the occurrence of a single flood appears to have less influence than a succession of increasingly severe floods in prompting higher levels of migration out of affected areas (Berlemann and Tran 2020), but overall, migration decisions in the wake of floods appears to be more heavily influenced by economic and other non-environmental factors (Bayrak et al. 2023). In the Okavango delta of Botswana flooding is an annual risk and rural households ordinarily adapt in a variety of ways that do not require migration, but severe floods in 2004 and 2009 caused such significant damage to household assets and community infrastructure that temporary relocations ensued (Motsholapheko et al. 2011).

Extreme floods due to unusually heavy seasonal rainfall in Nigeria in 2022 displaced over 2 million people across rural and urban areas, with floodwaters taking more than a month to recede in many areas (NBS/NEMA/UNDP 2023). The most commonly reported impacts were loss of livelihood assets, damage to homes, outbreaks of infectious disease, and increased food insecurity. Six months after the floods, roughly 40% of those displaced had been unable to return to their homes, with rural households more likely to be displaced for longer periods of time than urban ones. Approximately 88% of those displaced remained in or close to their home community, and 7.6% moved to a different community but remained within the same state. Only 2.8% of those displaced moved to a different state, reinforcing a recurrent theme throughout this chapter that most flood-related displacement occurs at local scales, and that longer-distance relocation is undertaken by relatively small numbers of those displaced.

In urban areas of low-income countries, locations close to flood prone waterways and steep slopes prone to precipitation-induced mudslides are often occupied by poor people living in self-built or informal housing, including people that have arrived from rural areas. Such locations can prove to be deadly. Local authorities may attempt to remove them, but the people will often return due to poverty and lack of alternative housing options. In Brazil, where heavy precipitation events are common across large areas, urban centers often have large numbers of people living in informal settlements (favelas) built in hazardous locations, and significant numbers of deaths and displacement often follow flash flooding events (Wink Junior et al. 2023, Marengo et al. 2023). In Rio de Janeiro, nearly a quarter of the population lives in favelas, with the government attempting for decades to relocate residents of the most hazard-prone favelas with mixed success. Severe floods and mudslides in April 2010 killed 67 residents and prompted authorities to forcibly remove 100,000 favela residents and destroy their homes, but limited financial assistance was given to those who were removed, and many ended up living in other locations also exposed to floods (Barbosa and Coates 2021).

Many urban centers across Africa and Asia face similar challenges as Brazil, having large numbers of low-income people living in flood prone locations, with urban sprawl removing natural features that absorb precipitation and slow the rise of flood waters. In Manila, roughly one-quarter of the city's population of over 12 million people lives in informal housing that is flooded so often that residents have developed a vernacular to describe how they cope with different levels of water (Akyelken 2020). In Dhaka, Bangladesh, a network of canals known (locally as khals) that had been built over centuries to help drain monsoonal water and prevent the city from flooding have been mostly lost due to rapid urbanization since 1990, increasing the flood hazard (Sakib et al. 2023). Despite the growing flood hazard, hundreds of thousands of people migrate into Dhaka's low-income neighborhoods and informal settlements each year seeking economic opportunity (Akther and Ahmad 2021). The poorest of them often end up living in low lying areas on the edge of wetlands, where their homes are regularly flooded each year. In both Nigeria and Ghana, floods regularly affect rural and urban populations, but densely populated cities with their large areas of informal settlements situated in flood-prone locations experience the highest rates of flood-related displacements (Echendu 2022). For example, 2018 floods in the Ghanaian cities of Kumasi and

Accra displaced approximately 34,000 people. In some informal settlements, such as those in Nigeria's Lagos region, flooding may occur once or even twice in an average year (Lawanson et al. 2023). A common theme across these and other examples from low- and middle-income countries is that flood risks and frequent displacements do not discourage migration into flood-prone locations, meaning that non-climatic, socio-economic factors are the dominant drivers of population movements and settlement patterns.

2.15 Case Study: Pakistan Floods

The geography and climate of Pakistan makes much of the country inherently prone to flooding. From its headwaters in the Himalayan ranges of western Tibet and Kashmir, the Indus River flows southward into Pakistan and bisects the country until it empties into the Arabian Sea, and with its tributaries drains over a million square kilometers of land in the process. River water levels begin rising in the month of May due to snowmelt in the mountains, with volumes increasing rapidly with the arrival of monsoon rains in July. When extreme floods occur, it can take months for the inundated areas to fully drain, preventing farmers from growing crops and increasing the risk of water-borne and mosquito-borne diseases (Atif et al. 2021). A network of dams is used to moderate flows and store water for irrigation and other uses during the dry season, but these structures can be overwhelmed and breached during exceptionally rainy periods or when snowmelt from upstream mountainous areas is rapid or high. Along many stretches of the river, especially in its upper reaches, dams, and embankments have aggravated the risk of wider scale floods by causing sediments to collect in the main channel and raising its bed so that it can hold less water (Gaurav et al. 2011). In both rural and urban areas of Pakistan, it is disproportionately poor households that occupy the most precariously situated floodplain land (Mustafa and Wrathall 2010).

There is growing concern among Pakistan-based researchers that heavy precipitation events and extreme floods have become more commonplace in recent decades, and that the socio-economic and human impacts of flood events are also rising (Atif et al. 2021, Manzoor et al. 2022). Severe floods in 2010 due to heavier than usual monsoon rains in the upper reaches of the watershed left nearly 2,000 people dead and destroyed an estimated 1.8 million homes, with cholera and lack of clean drinking water affecting millions more (Hartmann and Andresky 2013). Follow-up studies conducted by Kirsch et al. (2012) in the most affected districts of Balochistan, Khyber Pakhtunkhwa, Punjab, and Sindh provinces provide insights into the consequent displacement and migration patterns. Roughly 55% of homes in those areas were damaged by floodwaters, nearly half of which could not be repaired. Over 85% of households had to evacuate their homes for at least 2 weeks, and nearly half stayed in an internal displacement camp during this initial evacuation period. During the following 6-month period, two-thirds of households stayed in only one place, and the remainder had to move two or more times. After 6 months, 52% of rural households and 74% of urban households were able to return to their homes. Most people stayed in the same district, even if they were not able to return to their original homes; only 12.6% of displaced urban households and 18.4% of displaced rural households moved to a different district. This again illustrates how place-specific social, cultural and economic connections tend to keep people from moving to more distant locations after floods.

The decade following the 2020 flood saw Pakistan experience a number of notable flood events, with floods of 2022 eclipsing all previous recorded flood disasters in the country. In August of 2022, large areas of Pakistan received up to five times the normal amount of precipitation, with the resulting floods displacing 32 million people and causing over US$30 billion in damage to housing stocks, agriculture, livestock, and transportation and communications infrastructure (Nanditha et al. 2023). The ensuing humanitarian crisis was likely exacerbated by a slow and uncoordinated response by national and provincial governments (Bhutta et al. 2022). Roughly two-thirds of the country was affected, and 6 months afterward an estimated 1.8 million people were still living near stagnant, standing flood waters (Shehzad 2023). Rates of malaria and waterborne illnesses increased, and many farmers were unable to plant at the usual time following the rainy season, creating widespread food insecurity. Scientists that assessed the unusual rainfall intensity in Pakistan in August 2022 have found that up to 50% of the intensity was attributable to anthropogenic climate change (Otto et al. 2023). The government of Pakistan subsequently helped lead a coalition of low-income and small island state countries at the December 2022 COP27 UN climate negotiations that successfully negotiated the creation of a dedicated fund to provide them financial support for losses and damages experienced due to climate change (Tietjen and Gopalakrishnan 2023).

2.16 Case Study: Changing Flood Risks in China's Yangtze River Basin

The Yangtze River and its tributaries drain a massive area in the central part of China, flowing eastward from mountainous headwaters in Tibet through the central part of the country and discharging into the Pacific Ocean near the city of Shanghai. Over 440 million people live in the watershed, accounting for roughly one-third of the total population of China (Zhu et al. 2020). Flooding during the June–August monsoon rainy season has always been a hazard for people living in the Yangtze basin. The frequency of floods increased over the second half of the twentieth century, largely due to land use change in the basin (Yu et al. 2009). In

Box 2.12
What Is an "Atmospheric River"?

Distinct from other types of extreme weather events described in this chapter is a phenomenon known as an "atmospheric river," which describes a long and narrow corridor of low-elevation, horizontally moving moist air that can deliver large volumes of precipitation to coastal areas (NOAA 2023). They can be accompanied by strong winds and severe temperature swings. Atmospheric rivers are generally short-lived, typically originate in the tropics, and are associated with a cold front generated ahead of a cyclonic storm moving over the ocean. The finger-shaped mass of moist air (which can be several hundred kilometers wide) delivers large amounts of rain or snow in a short period of time when it is forced upward by mountains or by colliding with a warm air mass. The resulting precipitation can be truly staggering in volume, triggering floods, and/or landslides and causing severe damage to infrastructure. A 2007 atmospheric river transporting more than fifty times the amount of water that is discharged at the mouth of the Mississippi River hit the US state of Oregon, triggering a 1 in 500-year flood event in the Chehalis River and killing 11 people (Dominguez et al. 2018). Atmospheric rivers can occur anywhere in the world and are a relatively common phenomenon, although in recent years they have become notoriously associated with especially heavy precipitation events along the west coast of North America, where they are sometimes referred to as the "Pineapple Express." As with other extreme precipitation events, there is growing concern that anthropogenic climate change may amplify the severity of atmospheric rivers by increasing the amount of precipitation they can potentially deliver (Payne et al. 2020).

A massive atmospheric river event in February 2024 hit southern California, causing flash floods and landslides in many areas, and heaping extra misery on the estimated 50,000 homeless people living in tents and self-built shelters on the streets of Los Angeles. Emergency homeless shelters report being overwhelmed by unhoused people seeking refuge, with some homeless people taking the very dangerous step of hiding from the rains in storm drains (Miller 2024). The homeless are invariably the most vulnerable members of any community to extreme weather, and as homelessness increases in many cities in high-income countries, additional humanitarian challenges will emerge. Systematic reviews of existing research on the vulnerability of homeless people to a changing climate (e.g. Bezgrebelna et al. 2021) echo themes that have appeared throughout the present chapter with respect to links between displacement risks and poverty, living in precarious locations in substandard shelters, and institutional inability to address the root causes of vulnerability. This represents an important subject for future research.

the 1990s, the government initiated the massive Three Gorges dam project in the middle reaches of the river to generate hydroelectricity and protect large urban centers downstream from flooding. In doing so, an estimated 1.1–1.35 million people had to be relocated from areas that were flooded by the reservoirs created upstream, to new farms and/or new towns constructed by the authorities (Duan and Wilmsen 2012). In many cases, these new settlements were less attractive or viable, and many people, especially better-off urbanites, decided to relocate elsewhere of their own accord in search of better jobs or better housing (Wilmsen et al. 2021).

The two most extreme flood events in the Yangtze basin in living memory occurred in 1998 – just before the Three Gorges project was completed – and in 2020, after its completion. In both cases, above-average summer monsoonal precipitation was a direct cause of the floods. The 1998 flood killed an estimated 1,300 people and left 13.2 million homeless (Ye and Glantz 2005). Although precipitation levels were higher in 2020 than in 1998, the number of deaths and value of economic losses were significantly smaller, and an estimated 3 million were temporarily displaced (Jia et al. 2022). Among a number of factors that explain the different displacement outcomes between the two flood events, a key one is the massive water storage capacity of the Three Gorges complex.

An important question emerges, for which there is no easy answer: Was the deliberate displacement and relocation of over a million people justified, even if it meant 10 million fewer people were displaced during the next extreme flood event? Many of those moved to make way for the dams found themselves impoverished and dislocated from friends, family, and community, and it is likely cold comfort to them to know that their removal from their homes spared other people from losing their homes a decade later. China is not the only country where forced relocations have been or are being made to make way for dams, other examples including populations living along Southeast Asia's Mekong River (Kura et al. 2017), at the junction of the Tokwe and Mukorsi rivers in Zimbabwe (Mavhura 2020), and the Amazon River (Mayer et al. 2020), among others. In a changing climate, where hydroelectric energy is seen as an important alternative to fossil fuels, we can expect to see more dam construction – and more dam-related population displacements – in coming decades, especially in decarbonizing low- and middle-income countries.

3

Migration and Displacement Associated with Aridity, Drought, Heat, and Wildfires

3.1 Introduction

All the climatic hazards discussed in this book – and changes in their frequency, severity, and geography – are affected by rising global average temperatures due to anthropogenic greenhouse gas (GHG) emissions. The present chapter focuses on migration and displacement associated with events that are directly linked to hotter air temperatures and/or an associated lack of moisture experienced at local and regional scales: droughts, increased aridity, desertification, heat, and wildfires. With the exception of wildfires – which share many characteristics comparable to the rapid-onset extreme weather events assessed in Chapter 2 – the hazards assessed in the present chapter are gradual in their onset and impacts. They do not cause destruction to homes or built infrastructure, and so they rarely lead to immediate, involuntary displacement. Their impacts accumulate with each passing week, month, and/or year, steadily eroding the water, food, and/or livelihood security of households and communities. The slow rate of onset allows exposed populations an opportunity to adjust and adapt through means that do not require changes to existing mobility practices and patterns, sometimes referred to as *in situ* adaptation responses. It is only after hot and/or dry conditions persist beyond a particular threshold of duration and/or severity that *in situ* adaptations no longer prove to be sufficient and changes in migration decision-making and outcomes emerge. There is thus often a lag period of months or even years between the onset of the hazard and the observed change in migration patterns. The thresholds in question vary from one household, community, and region to another, and reflect differences in vulnerability, exposure, and adaptive capacity reviewed in Chapter 1.

In high-income countries, droughts, increased aridity, desertification, and hotter temperatures tend not to have a significant influence on migration, displacement, and/or population patterns today (but wildfires do). Most present-day migration

95

associated with such hazards occurs in low- and middle-income countries, particularly among small-hold farmers and pastoralists. In most cases, the observed migration occurs within the same country or flows between adjacent countries within the same geographical region, but there can be some instances where longer distance international migration rates increase. Migration outcomes can vary considerably from one country and rural population to another, even within the same geographical region and in response to similar climatic conditions, meaning that sweeping statements need to be avoided. A common thread among the examples given in this chapter is that heat, drought, or aridity in and of itself is rarely the only factor that leads to migration; rather, it is typically the combination of these climatic conditions with non-environmental factors such as poverty, inequality, political instability, economic crises, agricultural system structures, and conflicts that leads to changes in migration behavior.

3.2 Migration Associated with Droughts, Increasing Aridity, and Desertification

3.2.1 *Important Terminology: Aridity, Dryland, Desertification, and Drought*

The terms arid (or aridity), drylands, desertification, and drought are sometimes used interchangeably, but there are important distinctions between them that are relevant for discussing migration and displacement in a changing climate. Care is taken in this chapter to use each term in its technically accurate sense. To begin, the word *arid* is used to describe land areas where precipitation and soil moisture are so scarce that the growth of vegetation is limited. In bioclimatic terms, an *arid* area is defined as one where the supply of moisture measured as average annual precipitation (P) is significantly lower than the demand for moisture, as measured by potential evapotranspiration rates (PET). Land where the moisture supply is 65% or less of the moisture demanded is known as *dryland*, and such areas represent roughly 45% of the Earth's land surface (Berdugo et al. 2020) (Box 3.1). A simple Aridity Index developed by the United Nations Environment Programme is typically used to sub-classify the degree of aridity within drylands, using the ratio of P/PET. Where P/PET is less than 0.2, the area is said to be arid, and is hyper-arid (i.e. a desert) if the ratio is below 0.03. Areas with an Aridity Index score of between 0.2 and 0.5 are classified as semi-arid areas. As is shown in Figure 3.1, large areas on all continents fall within these classifications under the Aridity Index. Upon closer examination, the reader will see that the areas indicated in Figure 3.1 include sparsely inhabited lands (such as polar regions and large deserts, such as the Sahara and Arabian deserts) as well as heavily populated areas

Figure 3.1   Map of world's arid regions. Darker colors indicate increasing aridity, from semi-arid to hyper-arid (i.e. desert) lands. Source: FAO 2024.

(such as Pakistan, western India, southern California, and the Horn of Africa). This highlights a limitation of bioclimatic measures of aridity: They are not an especially good indicator of the suitability of land for human use or settlement. For monsoonal climates, such as India and Pakistan, the Aridity Index (which uses annual precipitation values) does not reflect very well the seasonality of precipitation. Whilst land in monsoonal climates can indeed become very dry in the late months of the dry season – and vegetation will often go dormant during this period – it can then become very moist and productive once the rains arrive.

There is some debate among climate scientists whether higher average global temperatures will lead to a global geographical expansion of arid or dryland areas in coming decades, or whether changes will vary at regional scales (Roderick et al. 2015, Asadi Zarch et al. 2017, Bonfils et al. 2020). This uncertainty in part reflects the ongoing challenges in modeling how precipitation patterns are likely to respond to warming temperatures. What is well established is that, as an area of land becomes more arid, it proceeds through a stepwise progression of ecological changes, beginning with a decline in vegetative cover and plant productivity, followed by decreasing soil fertility, and finally a threshold-type shift to a desert-like environment where many previously present plant and animal species disappear entirely (Berdugo et al. 2020). This process of reduction or loss of plant productivity in drylands is known as *desertification*, and can be driven not only by climatic variability or change but also by land use practices that cause a net loss of vegetative cover and/or a decline in soil fertility. According to the secretariat of the UN Convention to Combat

Desertification (UNCCD 2017), between one-quarter and one-third of land in the world's dryland regions is experiencing or at risk of desertification, with areas of particular risk including semi-arid and arid lands located along the southern border of the Sahara Desert, in the Horn of Africa, in Iran, Pakistan, and parts of Central Asia.

The terms aridity, drylands, and desertification refer to long-term conditions and processes, whereas the term *drought* refers to a temporary period of unusual dryness (Box 3.2). There is no universal definition of what constitutes a drought, for the conditions that give rise to a drought are specific to a given location or system. There are three standard ways in which scientists categorize and describe droughts: meteorological droughts, hydrological droughts, and vegetative (or agricultural) droughts (National Drought Mitigation Center 2024). A *meteorological drought* is one where precipitation levels fall markedly below observed averages for a given area for an established period of time, with local definitions and parameters typically set by state meteorological organizations. In Ireland, for example, a meteorological drought is declared after a period of 15 or more consecutive days with less than 0.2 mm rainfall on each day (Met Eireann 2021). In India, the Meteorological Department has both seasonal and annual definitions of droughts. There, a seasonal drought is said to occur when precipitation is less than 75% of expected amounts, and an annual drought (or "Drought Year") is declared if an overall monsoon rainfall deficiency of 10% or more covers 20%–40% of the country (Udmale et al. 2020).

Meteorological measures of drought do not always reflect the practical implications of precipitation shortfalls, hence two additional categories or definitions are also used. A *hydrological drought* refers to a situation where water levels in a particular region or catchment area drop below a critical, predetermined threshold that reflects the human demand for water, the way in which water is used, and/or the minimum volumes of water needed to maintain ecological integrity within surface water bodies. A hydrological drought is typically identified using measurements recorded by a network of surface water-level gauges placed throughout a catchment area, and these may be combined with estimates of groundwater levels (which tend to shrink when there is not enough precipitation to recharge them) in areas where groundwater is an important component of human use. The impacts of hydrological droughts can be many, including shortages of water for urban consumption, transportation, agricultural irrigation, and recreational purposes. A *vegetative drought* occurs when there is insufficient moisture in the soil to support healthy plant growth, and is specifically referred to as an *agricultural drought* when crops, fodder, and/or grazing is adversely affected. Some authors have recently suggested a fourth category of drought – *socio-economic drought* – for situations when there is a shortfall in supply of an important economic good other than agricultural products due to a weather-related shortfall in water supply, hydroelectric power being an example (National Drought Mitigation Center 2024).

The potential for any of the above types of drought to occur is greatest when shortfalls in precipitation coincide with extended periods of above-average temperatures, the latter causing evapotranspiration rates to rise. Farmers in western North America will sometimes refer to these as "hot droughts," to distinguish them from periods of low precipitation but normal temperatures that are less difficult to cope with. *Evapotranspiration* refers to the amount of moisture being released into the air from the soil through direct evaporation and through transpiration by plants as they photosynthesize. Even though the impacts of droughts are specific to local ecological and hydrological conditions and human water needs, standardized indices of drought risks have been developed to facilitate the monitoring of drought risks across large areas. The two most common of these are the Standardized Precipitation Evapotranspiration Index (SPEI), which can be calculated for timescales from 1 to 48 months and combines measurements of precipitation and temperatures with estimated evapotranspiration rates (which may in turn be obtained from global datasets [Hobbins et al. 2023]); and the Palmer Drought Severity Index (PDSI), which uses monthly precipitation and temperature values to estimate a running soil moisture balance that can be compared with long-term averages (Vicente-Serrano et al. 2010). The PDSI is useful for estimating longer term drought conditions (i.e. 1 year or more) in mid-latitude regions, whereas the SPEI can be used for any region and across a wider range of timescales. Many of the quantitative studies described later in this chapter have compared historical population data with historical SPEI data to search for associations between drought and migration at large spatial scales. In both the SPEI and PDSI, the combination of hot temperatures and low precipitation generates the strongest drought index values.

Box 3.1
Population Change in Dryland Regions

Approximately 2 billion people globally live in dryland regions, 90% of whom live in low- and middle-income countries (UNCCD 2017). Roughly three-fourth of drylands that are put to human use are rangelands where an estimated 50% of all livestock are grazed. The remaining area represents roughly 44% of all global cropland (Maestre et al. 2021). Rural dryland populations in South Asia, the Middle East, North Africa, and sub-Saharan Africa are expected to increase by as much as 50% by the middle of this century, through a combination of natural population growth and a geographical expansion of dryland areas due to a changing climate (Stavi et al. 2022). There are approximately 580 urban centers globally with populations greater than 300,000 that are situated in drylands (EC Joint Research Centre 2019). The rate of urbanization in dryland areas is expected to grow, but there will nonetheless continue to be large rural populations in dryland areas of low- and middle-income countries exposed to growing risks of drought in coming decades.

3.2.2 Established Migration Patterns in
Rural Dryland Regions

Compared with other ecological regions, countries with large dryland areas generally tend to have relatively higher rates of internal migration (Hoffmann et al. 2023). This is not entirely surprising, for drylands are challenging environments in which to gain a livelihood, and mobility is a useful tool for meeting such challenges. Even in years with average precipitation and temperatures, drylands tend to have low vegetative productivity and limited surface water availability. Rural populations in dryland regions must be able to adapt not only to chronic, long-term water insecurity but also the inevitable drought periods when water and soil moisture are acutely lacking. In dryland areas of high-income countries, such as the North American Great Plains, Mediterranean Europe, and Australia, rural adaptation has been achieved through decades of specialization and mechanization in agriculture, irrigation using both surface and groundwater, engineering crop varieties that use minimal water, creating crop insurance programs and government programs that provide farmers with income support in times of hardship, and dedicating the driest areas that lack irrigation water to livestock ranches. The need for labor in the agricultural economy is low, and so, too, are rural population densities; indeed, the majority of people in dryland areas of high-income countries (and growing numbers in low- and middle-income countries as well) live in urban centers.

By contrast, dryland areas in low- and middle-income countries often have relatively large rural populations, many unirrigated smallhold farms, and large numbers of people pursuing traditional livelihoods centered on keeping livestock herds. Basic government services and infrastructure are often limited or inaccessible in rural areas, and so communities, households, and individuals must adapt as best they can without much outside support. People must be flexible and diversify their livelihood strategies and income sources to the extent possible, and migration often becomes part of these strategies. There are three general types of migration strategies observed in rural dryland areas in low- and middle-income countries. One is *pastoralism*, a practice whereby people move their livestock over the course of the year to locations where fodder and water are most abundant in a particular season. Pastoralism is well suited to arid regions, as well as to areas with extreme temperature regimes (e.g. mountains) and areas with highly seasonal precipitation regimes. It may be practiced in a largely *nomadic* approach (i.e. the entire household moves continuously with its livestock throughout the year), a *transhumance* approach (whereby the household resides in a fixed place and an individual or individuals from the household move its livestock between pastures on a seasonal basis), or various permutations of these. It is estimated that up to 100 million people worldwide practice pastoral livelihoods, with the proportion of rural households engaged in it being especially large in countries in Sudano-Sahelian Africa and the higher altitude regions of Western and Central Asia (Manzano et al. 2021). The geographical destinations to which pastoralists and their herds move within a given region have often been established through generations of experience and practice. In many regions, the territories used by pastoralists are constrained or shrinking because of the encroachment of people establishing crop farms and by government authorities seeking to force nomadic pastoralists into permanent settlements (Dong et al. 2011). Climate change presents an additional set of risks to pastoralists by potentially altering the vegetative productivity of their grazing areas and/or the availability of surface water for livestock (Herrero et al. 2016).

Another common migration-centered livelihood practice of households in dryland regions of Africa, Asia, and Latin America is for young adult members to migrate elsewhere in search of wage labor during the dry season or at other times in the year when their labor is not needed at home. In rural West Africa, this practice is described using the term, "eating the dry season" (Rain 1999) and allows households to temporarily reduce the number of people who need food and water and to potentially gain remittances at a time when no income can be generated locally. These regularized, circular migrations can see migrants moving to other rural areas or to nearby urban centers, and depending on the culture may include women as well as men. Most such migration occurs across relatively short, inexpensively traveled distances within the country of residence or to nearby contiguous countries, assuming borders are easily crossed. Not all communities, groups, or households employ (or are able to employ) temporary or seasonal migration as part of their wider livelihood strategies, and its use is best described as strategic or situational (Hoffmann et al. 2022).

Rural households in drylands may also send members to more distant locales for an indefinite time to seek wages and opportunity. This is a speculative venture to seek economic gains in destinations more prosperous than those that are more easily accessible. Because long-distance migration is inherently risky, is financially expensive, and forces the household to forego the labor of the missing migrant at times when it is ordinarily needed, it tends to be undertaken only during times when the household has resources to spare, which is typically when climatic conditions are favorable, and not during times of drought. Specific examples of this distinction appear in examples further below.

Box 3.2
Projected Changes in Drought Risks

On a warming planet, some regions of the world will become drier and experience greater risks of droughts than they do today, while other regions will become wetter and less drought prone. The extent to which such changes will emerge depends heavily on the amount of additional global warming that occurs. The most recent IPCC assessment projects that, in general, most areas of North, Central, and South America will become progressively drier with additional warming, as will the Mediterranean region, southern Africa, Australia, and areas of eastern China (Figure 3.2). Conversely, Sudano-Sahelian Africa, the Horn of Africa, and Central and South Asia are expected to become wetter. However, many of the regions expected to become wetter are monsoonal climates, where the additional moisture is likely to be received in the wet season, increasing flood risks and not eliminating dry season drought risks. Recent studies suggest that the greatest increase in the frequency and severity of droughts in coming decades will occur in southern Africa, the Mediterranean region, Western and Central Asia, and the North American Great Plains (Meza et al. 2020, Tabari et al. 2021).

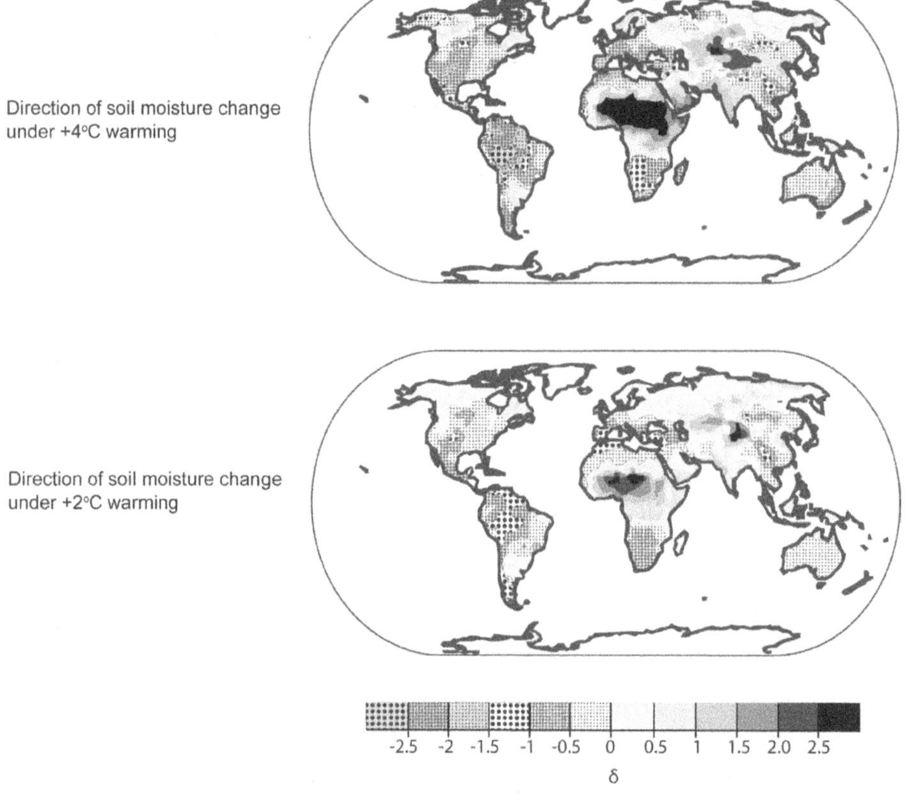

Figure 3.2   Projected changes in average soil moisture under increasing levels of global warming. Adapted from Figure TS.5 Panel (b) in IPCC, 2021: Technical Summary. In: *Climate Change 2021: The Physical Science Basis.*

Figure 3.2   Caption (cont.)
Contribution of Working Group I to the Sixth Assessment Report of the Intergovernmental Panel on Climate Change [Chen, D., M. Rojas, B.H. Samset, K. Cobb, A. Diongue Niang, P. Edwards, S. Emori, S.H. Faria, E. Hawkins, P. Hope, P. Huybrechts, M. Meinshausen, S.K. Mustafa, G.-K. Plattner, and A.-M. Tréguier, 2021: Framing, Context, and Methods. In: Climate Change 2021: The Physical Science Basis. Contribution of Working Group I to the Sixth Assessment Report of the Intergovernmental Panel on Climate Change [Masson-Delmotte, V., P. Zhai, A. Pirani, S.L. Connors, C. Péan, S. Berger, N. Caud, Y. Chen, L. Goldfarb, M.I. Gomis, M. Huang, K. Leitzell, E. Lonnoy, J.B.R. Matthews, T.K. Maycock, T. Waterfield, O. Yelekçi, R. Yu, and B. Zhou (eds.)]. Cambridge University Press, Cambridge, United Kingdom and New York, NY, USA, pp. 147–286, doi:10.1017/9781009157896.003. Reprinted with permission.

3.2.3 Drought and Migration Responses in the World's Main Dryland Regions

Migration patterns and behavior that are common in dryland regions – which might be referred to as "background migration" or "normal migration" – can undergo changes during times of drought. These changes, however, can vary considerably from one region or community to another and even from one drought event to the next, given the many potential intermediary social, economic, cultural, and other factors present at local and regional scales (Wolde et al. 2023). For example, smallhold farm households that are entirely dependent on rainfall for their crop success are much more vulnerable to the impacts of drought and more likely to exhibit changes in migration behavior and/or make migrations sooner than other households. Similarly, households that rely on local surface water and/or shallow wells have limited water security, and are more likely to leave their homes during droughts (Stoler et al. 2021). Even so, changes to migration patterns may not occur immediately; instead, there is often a lag time between the onset of drought conditions and observed changes in migration, reflecting attempts by households to cope through other, less disruptive strategies, such as planting more drought tolerant crops, eating less, drilling irrigation wells, drawing down savings, taking out loans, cutting alternative fodder, and selling off livestock, among other options (Mardy et al. 2018, Gidey et al. 2023). Only once these *in situ* strategies prove inadequate owing to the severity or duration of drought conditions do changes in migration patterns start to emerge. However, as is noted in examples from the world's main dryland regions summarized below, the poorest rural households, especially those without pre-existing migrant network connections, may lack the resources to move and can become trapped in place.

In dryland regions of Africa, migration responses to drought are highly variable across countries (Gray and Wise 2016). In some countries, migration rates increase during droughts, in others they decrease, and in still others there is no clear evidence of change. In those cases where migration increases from drought-hit areas, it tends to result in an increase in short-duration migration either to less hard-hit rural destinations or to urban centers within the home country or region, often following well-established migrant networks – but again, the actual observed outcomes can be varied and mixed. For example, in dryland areas of Tunisia, farmers grow olives, pomegranates, figs, dates, and other crops in oasis-style plantations in locations where there is supplemental ground water available for irrigation. Population growth has magnified the challenges of farming in these dry areas, requiring continuously higher levels of adaptation. In years with below-average precipitation, there is an increased amount of migration out of affected areas of short-, medium-, and long-term duration (Sobczak-Szelc and Fekih 2020). The aim of migrants is to acquire money to help the household get through the immediate period of hardship, with additional hopes of earning extra money that can be used to diversify and/or intensify the household's agricultural production, such as by digging additional wells, buying additional land to farm, and investing in other types of crop production.

In East Africa, the relationship between droughts and migration patterns appears to be subtle and complex. A broad statistical analysis of household survey data collected in four East African countries (Ethiopia, Tanzania, Malawi, and Uganda) between 2009 and 2014 found few observable changes in long-term rural out-migration during droughts but did observe a reduction in migration out of urban centers, which the authors associate with indirect impacts of drought on non-urban labor market opportunities (Mueller et al. 2020). Similar research for southern Africa found that excessively hot, dry conditions are associated with lower levels of migration in Botswana and Kenya, but with small increases in migration rates in Zambia, with labor markets playing a moderating role (Mueller et al. 2020). A detailed study of migration patterns in Zambia during droughts found that higher income households and those with strong migrant network connections were much more mobile than low-income rural households, and that these latter groups might be trapped in difficult circumstances (Nawrotzki and DeWaard 2017). In semi-arid parts of Malawi, drought and other climate hazards typically do not affect intentions or aspirations of rural people to migrate, but may in the short term reduce their ability to move given the temporary lack of resources available to the household (Suckall et al. 2017). In Uganda, short duration droughts are associated with changes in the movements of pastoralists and higher rates of temporary rural out-migration, especially among the poorest rural households (Twinomuhangi et al. 2023), but extended dry periods cause temporary migration to ebb and permanent out-migration to increase (Call and Gray 2020). In the Republic of South Africa, rural districts have higher rates of out-migration and lower rates of in-migration during short-term droughts, but rates of out-migration to urban centers fall when the drought conditions persist for a full year or more (Xiao et al. 2022). In Mozambique, rural out-migration rates increase significantly in the year immediately following a drought, with the effect tapering off with the passage of time (Yoo and Agadjanian 2024).

Ethiopia is the most studied of all countries globally in terms of the relationship between drought and migration (Xu and Famiglietti 2023), dating back to research on famine-related displacements of the 1980s (Ezra and Kiros 2001). Migration has long been an important method for smallhold farmers and landless rural households to cope with the seasonality of the climate (Dessalegn et al. 2023). Late-arriving seasonal rains and droughts have been observed to generate higher rates of short-distance migration in central Ethiopia (Etana et al. 2022) and in the highlands of northern Ethiopia (Asefawu 2022), particularly among young men (Gray and Mueller 2012). Migration does not automatically ensue, as farmers will often first try to cope by selling off livestock and seeking out local income opportunities in the informal sector (Hermans and Garbe 2019). Most people do not migrate during droughts, but an increased level of out-migration is nonetheless observed. This migration includes short distance, temporary moves as well as longer distance, longer duration moves, with different distances/durations of migration appearing to have different motivations. Migrants who stay within Ethiopia tend to move to urban destinations in search of wages to provide financial support to the household. People who migrate internationally (typically to Persian Gulf states) stay away longer and tend to remit money home less frequently. Their motivation appears to be driven by a desire for financial independence from the household rather than a desire to increase their contribution to the household. The contrasting migration motivations and outcomes suggests that drought influences migration decision-making but is not in itself the sole determinant of it, with cultural, social, and economic factors playing important roles (Groth et al. 2020). This conclusion is consistent with wider changes in internal and international migration patterns being observed in Ethiopia, where nomadic and semi-nomadic pastoralists are being forced to settle in fixed locations, a growing amount of migration is destined to urban centers, and the range of international destinations to which Ethiopians move is expanding (Schewel and Asmamaw 2021).

Similarly, in rural drylands of West Africa, where a large amount of migration occurs on an ongoing basis, drought is seen not as a primary motivation for moving, but as one of a mix of interacting factors that shape migration decisions and patterns (Gautier et al. 2016). Drought–migration outcomes play out differently across countries in the region (e.g. Box 3.3). In Senegal and Burkina Faso, droughts tend to lead to temporarily lower levels of international migration, which in contrast increase during highly favorable climatic conditions (Nawrotzki and Bakhtsiyarava 2017). The drought–migration relationship in those countries appears to vary depending on the nature of the local agricultural system and preferred crop types, with drought-related out-migration being higher in areas of Senegal where groundnut is the dominant crop. In dryland regions of Ghana, smallhold farmers employ a wide variety of tactics to adapt to general dryness, including crop and livestock diversification, adjusting the planting schedule and planting drought-tolerant crops, and seasonal migration (Antwi-Agyei and Nyantakyi-Frimpong 2021). During droughts, farm families will eat less, sell off livestock and non-farm assets, seek out local non-farm jobs, make

---

Box 3.3
Does Drought Cause Higher Levels of Migration to Europe?

Since the so-called "migrant crisis" of 2015, when over a million asylum seekers arrived in European Union countries from Africa, the Middle East and Western Asia (Crawley 2016), there has been ongoing concern among European governments and policymakers as to whether climate change will lead to higher future levels of irregular migration (Özdemir 2023). In terms of drought, such fears are not supported by the data available. Research conducted by Cottier and Salehyan (2021) examined statistics for irregular migration into the European Union between 2010 and 2015 and found no evidence of any association between drought conditions in migrant source countries (as measured by SPEI) and migrant arrivals from those countries. Indeed, the researchers found that in many cases, migration levels from some source countries actually went down during drought periods.

---

charcoal, draw upon resources of friends and relatives, and/or migrate temporarily. Drought-related migration in Ghana can be both rural-to-urban and rural-to-rural in nature, often from drier northern regions to wetter, southerly regions of the country, with community-level factors and pre-existing migrant networks playing important roles in destination decisions (Jarawura 2021). The out-migration of young adult women from rural areas can increase during droughts, particularly from areas where there are few local economic opportunities for them (Azumah and Ahmed 2023). By contrast, in northern Burkina Faso, women tend not to migrate during droughts, but remain behind in villages where they face an increased burden in helping remaining household members cope (Vinke et al. 2022). In Mali, rural areas experience lower rates of net migration during droughts, but out-migration from the country as a whole to international destinations tends to rise (Defrance et al. 2023).

Rural migration patterns in arid areas of Asia share commonalities with those in Africa, showing variations between one country and region to another, with migration responses to droughts being highly moderated by non-environmental factors operating at local and larger scales. A study of migration patterns in six Asian countries (China, Indonesia, Malaysia, Nepal, the Philippines, and Vietnam) found that periods of low precipitation are associated with generally lower levels of out-migration from affected areas, although the strength of this association and specific outcomes vary across and within these countries (Thiede et al. 2024). Some examples of observed drought-related migration outcomes from countries in Asia now follow, beginning with Iran. Increasing aridity, drought, and decades of surface water diversions for irrigation in Iran have caused lakes Hamun and Urmia to dry up (Shahi 2019), similar to the infamous Aral Sea disaster in the Central Asian Karakalpakstan region of the former Soviet Union (White 2013). This in turn has led to the abandonment of thousands of villages due to water scarcity, contributing to the much larger flow of rural migrants into Iran's swelling cities.

In Pakistan, dryland farmers use seasonal and longer duration migration as part of a range of activities to adapt to dry conditions. Meteorological evidence and accounts of rural people suggest that the severity and duration of droughts are increasing, particularly in Sindh and Balochistan provinces, creating severe shortages of surface water during the dry season and declining groundwater reserves in many rural areas (Ajani and van der Geest 2021, Dahri et al. 2021). During droughts, rural households will attempt a variety of *in situ* coping strategies, including eating less and spending less money on non-essential goods, children's education, social activities, health care, and building maintenance, but many will turn to migration, particularly households with lower levels of education, income, and/or access to government support (Ashraf et al. 2014). Unusually hot and dry conditions are much more likely to stimulate observable changes in rural out-migration than are floods, particularly among poor, smallhold farming households (Mueller et al. 2014). The duration and distance of migration varies, with observable increases in all types. Men are more likely to migrate than women in most regions, although migration rates of both men and women rise during high heat events, likely because of the adverse impacts on crops and associated income.

There is a surprisingly limited amount of recent research available on linkages between drought and migration patterns in dryland regions of India, although a number of studies conducted in non-arid regions in the east of the country provide useful insights, and are assessed later in this chapter. One of the most comprehensive available studies found that across India, rural districts in semi-arid and arid areas had higher rates of temporary out-migration during droughts, with roughly two thirds of that migration being to urban centers (Sarkar et al. 2022). Migration participation rates vary by household income, caste, gender of the head of household, and other socio-economic and demographic factors, with socio-economically marginalized groups being most likely to migrate during droughts. Similar results were reported by Sedova and Kalkuhl (2020), who compared household survey data collected in 2004 and 2012 with weather data and found that unusually low levels of precipitation and excessively hot conditions are associated with higher rates of rural–urban migration, originating especially from lower skilled agricultural households. Data for the decade of the 1990s show that higher levels of interstate migration in India originated from drought-affected agricultural states relative to others, particularly rural-to-rural migration, with drought frequency, duration, and severity all being relevant factors (Dallmann and Millock 2017). State-level research is generally consistent with the aforementioned country-wide analyses and adds further context. In rural Andhra Pradesh, smallhold farmers with no access to irrigation and no livestock are most likely to migrate during droughts (Pragathi and Anitha 2019). In Uttar Pradesh, rural out-migration rates increase in some districts during droughts but do not in others (Islam and Singh 2021), suggesting non-environmental factors play a heavily moderating role.

In neighboring Nepal, seasonal migration, especially of young men, has for generations been an important adaptation of rural households to the seasonality of precipitation and risks of droughts and other hazards (Nepal et al. 2021). Growing rural poverty and declining profitability of smallhold farming have led migrants to move away for increasingly longer periods of time (Gautam 2017). Evidence for the period 2011–2017 shows that during droughts there is an increase in male out-migration from rural areas to both destinations within Nepal and internationally, and an increase of female outmigration to Nepalese destinations (Epstein et al. 2022). At the same time, there can be an increase in return migration of migrants who were living and working away, especially those of an older age than the ones migrating out of the region – suggesting that during the drought, these incoming return migrants may be taking up roles caring for remaining family members and assisting with the household's livelihood practices during a time of need. Williams and Gray (2020) have made generally consistent findings that temperature and precipitation extremes in Nepal do not necessarily alter the total number of migrants, but can temporarily alter the distance and duration of migration in which people engage. Decisions to migrate – or to remain in place – during extreme weather of all types in Nepal are shaped strongly by non-environmental factors including cultural values, place attachment, local social capital, local adaptive capacity, and larger socio-economic factors (Bhusal et al. 2021).

Central and South America experience a similarly wide array of drought risks, rural vulnerability, adaptation, and migration responses as in Africa and Asia. In semi-arid northeastern Brazil, millions of people live on smallhold farms that lack irrigation and are thus highly exposed to drought risks (Marengo et al. 2022). Rural–urban migration is an ongoing phenomenon in this region. Where large-scale migration was once a common outcome of droughts, gradually increasing government social programs and support for farmers in the region in recent decades have lessened it (Marengo et al. 2020). Even so, observable changes in migration patterns continue to occur during droughts: Out-migration rates decline among the lowest income, poorest educated segments of the rural population, but increase for people with higher incomes and education and thus fewer economic barriers to moving (Delazeri et al. 2022). General aridity and drought risks are expected to grow considerably in northeastern Brazil in coming decades (Marengo et al. 2020), and econometric models assessing the potential impacts on inter-regional migration within Brazil forecast that the northeast will see considerably higher levels of out-migration in coming decades relative to other parts of the country (Oliveira and Pereda 2020).

In dryland Chile, poor rural households use temporary and longer-term labor migration as part of an ongoing set of strategies to cope with economic inequality and poverty, and although migration rates do increase during droughts, participants themselves frame it as an opportunity to gain economic security rather than being a deliberate drought adaptation (Wiegel 2023). Beginning in 2010, successive years of severe drought in central Chile led to higher rates of rural out-migration from affected areas (Aldunce et al. 2017). In recent years, Chile has received growing numbers of immigrants from lower income countries in the region, particularly Venezuela, Colombia, Bolivia, and Haiti, many of whom may have higher levels of vulnerability to environmental hazards including drought because of their lower socio-economic standing within Chile (Bronfman et al. 2021).

In Central America's semi-arid "Dry Corridor" of arid land that spans parts of Honduras, Guatemala, Nicaragua, and El Salvador, drought risks are increasing and projected to grow considerably in a changing climate (Depsky and Pons 2020). Large numbers of people move within and between these countries, and growing numbers of migrants from them are appearing at the US border, with drought often cited in popular media accounts and nongovernmental organization reports as being a main driver (e.g. Lustgarten 2020, Läderach et al. 2021). However, the amount of peer-reviewed scholarship on the subject is limited. In rural areas across this region, migration patterns vary considerably according to socio-economic status, with the lowest income households often having lower rates of migration than people with greater wealth and education, and low-income people are most likely to migrate over short distances or in circular patterns according to demand for crop harvesters (Huber et al. 2023). Young adults, both male and female, often move to cities within the region and/or to neighboring Costa Rica in search of wage labor opportunities. Qualitative research conducted in rural Honduras suggests that, as in many other regions, households view migration principally as being an option for escaping poverty and lack of economic opportunity, with drought being a secondary factor in decision-making (Cárdenas-Vélez et al. 2024). Longer distance migration from the Dry Corridor to the US is growing, and is most often undertaken by young adults from higher-income households (Huber et al. 2023). Researchers who examined US Border Patrol apprehension data from 2012 to 2018 found that, after controlling for reported rates of criminal violence and poverty, local areas within the Dry Corridor that experienced drought conditions had a 1.7 times higher rate of out-migration than areas not experiencing drought (Linke et al. 2023). The overall picture that emerges from this particular dryland region is that drought plays a potential contributory role to migration decision-making and outcomes, but its effects filter through a range of other factors, including changing agricultural practices (including the expansion of coffee plantations), endemic poverty, weak governments, geographical patterns of social networks, and US government immigration and border policies (Reichman 2022).

3.2.4 Case Study: Drought and Migration in Mexico

Large flows of migration between Mexico and the US have existed for the better part of two centuries. Under an 1848 treaty, a large area of northern Mexican territory was ceded to the US, creating an instant diaspora of tens of thousands of Mexicans within the southwestern US, their numbers swelling by hundreds of thousands following the 1910 Mexican revolution (Cano and Délano 2007). Between World War II and the 1960s, a temporary worker scheme called the Bracero program facilitated the documented and legal movement of hundreds of thousands of Mexican agricultural and railroad workers into the US, but by the time the program ended in the mid-1960s, millions of Mexican migrant workers and their US employers were bypassing the formal program. The number of undocumented Mexican workers moving to the US continued to rise steadily from the 1960s through to the 1990s (Durand and Massey 2019). During that period, the long US–Mexican land border was relatively porous, poorly patrolled, and had many easily accessible informal crossing points. As a result, many undocumented Mexican workers moved repeatedly back and forth between the two countries, hiring a guide for their initial crossings into the US, but making subsequent crossings without help as their knowledge and experience improved (Singer and Massey 1998). US enforcement of labor laws pertaining to undocumented workers was also lax, creating an exploitative situation in which undocumented Mexican workers were earning on average 40% lower wages than documented Mexican workers (Rivera-Batiz 1999). US border policies changed markedly following the September 11 attacks, initiating a steady increase in border surveillance by the US government that continues to this day and increasingly relies on a range of remote sensing devices, border barriers, and frequent patrols to keep undocumented migrants from entering. The relatively free flow of Mexican workers seeking to newly enter the US ebbed, and those who entered the US without documentation are not able to leave and re-enter easily as was formerly the case. Migration rates from Mexico dropped by more than two-thirds between 2005 and 2012 from the combination of tighter border enforcement, greater economic prosperity in Mexico, and the 2008 economic crisis that caused a drop in demand for workers in construction and other sectors once dominated by Mexican workers (Villarreal 2014). Large numbers of Mexicans continue to make undocumented crossings into the US in search of work – over 800,000 were interdicted by US border patrols in 2022, mostly single adults – but their numbers are now lower than third-country nationals from Central and South America and Caribbean countries (WOLA 2022).

Multiple studies of historical data on migration within Mexico and to the US find that droughts in Mexico have been an important driver of rural out-migration. For example, it has been estimated that up to one-third of all migration from rural Mexico to the US between 1970 and 2009 originated during times of drought, disproportionately from arid areas (Murray-Tortarolo and Salgado 2021). Two separate studies that looked at migration data for the late 1990s and early 2000s found that a 10% decrease in precipitation corresponded with a 2% increase in outmigration from the affected area (Feng et al. 2010), and that areas that experienced a 20% or greater shortfall in precipitation saw an average increase in out-migration of approximately 10% (Puente et al. 2018). Most observed drought-related migration to the US in decades prior to 2000 originated in arid- and semi-arid states (Nawrotzki et al. 2013), with rates being especially high among smallhold dryland farmers that have no access to irrigation (Dobler-Morales and Bocco 2021, Fishman and Li 2022). Drought-related migration to the US has historically not been instantaneous; rather, there is typically a lag effect (Hunter et al. 2013), with relatively modest changes in out-migration in the first year after a drought but growing to a peak in the third year, and thereafter returning to previous levels (Nawrotzki and DeWaard 2016). Droughts also lead to increased rural-to-urban migration within Mexico, with each additional month of drought increasing migration likelihood by 3.6% up until the end of the third successive year of drought, at which point rates begin increasing nonlinearly (Nawrotzki et al. 2017). Conversely, when precipitation rates are higher than normal, rural–urban migration rates within Mexico tend to decline.

However, it is worth noting that most of the preceding studies looked at data from periods when migration rates from Mexico to the US were generally high. A study that looked only at the period 2000–2010, when migration between the countries was slowing for reasons noted earlier, found that US-bound migration rates did not increase in most states except during especially severe drought conditions and only in areas where households had strong connectivity to migrant networks (Riosmena et al. 2018). In all other situations, drought may decrease the likelihood of migration to the US. This is again consistent with research in other dryland regions of the world that when households experience depressed income during droughts, they may lack the financial resources necessary to undertake longer distance migration.

It is expected that climate change will lead to increases in migration within Mexico and to the US due to declining crop productivity, more frequent drought-related crop failures, and changes in the availability of fodder and surface water for livestock, which will in turn lead to rural household income losses (Thalheimer et al. 2021). One study has suggested that by the 2080s, between 1.4 and 6.7 million Mexicans could have to migrate depending on future warming trajectories (Feng et al. 2010), but it is again worth noting that the underlying assumptions about drought–migration linkages were based on findings from data collected during the peak period of Mexican migration.

3.2.5 Case Study: Drought and Migration on the North American Great Plains

Precipitation patterns and periodic droughts have played a significant role in shaping rural migration patterns on the Great Plains since the early days of European settlement in the nineteenth century. The Plains (known in Canada as "The Prairies") are an enormous grassland ecosystem of over a million square kilometers dominating the central part of the North American continent. Although the soils are very fertile, much of the region is arid or semi-arid. There is considerable interannual variability in the amount of precipitation received, and there can be considerable spatial variability in the distribution of precipitation in any given year. Multi-year and even decade-long droughts are not uncommon. Indigenous people of the Great Plains, whose livelihoods were based heavily on hunting huge herds of bison, were forcibly displaced in the 1800s by governments and settlers to make way first for livestock grazing and later for crop farming. Beginning in the 1860s in the US and the 1870s in Canada, individuals or families could claim title to 160 acres (65 hectares) of unoccupied land for the purpose of farming under a system known as *homesteading*. The rate at which homesteads were established varied over time, with wheat prices and interest rates being important factors (McFerrin et al. 2012). Precipitation also played a role; periods with abundant precipitation saw an influx of settlers, but during dry periods, the number of settlers would slow (Mock 2000). When droughts occurred, homestead farms would often be abandoned, as was the case in western Kansas in the mid-1890s and in eastern Montana and southern Alberta in the late 1910s/early 1920s (Libecap and Hanson 2002, Jones 2002). By the early 1930s, rural population and farm numbers had peaked across most states and provinces of the Plains (Carlyle 1994, Johnson 2011).

The 1930s brought environmental and economic disaster to the Great Plains. The 1929 stock market crash triggered the Great Depression that would last until the onset of World War II in 1939. Prices for grain, cattle, and other commodities produced by Great Plains farmers collapsed, driving down household incomes. Farmers who had taken out farm improvement loans when interest rates were low in the 1920s found it difficult to repay, and many were foreclosed (Grant 2002). Between 1929 and 1937, extreme summer heat and severe droughts emerged across large swaths of the Plains, the period between 1932 and 1936 being especially dry and 1934 likely being the worst drought year on the Plains in the last thousand years (Cowan et al. 2017, Cook et al. 2014). Enormous dust storms formed as winds eroded the desiccated soil, the situation being especially bad in the "Dust Bowl" area of the southern Plains. Hundreds of thousands of people fled the Great Plains and adjacent farming areas for the Pacific coast, the Canadian parkland, and states and provinces to the east; other displaced people who lacked the means to leave gravitated to urban centers on the Plains, where large homeless camps formed on the outskirts (Gregory 1989, Worster 1979, Gray 2004, McLeman et al. 2013). Disproportionately represented among those who left the rural Plains were young families with transferable skills and migrant network connections elsewhere – what might be best referred to as the rural "middle class" – while those who remained consisted of financially stable owners of well-established farms and rural businesses who had strong local community networks, and large numbers of destitute and landless immobile people (McLeman et al. 2008, Gilbert and McLeman 2010).

Rural populations on the Great Plains steadily declined throughout the remainder of the twentieth century as the agricultural economy shifted toward larger, mechanized, specialized farms. Severe droughts occurred in the 1950s, 1970s, early 2000s, and early 2010s, but were not accompanied by massive migrant outflows as occurred in the 1930s. This is due to there being ever fewer people living in rural dryland areas of the Great Plains (meaning that few people are exposed to drought risks) and improved farm-level adaptation to drought risks through irrigation, crop selection and diversification, improved soil management practices, and other risk-reduction methods (McLeman et al. 2013). That said, subtle changes in rural population numbers have occurred in the post-Dust Bowl era in specific locations on the Plains at times when severe droughts have coincided with trying economic conditions – but the number of people on the move are too small to draw widespread attention (McLeman et al. 2022a).

3.2.6 Drought and Migration in Non-dryland Regions

Preceding sections focused on migration patterns in dryland regions, where drought is an expected hazard, and where exposed populations learn to adapt over time according to the economic, social, and other forms of capital available to them. In non-dryland regions, droughts can also create occasional risks, especially for rural populations, and the impacts may in some instances be greater specifically because of the infrequency of droughts and a lack of established methods to cope with them. In the case of monsoonal climates, where flooding and heavy precipitation are the more common hazards, a severe drought may prove to be a greater disruption to household well-being and potentially generate migration responses that might not otherwise occur. In such climates, the specific combination of weather conditions that gives rise to drought are a monsoon season that is shorter and/or delivers less rainfall than average followed by an unusually hot dry season. In the absence of water storage and/or groundwater access, rural households (or indeed urban centers – see Box 3.4) may find themselves running short of water in the later stages of the dry season.

Such conditions periodically occur in eastern India's West Bengal state and when they do, poorer rural households and farmers with small land holdings use short-term migration out of affected areas as an adaptation strategy (Debnath and Nayak 2022). There are few off-farm local jobs to work at during drought, so non-migration adaptation options tend to be limited to eating less, selling off livestock, and/or borrowing money. State and national governments are investing in dams, wells, and irrigation systems in the region and establishing programs to provide short-term employment for drought-affected households, but these institutional adaptation investments are often not sufficient. As a result, during severe droughts up to one-half of households in rural villages will send a male member of working age as a migrant. For those that own farmland, the duration of migration is typically less than a year, but migrants originating in landless households are more likely to stay away for longer periods. Areas with poorer quality soil tend to have higher out-migration rates during droughts – a phenomenon also observed in rural drought migration patterns on the Canadian Great Plains during the 1930s (McLeman and Ploeger 2012) (see Section 3.2.5).

Data and research from Bangladesh – again, not an arid country – show that droughts and extended periods of hot and dry conditions are often more likely to generate observable changes in rural migration patterns than floods, the latter being the more common hazard that rural households expect and adapt to. Drought conditions stimulate higher levels of short- and medium-term labor migration as farming households seek alternative income sources (Call et al. 2017, Carrico and Donato 2019). In northwestern Bangladesh, the Indigenous Santal people use a variety of water storage and other strategies to prepare for water shortfalls, which are complemented with a temporary out-migration of large numbers of young adults to nearby cities during dry periods, the migrants usually returning home with the arrival of the rains (Ahmed et al. 2019). In northern Bangladesh, short-term out-migration during droughts is highest among landless and impoverished households, especially those with larger family sizes (Kabir et al. 2018). Microcredit programs implemented in that area by governments and NGOs are often used as a short-term adaptation to droughts, but poorer households may be unable to repay, and end up becoming more likely to migrate in the future as their indebtedness grows.

A somewhat different set of drought–migration dynamics has been reported in Southeast Asia – a region with monsoonal climates but not arid ones. Short-term rural–urban migration is a common adaptation strategy among many communities; one study estimated that one-third of rural households in Vietnam's Mekong Delta region use it as a drought-coping strategy (Tran et al. 2021). Droughts are also generally associated with higher rates of migration from Cambodia to Thailand (Bylander 2016). However, when drought conditions become recurrent, rural out-migration rates may in some instances decrease (Quiñones et al. 2021). In this study, data from Thailand and Vietnam suggest a single, occasional drought event leads to a small increase in outmigration as households seek short-term sources of income, but repeated droughts reduce migration participation rates, especially for longer distance, longer duration migration. The reason appears to be that multiple drought years erode households' financial resources to a point where they cannot afford to send members on long migration journeys, poorer households being most affected.

There is not a great deal of published research on drought-related migration in non-arid regions in the Americas, suggesting an opportunity for future research (e.g. Box 3.4). One broad-based study of Latin America and the Caribbean – including non-arid locations within the region – found that rural out-migration of people in their late teens/early twenties increases during droughts, with participants typically moving short distances within their home regions/countries in search of income opportunities (Baez et al. 2017a). A study looking at rural migration patterns in eastern South Dakota, USA, in the mid-1970s observed that locally severe drought conditions and rising interest rates led young adults to move out of the region in search of employment, while simultaneously stimulating a return migration of older adults to assist on family farms no longer able to hire labor (McLeman et al. 2022b). Although South Dakota is generally considered to be a Great Plains state, eastern parts of the state are not arid, and flooding and waterlogging of soil are more common hazards than droughts. Most eastern South Dakota farms were not irrigated, and the farming practices common to the area at that time made them unusually vulnerable to droughts.

Box 3.4
Urban Drought Risks

Most of this chapter – and indeed most of the established research literature – focuses on the influences of heat and drought on migration patterns in rural areas. This is not entirely surprising, given there is clear evidence that farming livelihoods are inherently susceptible to variations in precipitation, which is often referred to as the *agricultural pathway* between climate and migration (Falco et al. 2018, Nawrotzki and Bakhtsiyarava 2017). Yet urban centers also require water to meet their residents' basic needs, support industry and commerce, and accept and transport human sewage waste (ideally only after that sewage has been properly treated). Somewhat paradoxically, urbanization rates have grown steadily in many of the world's most water-scarce regions, where cities are exposed to periodic drought hazards that can cause acute, short-term shortfalls in water supplies. Since the year 2000, 79 cities globally have experienced at least one drought event that necessitated an urgent water conservation and rationing response (Zhang et al. 2019). It has been estimated that roughly one-quarter of the world's largest cities (representing 233 million people) will by the year 2050 have greater local demand for water than available surface water supplies (Flörke et al. 2018).

There are no modern examples of large cities being entirely without water for any extended period of time, with the exception of those that have sustained severe damage to water infrastructure during wartime. There have also been instances where cities have experienced contamination of water supplies that required emergency measures, such as Flint, Michigan, where a 2014 government decision to switch water sources created a public health disaster (Pauli 2020). But, there have been no instances where a city's water supplies ran out because of drought. The city that has come the closest to doing so in recent decades is Cape Town, South Africa. That city has a Mediterranean climate – generally dry, with mild temperatures throughout the year, and a distinct rainy season running from late May to early October. The city's 4.3 million residents rely on water stored in reservoirs to get through the dry months. Between 2015 and 2017, seasonal rainfall levels were below average, and by the end of the 2018 dry season, the city's reservoirs had almost completely run out of water. Local government warned of an impending "day zero" – a point at which the city would shut off water supplies to households and businesses and would allocate only 25 liters of water per person per day, to be collected at communal standpipes (Calverley and Walther 2022). City officials employed a variety of conservation measures, public awareness campaigns, and temporary restrictions on large water users that successfully enabled residents to avoid day zero, and heavy rains in the 2018 rainy season allowed for reservoirs to be successfully recharged to near capacity.

The Cape Town case serves as a warning for other cities with precarious water supplies and high rates of in-migration (Rodina 2019). It is worth noting that the shortfall in precipitation in the Cape Town region during the drought years was not especially large in percentage or absolute terms, but the city's water storage infrastructure had not kept pace with its rapid population growth and the increasing per capita water use rates of residents. The combined effects of climate-related precipitation change and urban population growth is expected in coming decades to generate particularly high drought risks and consequent water supply–demand crises for cities in India, Pakistan, the southwestern US, eastern Brazil, and western Africa (Stolte et al. 2023).

3.2.7 *Links Between Droughts, Conflict, and Migration/Displacement*

Since the 1980s, a debate has simmered among researchers regarding potential linkages between climatic hazards (particularly droughts), violent conflicts, and refugee movements. Two general types of connections or pathways have been suggested by those who study the topic:

1. Droughts (and other environmental hazards) can generate migration out of affected areas, with conflicts emerging in destination areas as new arrivals compete for scarce resources with established populations (hereon referred to simply as the *drought–migration–conflict* pathway, or D-M-C).
2. Droughts can generate conflicts between groups competing for scarce resources, which in turn produces refugees who flee the conflict that has emerged (i.e. the *drought-conflict-migration* pathway, or D-C-M) (Figure 3.3).

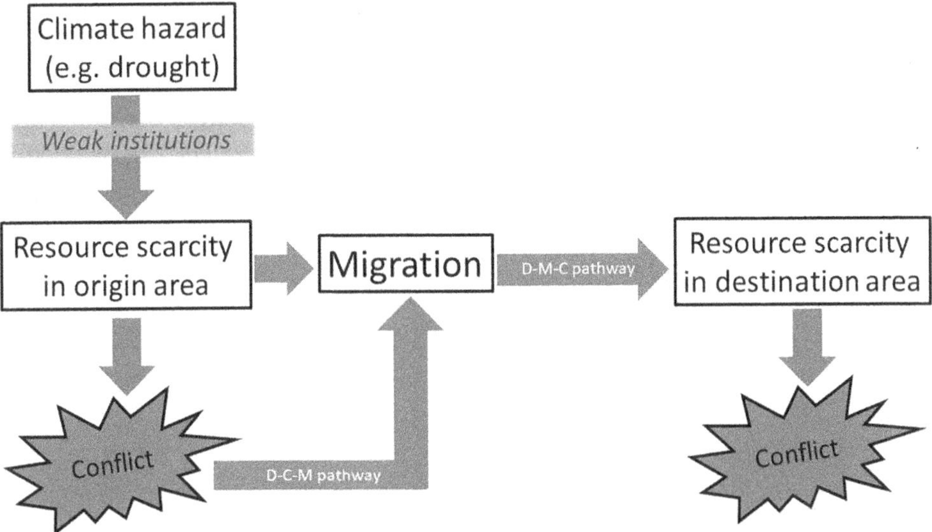

Figure 3.3 Schematic diagram showing two possible pathways through which droughts or other climate hazards may lead to conflict and/or migration.

The D-M-C pathway made its first notable appearance in a high-profile 1985 report on environmental refugees released by the UN Environment Programme (El-Hinnawi 1985), which described 1984 conflicts between drought-displaced Afar herders and farmers residing in Ethiopia's Borkena Valley. The D-M-C pathway gained notoriety nine years later in an *Atlantic Monthly* essay and a subsequent book with the same provocative title, "The Coming Anarchy" (Kaplan 1994), which described how droughts and environmental degradation were causing conflicts and refugee flows in sub-Saharan Africa, with more likely to arise in future decades. Another UN Environment Programme report, issued in 2008 described how drought and desertification (see Box 3.5 for more on desertification) led to violent intergroup conflict in northern Darfur, triggering large flows of refugees out of the region and into neighboring countries, seemingly confirming Kaplan's predictions and providing further evidence for the D-M-C pathway (UNEP 2008). This latter report gained considerable attention in the popular media, leading to stories with headlines such as, "Darfur conflict heralds era of wars triggered by climate change, UN report warns" (Borger 2007).

Although both pathways and the examples listed above seem plausible, caution needs to be taken before accepting them at face value for several reasons. First, there is limited scientifically validated, empirical evidence for either pathway: There are far, far more examples in Africa and around the world where droughts have not triggered conflicts and/or refugee flows and where drought-related migration has not led to violence than there are examples where such things have actually occurred. A second problem is the conflation of association with causality: The fact that a drought and a conflict have occurred in the same country or region at approximately the same time does not automatically mean that the drought caused the conflict (just as one would not assume the conflict caused the drought). A third problem is the underlying assumption that when times get difficult and resources become scarce people become more inclined to act violently toward one another, an assumption for which there is especially little empirical evidence. Indeed, the evidence suggests that resource scarcity may promote greater levels of cooperation rather than competition or conflict depending on the circumstances (Matthew and Gaulin 2001).

There are no examples in modern times where drought has been the direct cause of a violent conflict between countries. There is, however, competing evidence with respect to the effects of droughts and other climatic conditions on conflicts within countries. A meta-analysis of 45 pre-existing studies of conflicts since World War II suggested that globally there is a significant statistical association between hot, dry conditions and inter-group conflicts within states (Hsiang et al. 2013). However, a global analysis by Cattaneo and Bosetti (2017) found no evidence of any statistical association between conflicts and climate events of any kind between the years 1960 and 2000. Which is to be believed? The answer probably lies somewhere between.

Multiple studies that have directly compared data on conflict with climatic conditions at regional scales have identified a limited number of examples where a positive association exists. For example, in a detailed statistical analysis for the entire continent of Africa for the period 1989–2014, von Uexkull et al. (2016) found no instances where a new conflict started specifically because of a drought, and found that in most areas of Africa there was little or no statistically significant association between drought and conflict. The one exception is that among agriculturally dependent groups in some very low-income African countries, droughts may have increased the likelihood that pre-existing conflicts would be prolonged. A study of conflict onset in sub-Saharan Africa between 1962 and 2006 by Bell and Keys (2018) found that severe droughts are no more likely to affect conflicts in socio-politically unstable countries than in more stable countries. In a study comparing conflicts and rainfall patterns in East Africa between 1997 and 2010, Raleigh and Kniveton (2012) found that in locations where civil conflicts have been known to occur, the frequency of violent events can increase during times of drought. However, they also found there was a higher statistical chance of this happening during times of high rainfall. Research using a similar methodology found no correlation between conflicts in Asia and droughts or any other climate-related hazards (Wischnath and Buhaug 2014). A study that examined the origins of asylum seekers in Europe for the period 2006–2015 found no evidence of an association between drought in migrant source countries and the number of claims originating from them, except for a period between 2010 and 2012 when countries in North Africa and the Middle East that experienced "Arab Spring" uprisings – widespread anti-government protests and violence during a period of above-average food prices and droughts in the region – saw an increase in asylum seekers moving to Europe (Abel et al. 2019).

Featuring prominently in the "Arab Spring" period is the most studied and cited example of the D-M-C pathway: The civil conflict that emerged in Syria in 2011, where a severe drought that occurred between 2007 and 2010 is believed to have led to increased rural–urban migration, food insecurity, and high unemployment, which in turn is said to have intensified pre-existing political strife and helped fuel the ensuing violence (Kelley et al. 2015, Ide 2018). However, there is considerable debate among scholars as to the causal influence of the drought relative to non-environmental causal factors such as inefficient agricultural production systems, attacks by government and rebel forces in rural areas that were deliberately staged during the growing season, and even upstream diversions of water from the Euphrates River by other states (Gleick 2014, Karnieli et al. 2019, Linke and Ruether 2021). Some scholars have found little evidence that the impacts of the drought on rural populations were as severe as claimed or that the scale of rural–urban migration was as large as claimed in other studies (Selby et al. 2017). The most that can be said based on the evidence is that drought conditions did not cause the civil conflict in Syria, but their occurrence likely made an already bad situation worse. This is an example of what is often referred to in security research as the *threat multiplier effect*, a situation where an environmental event or condition works in conjunction with other factors to precipitate the emergence or prolongation of crises (Brown and McLeman 2009). Although it is a convenient conceptual way of framing the relationship between drought and conflict, it is inherently difficult to validate or quantify.

There is even weaker empirical evidence for the D-M-C pathway's most cited example, the eruption of violence in the Darfur region of Sudan in 2003. As noted earlier, situation reports from the UN Environment Programme suggested that droughts and increasing aridity caused by climate change were the root causes of the intergroup violence and subsequent displacement of 2 million people. Multiple studies by scientists have found no significant, identifiable changes in precipitation patterns or desertification in the Darfur region in the years leading up to the conflict (Kevane and Gray 2008, Brown 2010). Indeed, the framing of the conflict as an outcome of climate change diverted international attention from the fact that the Sudanese government was deliberately arming one particular group and encouraging it to attack others, which was the actual root of the conflict (Brown et al. 2007).

To conclude this section, it is worth noting there are few examples where people who moved because of drought or other environmental reasons have deliberately initiated violent interactions with residents of the locations to which they move. Research conducted in Vietnam and Kenya by Spilker et al. (2020) found that residents of receiving areas often perceive environmental factors as being legitimate reasons for migrating, and harbor no different attitudes toward environmental migrants than they do toward any other group. Environmental migrants in

Kenya were found to have a high likelihood to join social movements that promote migrants' rights and social integration, but they are more likely to experience violence and hostility than to perpetrate it in the areas they move to (Koubi et al. 2021). In short, no presumptions should be made as to how environmental migrants and receiving communities will interact with one another – amicably, with hostility, or in some other way – and the eventual outcomes will inevitably depend upon a wide range of factors specific to the case in question (McLeman 2020a).

Box 3.5
Does Desertification Cause Migration?

Desertification refers to an ongoing process of land degradation in dryland areas that results in reduced vegetative cover and a trend toward permanently drier conditions, and is distinct from droughts and general aridity. Challenges in understanding its impacts on migration include a lack of standardized measures of desertification and a common habit among migration researchers to use the term loosely and interchangeably with drought and aridity. Identifying potential associations between desertification and migration is a multi-step process, which relatively few studies have undertaken. The first step is to identify in a systematic way locations that are experiencing desertification, an efficient way being to locate dryland areas that have experienced a decline in vegetative cover over time using satellite or other remotely sensed data (Hermans and McLeman 2021). The next step is to then identify migration patterns particular to those locations during the period in which desertification is experienced, and then ideally to compare them either with migration patterns in similar areas in the same region that are not experiencing desertification and/or in the same locations but during periods when desertification was/is not occurring. Doing so does not prove that desertification is in itself a primary cause of the observed migration patterns, but it allows for at least a more confident identification of the potential association between the two variables, which can be further teased out through qualitative field research in affected areas.

Research that has systematically tried to isolate desertification from drought, heat, and precipitation shortfalls as a potential causal factor in migration include Call and Gray's (2020) study of migration in rural Uganda, Neumann et al.'s (2015) study of rural outmigration patterns in Burkina Faso and northeastern Brazil, Nebie and West's (2019) comparative study of two rural districts in Burkina Faso, and Hermans-Neumann et al.'s (2017) study of the northern Ethiopian Highlands. In the Uganda study, the researchers found very little evidence of land degradation or loss of vegetative cover having any significant influence on migration, but found that droughts and extended heat waves are significant factors. By contrast, land degradation does appear to be an important contributor to rural out-migration observed in Burkina Faso and northeastern Brazil. In Burkina Faso, institutional initiatives to address land degradation in the Bam district lowered outmigration starting in the 1990s. In Ethiopia, the effects of land degradation on migration appear to be compounded by uncertain land tenure and lack of institutional support for farmers.

3.3 Extreme Heat Events and Hotter Average Temperatures

There is no evidence that very short heat waves have any significant, direct effect on migration and displacement, although they can lead to higher numbers of doctor visits and hospital admissions, and can even lead to fatalities, particularly of older people and people with pre-existing health conditions (Song et al. 2021). These types of health impacts could have indirect implications for the future mobility and migration of such high-risk groups. The frequency and severity of heat waves are rising on all continents, as is the number of heat-related deaths, and it is estimated that over one-third of this increase since 1990 is directly attributable to anthropogenic climate change (Vicedo-Cabrera et al. 2021).

The influences on migration of extended periods of extreme heat or the elevation of average temperatures in regions that are already hot are poorly understood, reflecting a lack of empirical evidence on the possible causal pathways between the hazard, its impacts, adaptation options, and migration responses. A number of quantitative studies have compared large temperature datasets with population and migration data and identified a positive association between heat and migration – in particular higher levels of rural out-migration – but in many such studies, heat data in the form of temperatures above a selected threshold are being used as a proxy for drought instead of SPEI or PDSI data. Such studies have been useful for comparing temperature-related hazards with other types of hazards to see which may have the strongest associations with observed migration, and provide a jumping-off point for qualitative research into the causal mechanisms. For example, in rural Pakistan, higher levels of out-migration, particularly by men, are observed when extreme heat events occur during the November–April post-monsoon growing season (Mueller et al. 2014). The authors note that rural out-migration rates during heat events are highest among poor rural households, with the likely implication being that as their livelihoods become compromised, male household members move in search of income, in processes similar to those described in Section 3.2 above. Other studies have observed that migration rates of young, unskilled or lower-educated adults into urban centers in Central and South America and the Caribbean generally increase during periods of extreme heat, again probably reflecting a search for wage labor and economic opportunity (Baez et al. 2017b, Thiede et al. 2016). Such effects are not universal across the region, however – migration rates in Bolivia drop during periods of extreme heat – providing another reminder that the pathways between hazard and migration outcomes are always specific to local contexts. A broad-based study of the effects of extreme heat (and wildfires – see Section 3.4) on migration rates within the US found that rural counties that have an economy closely tied to outdoor recreation and environmental amenities may receive fewer in-migrants in the year following a year with an extended period of high heat (i.e. more than ten consecutive days with average daily temperatures above 90°F/32°C) (Winkler and Rouleau 2021). The authors suggest the reason is that while reliably warm temperatures are an amenity that migrants often seek out (Box 3.6), an area that becomes perceived as being too hot will become less attractive.

Box 3.6
Migration in Search of Hot Weather

Although the focus of this chapter is on cases where drought and heat events push people to move out of affected areas, it is worth recalling that there is also in many parts of the world *amenity-seeking migration*, where people deliberately seek out hot, dry, and (in some cases) water-scarce locations. This migration can take various forms and engage people across a variety of age groups.

The best known and possibly most numerous are middle- and higher-income retirees moving on a seasonal basis between locations with cold winters and sunnier climes, often referred to in English as *snowbirds* (in Chinese a similar term is used, *houniao* (Chen 2020)). This migration can be international (e.g. from the UK and northern Europe to Mediterranean countries, from Canada to the southern US, from the northern US to Mexico and the Caribbean) or internal (e.g. from northern China to southern China, from the northern US to the southern US), depending on the geography of the countries in question. Recent decades have seen growth in longer-distance international retirement migration, with European retirees, for example, traveling in increasing numbers to Southeast Asia, where the length and costs of travel are offset by the lower cost of living at the destination (King et al. 2021). Although there are numerous published studies on flows of retirement migration within and between particular countries, reliable global estimates of retiree migrants are scarce, but it is reasonable to assume they number in the tens of millions. As one example of the scale, a 2005 study estimated that the US state of Florida alone hosted 800,000 temporary elderly migrants (Smith and House 2006), a figure that likely underestimates the present day number. Some studies suggest that up to a million Canadians of retirement age travel to the US for at least a few weeks each winter (Kremmidas 2024), with between 300,000 and 375,000 staying for a longer period (Kelly 2023). The southern Chinese city of Sanya receives an estimated 400,000 seasonal retiree migrants each year (Kou et al. 2018). Temporary snowbirds that can afford to do so may decide to permanently relocate to their preferred warm weather destinations, depending on factors such as familiarity, satisfaction, and other subjective perceptions of the destination (Chaulagain 2023). In many cases described in the literature, the local economy of receiving areas is economically dependent to a significant degree on providing services for snowbirds.

A transformative technological innovation that was a precursor to snowbird migration and has been responsible for much larger scale, long-term shifts in migration and population patterns in many countries has been air conditioning. Until it became widely available and affordable, hot-and-humid and hot-and-dry locations were not typically seen as being desirable destinations in terms of amenity migration. They were also not practical places in which to establish manufacturing facilities or other types of non-agricultural enterprises that are not tied to any specific location and that might otherwise attract wage labor-seeking migrants from other areas. Thanks to air conditioning, the last fifty years have seen a massive expansion in economic activity in hot climates – think of the many automobile assembly plants in the southern US and northern Mexico, large industrial zones of southern China, and garment factories found throughout low- and middle-income countries of the tropics and sub-tropics. Tremendous population growth has followed that economic expansion, and explains why, for example, the fastest growing states and cities in the US over recent decades have been disproportionately situated in the southern part of the country (Gutmann and Field 2010).

An entirely different approach to studying the effects of heat on migration and displacement patterns is one that looks at the long-term *habitability* of regions that experience hot temperatures all or part of the year (Gilmore et al. 2024). Physiologically, the human body requires a skin temperature lower than 35°C to cool itself by radiating heat outward. At higher skin temperatures, the body relies on the evaporation of sweat to provide an additional latent cooling effect. However, if the air is too humid, sweat ceases to evaporate, the body can no longer cool itself and it begins to overheat, leading to a serious and potentially fatal health condition known as hyperthermia (Bach et al. 2024). Older people and people with pre-existing cardiovascular, respiratory, and other health conditions are especially vulnerable to hyperthermia (Cissé et al. 2022). To estimate the weather conditions at which hyperthermia can occur, scientists use a wet-bulb thermometer – one that is wrapped in saturated cloth and has moving air passed over it, so that the cooling effects of evaporation are captured. If the air surrounding the wet-bulb thermometer is extremely humid (i.e. approaching 100% humidity), there will be no evaporation, and the wet-bulb thermometer will register the same temperature as would a dry-bulb thermometer with which most people are familiar. At lower levels of humidity, the temperature recorded by the wet-bulb thermometer will be lower than the ambient air temperature. In other words, this thermometer registers the combined effects of heat and humidity. For scientists studying the relationship between climate conditions and hyperthermia risks – and therefore the habitability – of a given location, a wet-bulb temperature reading of 35°C or higher represents a physiological threshold where sweat ceases to evaporate. Exposure to temperatures higher than this for six or more hours per day creates a situation where hyperthermia-related health impacts can be expected to emerge even in healthy adults (Sherwood and Huber 2010).

A number of studies have used models of where 35°C wet bulb temperatures are expected to occur in a changing climate to project future habitability of geographic regions. At present, daytime high temperatures only occasionally surpass this threshold, even in the hottest inhabited parts of the world, and minimum nighttime temperatures are almost never higher than it (which is likely the absolute threshold for habitability). Depending on the SSP/RCP scenarios used, studies forecast that coastal areas of the Persian Gulf region and large areas of India and Pakistan will be among the first to regularly experience daytime wet bulb temperatures higher than 35°C in at least some months of the year, starting in the second half of this century (Raymond et al. 2020, Horton et al. 2021). A group of health researchers has suggested that wet bulb temperatures of greater than 35°C for even brief periods of time make it difficult for people to carry out even moderate physical work, and that exposure to wet bulb temperatures of greater than 40°C over the course of even a few days can be fatal (Andrews et al. 2018). Using these distinctions, large areas of tropical, sub-tropical, and even temperate regions will experience a significant increase in the number of days each year where outdoor work becomes physiologically difficult to perform, and at global warming levels of +2.5 to +3°C between 20 and 50 million people worldwide would be living in locations exposed to temperatures that would be too hot to survive. Other researchers have suggested that the 35°C wet bulb threshold is much too high, that observable health impacts occur at significantly lower wet bulb temperatures, and that studies that use this threshold underestimate the true risks of a warming climate (Vanos et al. 2023). When the habitability threshold is reduced to a minimum of one day per year when wet bulb temperatures reach 33°C, an estimated 789 million people would be exposed to such risks with 2°C of warming, and 1.22 billion people with 3°C of warming (Li et al. 2020).

All of this begs the question, if average temperatures in a given area start to routinely exceed a threshold where one's health or very life is at risk – whatever the absolute wet bulb temperature value might be – will people move away from such areas? Will the populations of such areas fall in terms of either absolute numbers or density? Will people in hot climates start to move out of or avoid moving to cities, which are invariably hotter than surrounding rural areas due to urban heat island effects, and will likely cross the habitability threshold earlier than other locations in the same country? (Tuholske et al. 2021). At present, there is no evidence to suggest that population numbers in a given area decline when exposed to excessively hot temperatures, although the rate of increase may slow down, especially in densely populated tropical and subtropical locations (Gray and Call 2023). There is also no evidence from recent experience that long-distance migration rates – which would be necessary to escape a region with dangerously hot temperatures – increase significantly during periods of excessive heat. For example, a comparison of changes in average temperatures and international migration rates for the period 1960–2000 found that international migration out of middle-income countries increases slightly with rising temperatures (typically destined to other middle-income countries but not to high-income countries) and that migration out of low-income countries decreases (Cattaneo and Peri 2016). However, such evidence represents observed population and migration responses to heat that have occurred without crossing the habitability threshold, and evidence elsewhere has shown that thresholds in environmental exposures and human adaptation can result in rapid and unexpected changes in behavior once crossed (McLeman 2018).

What seems most likely is that, as temperature-related habitability thresholds are approached and surpassed in specific locations, the potential for large-scale population movements out of those locations is likely to be heavily moderated by non-environment social, economic, and political factors, as is the case with other types of climate hazards. Many of the necessary adaptation responses to higher temperatures will take place at the household level, through such things as acquiring air conditioning, seeking out housing that is ventilated and insulated so as to minimize indoor temperatures in the hottest part of the day, and pursuing occupations and livelihoods that do not require laboring in full sun. Such options may not be affordable or available to low-income and agriculturally dependent households, and uneven access to such adaptations may create a widening economic gulf between households within a community and contribute to unequal economic growth across countries (Pavanello et al. 2021, Davis et al. 2021). Especially in low- and middle-income countries, governments and institutions will need to assist households in improving their homes to be more heat-resistant and ensure the electrical infrastructure can support the extra demand for air-conditioning during heat waves, something that infrastructure even in high-income countries often struggles with (Zuo et al. 2015). This in turn raises the question of whether the additional energy demands can be met without increasing consumption of fossil fuels; if not, a self-reinforcing feedback loop emerges where hotter temperatures lead to higher GHG emissions which in turn leads to even hotter temperatures. For heat risks in urban centers, other adaptation options that require institutional leadership include building temporary cooling centers for low-income residents and increasing the amount of greenspace and tree cover within cities so as to reduce the amount of heat trapped locally and increase the amount of available shade (Fernandez Milan and Creutzig 2015). These again tend to be options better available to cities in high-income countries than in low-income countries.

The impacts of heat on livelihoods and incomes of agricultural households, fishers, and a whole range of outdoor occupations may present even greater challenges for exposed populations than the immediate physiological need for cooling, and will hit low-income households and countries disproportionately. Compounding such risks would be the impacts on short- and long-term food prices that can be expected to occur due to heat-related disruptions of crop yields and livestock health, the financial impacts of which would again fall most heavily on low-income households and on low-income countries (Haile et al. 2017). For households that lack the means to change or diversify livelihood activities to reduce their exposure to the economic risks of hotter temperatures, there is a serious risk of their being trapped in locations that become increasingly uninhabitable, as they would also likely lack the means to relocate over long distances without external assistance. Being connected to geographically extensive migrant networks might help some low-income groups and individuals migrate away from dangerously hot locales, but people lacking such connections would seem destined to become trapped. At larger scales, it has been estimated that extreme heat events attributable to anthropogenic climate change are already generating billions of dollars in economic losses each year, and that the economic burden on low-income countries as measured by percentage of lost GDP is more than twice as large as for high-income countries (Callahan and Mankin 2022).

3.4 Migration and Displacement Associated with Wildfires

Wildfires occur on every inhabited continent, in all types of vegetated environments (i.e. grasslands, forests, etc.), and are typically associated with extended periods of high heat and low precipitation, referred to as *fire weather* by scientists. In many ecosystems, fire is a natural and regular ecosystem process that helps rejuvenate and revitalize vegetative cover. Some plant species are so well adapted to periodic fires they actually require fire to regenerate. For example, in North America, jack pine (*Pinus banksiana*) trees produce serotinous seed cones that are glued tightly shut with resin and will only open and disperse their seeds after being heated by fire. This allows their seedlings to germinate after a fire has removed competing vegetation and opened the forest canopy (Natural Resources Canada 2020). The lodgepole pine (*Pinus contorta*) is even more highly adapted in that individual trees will produce either serotinous or non-serotinous cones depending on the general frequency of fires in the local environment. Other tree species, such as trees in the birch (*Betula*) and aspen (*Populus*) genus, as well as most grasses and forbs in grassland ecosystems, are able to regenerate from their unburned roots after fires.

Although natural systems are highly adapted to periodic wildfires, human systems typically are not. Wildfires are not as common a driver of migration and displacement as compared with other environmental hazards – wildfires were responsible for 435,000 of 20.3 million weather-related displacements recorded globally in 2023 (IDMC 2024) – but in certain environments and regions they pose a disproportionately large risk to human settlements. Wildfire-related displacement risks are particularly great in Mediterranean-type climates with large populations, such as southern Europe, southern California, and southeastern Australia. Wildfires are a regular occurrence in the world's largest contiguous forests – the North American boreal forests and the Eurasian taiga – and tropical forests of South America and Southeast Asia may also experience large fires during unusually dry periods, but in these environments human population densities are relatively low, and so fewer people are at risk of displacement. In addition to the destruction caused to homes, commercial buildings, and infrastructure, the immense amounts of smoke and air pollution generated by wildfires can have serious health implications for people living across a much wider area, and there is some evidence that exposure to wildfire smoke can potentially have a moderate effect on some individuals' migration decision-making (Berlin Rubin and Wong-Parodi 2022).

As rising average global temperatures and shifting precipitation patterns create more favorable conditions for fire weather, the frequency and/or severity of wildfires is expected to increase in southeast Australia (Di Virgilio et al. 2019), Canada (Gaur et al. 2021), Europe (Dupuy et al. 2020), Russia (Shvidenko and Schepaschenko 2014), and the US (Heidari et al. 2021, Zou et al. 2021). In California, the number of fire weather days in autumn (the most dangerous season for large wildfires in that state) has doubled since the 1980s, and this is directly attributable to anthropogenic climate change (Goss et al. 2020). In addition to the expanding amount of fire weather due to climate change, fire risks have grown in many regions due to authorities having worked actively for decades to suppress wildfires and extinguish them quickly, allowing for a buildup of flammable material in unburnt forests (Parisien et al. 2020). This becomes especially problematic at the interface between wildlands and settled areas, where rapid population growth has been occurring in many wildfire-prone countries. For example, in the US, it is estimated there are approximately 50 million homes in the wildland–urban interface and this number increases by one million houses every three years (Burke et al. 2021). Another factor that elevates future wildfire risks is that climate change is expected to reduce the resilience of forests to fires, as drier precipitation regimes impede the post-fire regeneration of trees, which in turn increases the risk of more frequent future fires (Stevens-Rumann et al. 2018).

The dynamics of wildfire displacement are similar to those of tropical cyclones reviewed in the previous chapter. When weather conditions in a given area create elevated fire risks, authorities issue warnings to avoid any activities that might unintentionally spark a fire and, if fires are ignited, for residents to prepare for evacuation if necessary. Authorities monitor the emergence of new fires and, where resources and terrain allow, they dispatch firefighting crews to try and contain and extinguish the fire and/or to create firebreaks (i.e. areas cleared of vegetation) between fires and settled areas. When this proves inadequate, authorities evacuate areas in the fire's immediate path. Only after the fire has passed do returning residents learn if their homes, businesses, and livelihood assets have been damaged or destroyed. For those who have lost homes, decisions then need to be made about whether to re-establish residence in the affected community or relocate elsewhere. There is some empirical research on reconstruction and community-level recovery following wildfires for many countries, but surprisingly few studies on the characteristics of people that do not return, the determinants of why they do not return, and where they go.

For example, a massive wildfire in May 2016 required the evacuation of the city of Fort McMurray in Alberta, Canada, and an estimated 2,400 homes were destroyed or damaged (Public Safety Canada 2024). For several years, local officials made anecdotal reports that some residents did not return, using school enrollment figures as their primary data source, but no systematic attempt was made to identify who did not return, and why not. The next census count was held in 2020 and published the following year, and showed that the population of the community had experienced a modest increase from 71,594 to 72,326 since the 2015 Census count. However, these data do not indicate how many 2020 residents moved to the community after the fire, how many 2015 residents did not return after the fire, nor where the latter may have gone. In other words, no systematic comparison was done between residents who were temporarily displaced versus those that were permanently displaced. The failure to systematically account for and take stock of long-term displacement is a common occurrence not just for wildfires but for all types of climate-related events and warrants greater efforts on the part of governments and academic researchers alike.

Most of what we currently know about temporary versus long-term displacement after wildfires comes from research done in the US. In a broad-based study that compared extreme heat events and wildfires with county-level US census data for the period 1990–2015, both wildfires and extreme heat events were associated with increased out-migration from affected areas and lower levels of in-migration to those areas (Winkler and Rouleau 2021). The net migration rate of counties that experienced a wildfire in the previous year was reduced by an average of 32 net migrants per 10,000 as compared with counties that did not experience a fire and did not have a neighboring county experience a fire in the previous year. This methodological decision to include the effects of fires in neighboring counties is an astute one, as sub-national political boundaries such as county lines often do not reflect actual settlement patterns, transportation routes, and other socio-economic linkages in a given area. This is seen in the study's additional findings that counties that did not experience a fire in the previous year but had a neighboring county that did experience a fire also experienced a reduction in net migration rate, of an average of 18 net migrants per 10,000 as compared with counties that had no experience with fire and no neighboring counties that experienced a fire in the previous year.

A particularly interesting finding of the same study is that counties that experienced a fire <u>and</u> had a neighboring county also experience a fire saw a reduction of 15 net migrants per 10,000 as compared with other counties. The fact that the net migration rate declined, but not by as much as the other two situations, is because the decline is driven by a drop in in-migration rates to the fire area. By contrast, in situations where a county experiences a fire but its neighbor does not, the decline in net migration is due to higher rates of out-migration. And, in the case where only a neighboring county experiences a fire, the drop in net migration is due to a combination of higher out-migration and lower in-migration. If this sounds confusing (and it does), what does it actually mean? It means that most people who experience a wildfire either return to their home community or, if they relocate elsewhere, choose a nearby destination in the same geographic region.

This last conclusion is borne out by evidence from the 2010 Fourmile Canyon fire in Colorado, which destroyed 169 homes in Boulder County. Surveys of over 400 residents conducted on two different occasions after the fire found that only about 4% intended to move out of the immediate area, and almost all of them were resettling or planning to resettle in the same county (Nawrotzki et al. 2014). Similar results emerged from research conducted in the wake of large wildfires that destroyed nearly 7,000 homes in Sonoma County, California, and surrounding areas in October 2017 (Sharygin 2021). Out of nearly 7,800 persons displaced from the main area burned by fire, the researchers estimated that fewer than 1,000 changed neighborhoods afterward, and fewer than 500 moved out of the county.

Home ownership (or lack thereof) appears to be an important factor in determining which people return to an area that experiences a wildfire and which ones move elsewhere. A study by McConnell et al. (2021) investigated in-migration and out-migration patterns following wildfires in the US by comparing census data for 1999 to 2018 with counts of damaged buildings and changes in homeownership, consumer credit usage, and financial distress among people whose neighborhood suffered damaging fires, using Federal Reserve Bank of New York/Equifax Consumer Credit Panel datasets. They found that census areas that experienced a wildfire then saw significantly higher out-migration rates but few changes in in-migration rates, consistent with the Winkler and Rouleau (2021) study mentioned earlier. The authors did observe a significant drop in homeownership among those who experienced major fires, particularly among people over the age of 60. Overall, the study found that while the effects of wildfires on migration and on borrowing are measurable, the impacts are not as large as for hurricanes or other extreme weather events. Evidence from a 2018 wildfire that destroyed 90% of the

14,000 homes in the town of Paradise, California, indicates that people who were renting their homes were often unable to return because the rebuilding and recovery effort was so heavily focused on property owners (Chase and Hansen 2021)

In southern Australia 2019 and 2020 during what was known as the "Black Summer," large wildfires destroyed thousands of homes and displaced an estimated 65,000 people at least temporarily (Ahmed and Ledger 2023). Although most people who were evacuated are believed to have returned to their homes fairly quickly, researchers estimated that it would take up to 4 years after the event for people whose homes were destroyed to resolve insurance claims and be in a position to finalize decisions about whether to return or relocate permanently elsewhere (du Parc and Yasukawa 2020). Factors cited by people who were hesitant to return to their former homes included a general feeling of no longer being safe in that area, as well as more practical reasons related to employment prospects and reduced access to basic services. As with wildfire-related displacements in other countries, specific data on the number of temporary versus long-term displacements remains lacking. Interestingly, a study conducted just before the Black Summer found that for most Australians, natural hazards are not an important consideration on where to live or migrate to, except for wildfires (Zander et al. 2020). Follow-up research to see if such perceptions have changed or are beginning to change would be very welcome.

In one of the few available studies of wildfire displacements in low-income countries, researchers working in the Imizamo Yethu informal settlement in Cape Town, South Africa, investigated why people would move there and/or remain there given that it is in a location that has experienced repeated wildfires (Christ et al. 2023). Their findings were that, despite the dangers, the settlement continued to be attractive because it offered close proximity to employment opportunities, residents had strong networks of extended family and friends close at hand, and some people had strong feelings of place attachment. The effects of wildfires on migration and displacement in low- and middle-income countries is a subject where more research is strongly encouraged and presents an opportunity for early-career scholars looking to make a mark on environmental migration scholarship.

4

Migration and Displacement Risks Associated with Mean Sea Level Rise

4.1 Introduction

The world's oceans and the atmosphere are *coupled systems*, meaning that variations and changes in the function or characteristics of one have impacts on the other (Xue et al. 2020). As average temperatures rise due to anthropogenic greenhouse gas (GHG) emissions, oceans experience changes in sea surface temperatures and in the salinity and the acidity of seawater. These changes in turn have effects on regional and global weather patterns, on the health of coral reefs and marine organisms on which human livelihoods depend, and, potentially on the circulation of seawater and ocean currents in future decades, which would have disruptive impacts on weather patterns and impacts in some regions of the world (see Section 7.6). The focus of the present chapter is on the migration and displacement implications of measurable changes in mean sea levels, which are currently rising at an average rate of 3.3–3.7 mm per year (European Environment Agency 2024). Climate-related mean sea level rise (MSLR) is caused primarily by the combined effects of thermal expansion of ocean water and the melting of glaciers and ice caps on Greenland and Antarctica. Although 3.7 mm may seem like a negligible amount, the accumulated year-over-year rise of sea levels relative to the land presents serious hazards for human settlements in low-lying coastal areas and on small islands. Further, the rate at which sea levels are rising is accelerating in conjunction with the continued rise in atmospheric temperatures (Sweet et al. 2022), and by the end of this century, tens of millions of people (or more) globally may find that locations that are currently habitable no longer will be without enormously expensive coastal protection infrastructure (Lincke and Hinkel 2021, Hauer et al. 2020). Already, small communities in many parts of the world are having to implement relocation plans and move exposed homes, businesses, infrastructure, and livelihood assets to higher ground, a process that is disruptive, complicated, and very expensive (McMichael et al. 2021).

In this chapter we:

- review the physical processes that affect the elevation of coastal settlements; relative to the sea, and identify current and projected rates of change;
- describe the impacts of MSLR on coastal settlements and on small island states;
- provide rough estimates of the number of people exposed;
- identify options for *in situ* adaptation;
- describe common challenges in implementing planned relocations of communities at risk, with case studies from the Carteret Islands and Fiji;
- conclude by reviewing the cascading risks faced in Bangladesh, a country that experiences all of the climatic hazards reviewed in Chapters 2, 3, and 4.

4.2 Drivers of Sea Level Rise, Observed Changes in Mean Sea Level, and Projected Future Changes

4.2.1 Physical Process That Causes Sea Levels to Change

The level of the world's oceans relative to land is continually changing. This is due to geological and geomorphological processes that affect the elevation of the land relative to the ocean, and to climate-related processes that affect the volume of water in the oceans. Geologic processes affecting the elevation of land in coastal areas operate across a range of timescales and include:

- the gradual movement of the continental plates on the Earth's mantle, typically measured in timescales of millions of years or more (Conrad and Husson 2009);
- sudden-onset geological events such as volcanoes and earthquakes that can cause vertical changes to land in coastal areas (Denys et al. 2020), and
- glacial isostatic adjustment of northern regions of the North American and Eurasian land masses as they decompress after having been weighed down by large amounts of snow and ice during the last Ice Age, which ran from approximately 21,000 to 6,000 years ago (NASA Jet Propulsion Laboratory 2024). At higher latitudes, land that was under the greatest mass of ice and snow is presently rising by 10 mm or more per year, and more southerly land that was only lightly glaciated, or was adjacent to glaciated areas – such as the US mid-Atlantic and south Atlantic coasts and the Gulf of Mexico – is sinking, a phenomenon known as forebulge collapse.

The collective effects of these geological changes influence the positioning of coastal lands relative to the sea, and at any given location may be pushing the land upwards, downwards, or leaving it relatively stable. Added on top of these are the effects of geomorphological changes such as the ongoing erosion of the coastal lands by heat, weather, chemical and mechanical processes. These convert rock to sand and smaller particle sizes, with gravity, wind, and moving water transporting and depositing these along shorelines. When deposited at or near sea level, such as in river deltas or coastal plains, these sediments may temporarily raise the level of the land, but over time will gradually compact and compress under their own weight, causing the area of deposition to subside. These areas of deposition – beaches, deltas, salt marshes, and other coastal landforms – are then subject to further erosion by water currents, tides, and storms. Human activities, such as the creation of artificial drainage and extraction of groundwater or fossil fuels can increase subsidence rates in such environments (Brown and Nicholls 2015).

A second key factor in determining the position of oceans relative to adjacent land is that the volume of water in the world's oceans is continually changing over time. Sea levels today are at least 6 m lower than they were 125,000 years ago, the warmest part of the interglacial period between the two previous ice ages, when average global temperatures were between 3°C and 5°C warmer than they are today (Kopp et al. 2009). After the most recent ice age, global sea levels rose by approximately 130 m, and stabilized about 2,000–3,000 years ago close to the levels we are familiar with today (NASA Earth Data 2024). Since the mid- to late- nineteenth century, sea levels have risen globally at an average rate of 1.7 mm per year, with each ocean basin and coastal areas alongside each ocean basin experiencing varying rates of observed change owing to the combination of physical processes specific to each. The globally averaged change is referred to as *mean sea level rise* (MSLR). Although the average rate of MSLR for the period 1860s to the present is an estimated 1.7 mm, the observed rate has accelerated in recent decades, to an estimated rate of 3.3 mm per year ($\pm$ 0.3 mm) since the 1990s (Guérou et al. 2023). Changes in MSLR are measured through data collected from tide gauges maintained at approximately 2,000 sites worldwide, some of which have records dating back a century or more, and through satellite-collected altimetry data, which date back several decades (Church and White 2011).

Three key phenomena account for MSLR observed since 1900 (Frederikse et al. 2020). The first of these is the melting of ice and snow on land due to warming temperatures, the main sources being ice caps on Greenland and Antarctica and glaciers at high altitudes. Meltwater from the Greenland ice sheet is one of the largest single contributions to current MSLR, and it will contribute an additional 320–950 mm of MSLR over the course of this century depending on future atmospheric warming trajectories (Goelzer et al. 2020, Box et al. 2022). The global contribution of meltwaters from mountain glaciers to MSLR in recent years has been estimated at approximately 1 mm per year, with the largest amounts originating in Alaska, Antarctica, and Asia (Zemp et al. 2020). The exact contribution of Antarctica to MSLR is less easily measured, and there is additional uncertainty regarding the net loss of ice and snow from that continent in future decades owing to the offsetting effects of more rapid melting versus additional accumulation of snow due to shifting precipitation patterns. Based on current and projected warming trends, an additional contribution of up to 450 mm toward MSLR can be expected from Antarctica by the end of this century (Pattyn et al. 2020).

A second contributor to MSLR is a process called *thermal expansion*, whereby the volume of seawater increases as its temperature rises, thereby leading to an increase in sea level relative to adjacent land areas. Over the course of the past 150 years, annual sea surface temperatures have increased by a global average of 0.88°C, with the observed amount of change varying by ocean basin (Fox-Kemper et al. 2021). These increases, as well as a greater range of annual sea surface temperature variability within ocean basins, are directly linked to anthropogenic warming of the atmosphere (Shi et al. 2024). An estimated 38% of observed MSLR since 1900 is attributable to thermal expansion, which in recent decades has become an increasingly dominant driver, with melting from glaciers and ice caps being the most clearly dominant factors earlier in the twentieth century (Frederikse et al. 2020). The contribution of thermal expansion to MSLR is expected to accelerate as oceans accumulate growing amounts of heat from the atmosphere in the coming decades.

A third factor that influenced MSLR in past decades was the large-scale construction of dams and reservoirs in many regions of the world, especially in the late 1960s and 1970s, which temporarily decreased the amount of water drained by major rivers into the world's oceans (Frederikse et al. 2020). In the absence of such dams, the observed rate of MSLR might actually have been 0.2 mm higher in that period (Fiedler and Conrad 2010). Some researchers have suggested that constructing a series of dams around Greenland might be a way of slowing the future release of meltwaters from the icecap to the ocean (Hunt and Byers 2019).

It is worth noting that, although Arctic Ocean ice sheets are shrinking rapidly due to global warming (Fox-Kemper et al. 2021), this has no direct effect on MSLR. Because the ice is already in the ocean, its melting does not change the volume of water in the ocean, just as the melting of ice cubes in a full glass of water will not cause it to overflow. However, sea ice is highly reflective, and much of the solar radiation that shines upon it during the daytime is reflected back into space, preventing it from adding to heat accumulating in the atmosphere. As sea ice retreats, it is replaced by open water that is less reflective and absorbs incoming solar radiation, causing a summertime buildup of heat in the ocean that is later released to the surrounding air in the fall and winter (Mengis et al. 2016). The result is a positive feedback loop in which Arctic air temperatures are warming at a faster rate than the global average, which in turn accelerates the melting of ice. It is expected that by mid-century the Arctic will often be entirely ice free in the summer months (Fox-Kemper et al. 2021).

4.2.2 Future Projections for Sea Level Rise

It is virtually certain that sea levels will continue rising through the end of this century and beyond, the only question being by how much, which will in turn depend to a large extent on whether collective action is taken rapidly to reduce GHG emissions and thereby curtail additional atmospheric warming. The Intergovernmental Panel on Climate Change (IPCC) projects that if GHG emissions were to be immediately curtailed and then decline rapidly, there would be an additional 150–230 mm MSLR by the year 2050 and an additional 280–550 mm MSLR by 2100 from current levels (Fox-Kemper 2021). The continuation of MSLR even under low emissions and no additional warming reflects the lag effect of heat accumulation and melting of ice caps that is already under way (referred to as *committed warming* and *committed MSLR*). Under high emissions scenarios, an additional 200–290 mm MSLR can be expected by 2050, and 630 to 1,010 mm MSLR by 2100. The IPCC warns, however, that these estimates of the rate and amount of MSLR could be too low should there be a more rapid deterioration of ice caps on the perimeter of Greenland and Antarctica than what is currently foreseen. Indeed, many researchers believe the IPCC's estimates of future MSLR to be much too conservative, and that the IPCC is underestimating the chances of MSLR outcomes being at or higher than the upper range of future forecasts (Siegert et al. 2020). A survey of 106 scientists who work in this field showed their collective estimates for 2100 to range from 300 to 650 mm of additional MSLR under low emissions scenarios and from 630 to 1,320 mm under high emissions scenarios (Horton et al. 2020).

For the purposes of hazard planning – regardless of the type of hazard – a precautionary approach that uses plausible high-end scenarios is the most sensible course of action. For MSLR, experts suggest that for the year 2100, with low GHG emissions pathways, a plausible high-end scenario would be an additional 900 mm of MSLR from present levels, and under high emissions scenarios, a plausible high-end scenario would be 1,600 mm of additional MSLR by 2100 (van de Wal et al. 2022). It is also important to keep in mind that sea levels observed at any given coastal location fluctuate according to tides, and there can be additional temporary increases as a result of storm surges (see Chapter 2) and heavy wave activity, so mean sea level is only a rough indicator of the risk to which coastal populations are and will be exposed. The coincidence of a high tide with big waves and a severe storm surge creates on a temporary basis what is known as an *extreme sea level* (Jevrejeva et al. 2023). Adding extreme sea level risks on top of a high-end scenario for MSLR would create by the year 2100 an average extreme sea level risk of 4,200 mm globally, with some coastal areas of areas of East Asia and northern Europe facing extreme sea level risks of between 9,000 and 10,000 mm. The statistical likelihood of such genuinely catastrophic scenarios happening is low, but not zero.

As noted earlier, the relative amount of sea level rise will vary from one coastal area to another depending on local oceanic and geomorphological processes. In some parts of the world, such as along Hudson Bay and the northern Baltic Sea, the rate of post-glacial isostatic rebound is high, and so the relative change in sea levels experienced by settlements along those coasts will be lower than the global average (Nicholls and Cazenave 2010). Conversely, the land in many densely populated deltas is subsiding from a combination of natural processes and human activities (such as the depletion of groundwater through pumping for irrigation and direct consumption), and so the relative increase in sea levels experienced in those locations can be expected to be much higher than the global average (Syvitski et al. 2009). In the case of a country such as the US, which has coasts on the Atlantic, Arctic, and Pacific Oceans, SLR risks will vary considerably by coast. For example, the US Pacific coast is expected to see a relative SLR of 100–200 mm by the year 2050, but the US Atlantic coast will see 250–350 mm relative SLR and the Gulf of Mexico coast 350–450 mm relative SLR by 2050 (Sweet et al. 2022).

4.2.3 Risks Particular to Inhabited Atolls

Tropical regions of the Pacific and Indian Oceans contain many inhabited atoll reef islands, which have unique tectonic and geomorphological characteristics that make them particularly exposed to MSLR risks. Every atoll is distinctive in its physical characteristics. Generally, atolls develop from extinct oceanic volcanoes that have protruded above the sea surface, and which host reef-building corals around their fringes (Woodroffe 2008). As the volcano subsides lower into the sea, the fringing reef builds itself upward to maintain its position relative to the sea surface. Eventually the volcanic material sinks below the surface, leaving only a characteristically ring-shaped crown (reflecting the rims of the original volcanic caldera) protruding a meter or two above the ocean surface, encircling a central shallow lagoon. Underneath the atoll is typically found a small lens of fresh water, which is filled and recharged by rainwater. The process of building an atoll can take hundreds of thousands or millions of years (National Ocean Service 2024). So long as the underwater coral reef remains healthy, the crown of the island will remain at or above sea level.

Reef-building corals are very sensitive to changes in ocean temperatures, ocean acidity, and human disturbance. Reefs are built by living organisms – polyps – that live in a mutually beneficial relationship with algae that they host within their tissues. Healthy reefs are colorful compositions of many organisms, but when under pressure, the corals expel their algae and become ghostly white – an event known as *coral bleaching* (National Ocean Service 2024). Bleached corals do not immediately die, but they stop growing and will die if conditions do not change. The number of bleaching events in recent years has been growing, and it is expected that most coral reef ecosystems will be under increasing pressure and become degraded in coming decades due to warmer temperatures and more acidic seawater due to climate change (IPCC 2019).

Scientists have raised concerns that even small changes in ocean conditions can stimulate coral bleaching events which, if persistent, can cause reef die-offs (Pandolfi et al. 2011). Contamination of an atoll's underwater lens by saltwater intrusion, which can be caused by human overuse or by events like tropical storm surges, can render the groundwater unusable (Werner et al. 2017). It remains uncertain how MSLR will affect freshwater lenses underlying atolls. While atolls are widely believed to be especially at risk of inundation and erosion under MSLR projections, there have been only small observed changes in the size and overall area of atolls in the Indian and Pacific oceans since the year 2000, apart from areas where people have deliberately reclaimed land to increase their size (Holdaway et al. 2021). However, there is concern that as the rate of MSLR accelerates, critical thresholds may be crossed where the natural processes that maintain the physical stability of atoll islands, their ecology, and freshwater resources become overwhelmed (Kane and Fletcher 2020).

In some atoll nations, urbanization is accelerating, which creates additional risks and challenges associated with having dense populations and so much infrastructure concentrated on specific atolls. Malé, the capital of the Maldives, is an extreme example, with over 100,0000 people living on a land area of just under 2 km^2, none of which is more than 2 m above mean sea level (Brown et al. 2023). Engineering work is already underway to raise and expand the amount of land available to residents of Malé (additional details follow in Section 4.5).

4.3 Impacts of Sea Level Rise on Human Settlements, Land Use, and Infrastructure

The potential physical impacts of MSLR on coastal areas fall into two broad categories. The first category is higher frequency and/or severity of flooding in low-lying coastal areas, with the lowest lying ones likely becoming permanently inundated over time. Higher sea levels allow high tides and storm surges to penetrate farther inland, and when rivers flowing into coastal areas are at high water levels, they can overspill their banks and flood surrounding areas at locations farther inland. At present, less than 25% of coastal locations around the world experience an average of ten or more minor floods annually (Hague et al. 2023), which might be referred to generally as "flood-prone" locations. With MSLR of 200 mm – which has a high potential of occurring by 2050 given current warming trends – the number of flood-prone locations grows to 69%, with 11% experiencing 50 or more minor floods on average annually. With MSLR of 1,000 mm – a very real possibility for the end of this century – over 90% of coastal locations are projected to experience more than 20 days each year of major floods (i.e. water levels roughly 50% higher than those of "minor" floods for the same location/tidal range). Unlike floods in river valleys described in Chapter 2, coastal floods are affected by the diurnal tide range, which varies considerably from one location to another. It should be noted the definitions of "minor" and "major" floods used by Hague et al. (2023) come from a standardized method of categorizing location-specific coastal floods that takes tidal gauge measures of the height of water relative to the mean low water average and adjusts these by the local diurnal tidal range.

Other studies warn that MSLR can cause the risks of flooding to grow exponentially, not incrementally. Researchers studying the effects of MSLR on the odds of given US coastal locations experiencing extreme flooding (i.e. the type of flood that has a statistical probability of occurring only once every fifty years) by the year 2050 found that, under high-emissions scenarios the odds increase by a factor of 100 or more (Taherkhani et al. 2020). Even under low emissions scenarios and at locations with relatively slow rates of local SLR, the odds of extreme flooding are likely to double several times between now and 2050. Under high-emissions scenarios, by the year 2100, the total area of the world exposed to coastal flood risks in general is projected to increase by up to 48%, putting over 1 million square kilometers of coastal lands at risk (Kirezci et al. 2020). This risk is not evenly distributed globally; over half the expected increase in coastal flooding due to MSLR will be experienced by seven countries: India, China, Bangladesh, Vietnam, Indonesia, the Philippines, and the USA (Tiggeloven et al. 2020). The Netherlands would also fall into this group were it not for the heavy investments already made and that will likely continue to be made in protective coastal infrastructure. This latter point about the effects of investments in protective coastal infrastructure is important, as the expected annual damages from coastal flooding are highly contingent upon assumptions about proactive adaptation; we return to this point in Section 4.5.

A second set of impacts of MSLR is on erosion rates, vegetation, groundwater, and soils in coastal environments. The specific nature of these impacts will vary from one location to another and will depend on the type of coastal environment and human use of it, in addition to the locally experienced amount of relative SLR. For example, there is considerable concern that sandy beaches are threatened by higher rates of erosion due to MSLR, with some researchers (but not all) projecting sandy beaches may all but disappear over the coming century (Vousdoukas et al. 2020). Researchers are similarly concerned – and not entirely certain – about how marshy coastal areas will fare in the face of MSLR, if they will persist or erode away (Fagherazzi et al. 2020). The potential loss of coastal marshes and mangroves (see Box 4.1) is problematic as they play important roles in moderating the impacts of high tides, heavy waves, and storm surges in low-lying coastal areas, and often protect landward human settlements from storm-related damages. In Arctic regions, erosion of coastal permafrost areas during the ice-free summer season has accelerated in recent decades (Irrgang et al. 2022). Low-lying coral atoll islands are continuously undergoing physical changes due to waves, tides, and storm actions, and there is evidence they can grow vertically and change shape laterally even as sea levels rise (Masselink et al. 2020). However, even if an atoll island persists, human-built homes, and infrastructure may be lost or damaged, and the groundwater on which people depend for drinking and agriculture may become salinized, making the island less habitable (Storlazzi et al. 2018).

Box 4.1
What Are Mangroves, and Why Are They Important?

The term mangrove refers to roughly 80 species of shrubs and trees that grow along shorelines in the tropics and sub-tropics. They are highly tolerant of salty and brackish shallow waters, and are easily recognized by their stilt-like latticework of roots. Mangrove roots provide a protective nursery for small fish and other marine organisms, trap sediments being moved by tides, and absorb the force of incoming waves and storm surges, providing a natural barrier to protect the landward coast from erosion (National Ocean Service 2024). It has been estimated that 15 million people globally live in locations where mangroves prevent them from experiencing periodic flooding, and that the economic value of this flood protection is approximately US$250 million annually (Menéndez et al. 2020). The area of coastline protected by mangroves has in recent decades been declining at a rate of 0.13% per year, the majority of this loss being caused by people removing them to make way for such things as aquaculture ponds and agricultural crops (Goldberg et al. 2020). However, there is growing recognition of the valuable role mangroves play in protecting human settlements, with many governments and communities working proactively to protect and restore them (Hagger et al. 2022) (Figure 4.1). It is estimated that mangroves are able to maintain their stability and vertical growth up to an MSLR rate of between 6 and 7 mm per year, after which they begin to decline (Saintilan et al. 2020). At high rates of GHG emissions, this threshold will be crossed by mid-century.

Figure 4.1   Young mangroves growing in a protected area in the city center of Tampa, Florida. Photo by R. McLeman.

Concerns about saltwater intrusion into coastal groundwater reserves due to MSLR are not limited to small islands but are of concern in many continental coastal areas as well, including locations with large urban populations that depend upon groundwater for potable drinking water (Jasechko et al. 2020). Mean sea level rise is also expected to cause underground water tables to rise in many coastal areas, which can have the effect of settlements being flooded from below, the rising groundwater causing damage to building foundations and buried infrastructure, which can include such things as underground sewage pipes, electrical cables, and subway train tunnels (May 2020). The productivity of agricultural soils in many coastal areas may also be affected by increasing salinity as sea levels rise (Corwin 2021), particularly deltaic areas such as Asia's Mekong and the Ganges-Brahmaputra (Eswar et al. 2021) (see Bangladesh case study below).

Collectively, the aforementioned physical impacts can be expected to exact a staggering economic toll on human systems in coastal regions and will increase with each additional degree of global warming. Under high emissions/low adaptation scenarios, the global financial costs of coastal flooding due to MSLR by 2100 have been estimated at over US\$14 trillion, and annual losses could equal more than 9% of global GDP (Kirezci et al. 2020, Brown et al. 2021, Hinkel et al. 2014). A study of Europe under similar scenarios projects financial losses of up to EUR 1.27 trillion (USD 1.38 trillion) by 2100, but this would fall to EUR 0.2 trillion under a low emissions scenario (Vousdoukas et al. 2020). For China, there is an estimated US\$2 billion difference in annual economic damage costs from coastal flooding in the year 2100 under a warming scenario of +1.5°C versus +4.0°C (Brown et al. 2021). For obvious reasons, the economic costs of MSLR will be disproportionately highest for countries with long coastlines and for coastal areas within countries; one European study projected that the total economic impacts of MSLR averaged across all of the European Union and the UK would be approximately 1.25% of total GDP, but in coastal areas/countries with long coastlines could exceed 20% of total GDP (Cortés Arbués et al. 2024).

4.4 How Many People Will Be Affected by Sea Level Rise?

Estimating the number of people directly exposed to risks associated with MSLR is not a straightforward task, and this is reflected in the large range of estimates in existing studies, from 88 million to 1.4 billion people (Hauer et al. 2021). Key factors that affect the estimation process include:

- the time period in question;
- the GHG emission scenario(s) used (i.e. Representative Concentration Pathways (RCPs)) and consequent rate of warming and MSLR that would occur;

- the relative rate of local SLR along heavily populated coastal areas under the selected MSLR scenario, given adjustments for subsidence and other local factors;
- the definition of the area believed to be exposed to risks, sometimes referred to as the *low elevation coastal zone*;
- the definition of the risk in question (e.g. is the risk occasional flooding, complete inundation, or some other variable?);
- current and projected number of people living within the selected areas of exposure (typically derived either from Shared Socio-economic Pathways (SSPs) or demographic projections from the UN Department of Economic and Social Affairs [DESA] Population Division).

The first three of the above factors are relatively straightforward in methodological terms, with researchers able to draw upon a growing number of data sets to select the GHG emissions, global warming, and MSLR scenarios of interest for the period up to and beyond the year 2100. Until fairly recently, a methodological challenge was that different climate models would generate different MSLR projections for similar warming scenarios, especially for decades in the more distant future (Brown et al. 2016). Recent initiatives in the climate modeling community to build standardized climate model ensembles through the Coupled Model Intercomparison Project (CMIP) have alleviated this and provide more consistent predictions on which to base estimates of future MSLR risks (Hermans et al. 2021). The fourth factor – where to define the low elevation coastal zone – requires subjective decisions to be made by the researcher. The actual level or frequency of exposure to the risks identified in Section 4.3 varies considerably from one location to another, even over relatively short distances. Elevation above sea level is an important factor. For example, people living in houses built at an elevation of 1 m or less from the mean high water level (i.e. the highest point of water during the average diurnal tidal range) are obviously much more at risk of being flooded by storm surges than people in the same settlement whose homes are built 5 m above the mean high water level, and people with homes 10 m higher being at even less at risk. All homes within this same settlement would be potentially at greater risk of at least occasional flooding under a high GHG emissions/ MSLR scenario, but only some of them (probably the ones built at an elevation below 5 m) would have a higher flood risk under a low-emissions scenario. A researcher using elevations to define boundaries for the area of exposure might therefore use multiple markers, one being changes in the mean high-water level, another at 1 m above it, another at 10 m, and possibly even an intermediate elevation of 5 m.

As described in Section 4.2, local geomorphological processes and the physical nature of the coastal environment have important influences on actual flood risk at a given location. So, instead of using elevation to delineate the area of exposure – which ignores the local physical environment – a researcher might instead start by using maps of existing flood risks in coastal areas. As described in Chapter 2, such maps are typically created using statistical probabilities of flood occurrence, such as a 1% or 2% chance of flooding in any given year (which then become referred to as the 100-year flood plain and 50-year flood plain respectively). The researcher can then calculate changes in the probability of flooding within these boundaries under MSLR scenarios and/or generate new projected boundaries for the 100-year floodplain, 50-year floodplain, and so forth.

Once the researcher has determined what spatial parameters to use in defining the low elevation coastal zone, the researcher must decide what type of hazard they specifically want to focus on. Is it simply coastal flood risk? Storm surge damage to infrastructure? The potential for complete inundation? Or some other considerations? If the risk is potential exposure to occasional floods, the total number of people at risk will, by design, be many times greater than if the risk is permanent inundation. Many early studies of exposure to MSLR risks took the current population in areas of exposure as being fixed, and simply focused on estimating future changes in MSLR and changes in the spatial distribution of risk. More recent studies now include projections of future population change in coastal areas, and this has generally led to global estimates of the number of people exposed to MSLR risk being revised upwards, as many coastal areas, especially in Asia, are expected to see high rates of population growth for many decades to come. The many methodological options for defining low elevation coastal zone boundaries, the many possible scenarios for GHG emissions and MSLR, and the multiple choices of hazard options on which to focus explain why there is such a wide range of estimates regarding the number of people exposed to the direct impacts of MSLR (McMichael et al. 2020, Hauer et al. 2020). The remainder of this section provides a small sample of predictions from the dozens of studies that have been completed to date.

One of the first studies to estimate how many people currently live in coastal areas at risk of flooding was done by McGranahan et al. (2007), who used year 2000 population data to estimate that 634 million people were living at an elevation of less than 10 m above mean sea level, three-quarters in Asia. By these calculations, China had by far the largest number of people living in low elevation coastal zones (then estimated at 143 million), followed by India, Bangladesh, Vietnam, and Indonesia. In five small island states – the Maldives, Marshall Islands, Tuvalu, Cayman Islands, and Turks and Caicos – over 90% of the population lives less than 10 m above sea level. Residents of lower- and middle-income countries are disproportionately represented in the global population living near sea level. A study by Dasgupta et al. (2009) estimated that 56 million people in developing countries live within 1 m of mean sea level, the greatest numbers being in Vietnam, Egypt, Mauritania, Suriname, Guyana, French Guiana, Tunisia, the United Arab Emirates, the Bahamas, and Benin.

Using existing 100-year floodplains to describe the low elevation coastal zone, Neumann et al. (2015) estimated that 625 million people in the year 2000 lived in areas exposed to MSLR hazards, 83% of them in low-income countries (figures not dissimilar from the McGranahan et al. [2007] study). Combining demographic projections from the UN DESA Population Division with standardized development scenarios constructed for a World Bank project, the authors then estimated that the number of people living in exposed areas will grow to between 879 and 949 million people by 2030 and to between 1 and 1.6 billion by 2060. Just under three-quarters of the global population living in exposed areas in the coming decades will be in coastal areas of Asia. These estimates are notably higher than ones made in a different study by Nicholls et al. (2021), who project that by the year 2050 between 300 and 350 million people globally will be living in locations highly exposed to MSLR risks under mid-range RCP scenarios. Yet another study, by Kirezci et al. (2020), came up with an even lower estimate, finding that under current emissions pathways, up to 287 million people globally will be living in areas exposed to episodic coastal flooding. However, their study made no attempt to estimate future population change in exposed areas, so they have likely underestimated the actual future total.

There have also been regional and national estimates of the number of people exposed to MSLR hazards. For example, Hauer et al. (2021) estimated that in the year 2000 just over 600,000 people in the US lived in coastal areas at risk of an annual flood event, 150,000 lived in locations below the high-tide line (and therefore protected by some form of built infrastructure), and 2.4 million Americans lived in the 100-year flood plain. Under mid-level emissions and development scenarios (i.e. RCP 4.5 + SSP2 [see Chapter 2 for definitions of RCPs]), the authors projected that by the year 2100 over 4 million Americans will live in areas where an annual flood event may occur, approximately 1.2 million people will be living in areas below high tide, and 9 million will be living in the 100-year flood plain. The authors also note that in every US coastal county, the increase in the number of people exposed to MSLR risks will be driven primarily by changes in the sea level and not by increases in population.

A different study looked at MSLR risks in 32 American cities and estimated that by 2050, between 55,000 and 273,000 people would be at risk of being displaced by MSLR and between 31,000 and 171,000 properties would be damaged (Ohenhen et al. 2024). Yet another study identified the American cities of Boston, Miami Beach, New York, Atlantic City, and Stockton as being especially at risk under high-emissions scenarios (Kulp and Strauss 2017). In Europe, the number of people living in coastal areas likely to experience flood events by the year 2100 would be just over 1.6 million in a moderate emissions scenario and just under 3.9 million under a high emissions scenario (Vousdoukas et al. 2020). In Asia, urban areas in coastal deltas (e.g. Bangkok, Dhaka, Ho Chi Minh City, among others) will become hotspots for MSLR risks in coming decades given their rapid population growth in terms of absolute numbers and density, even though these areas make up a tiny fraction of the continent's low elevation coastlines (McGranahan et al. 2023).

4.5 Adaptation Options for Mean Sea Level Rise

Estimates of the number of people exposed to MSLR risks in future decades are typically made with the understanding that the actual number of people affected will depend upon the extent of future adaptation. Recalling from Chapter 1, adaptation takes on a variety of forms and, when done well, lowers people's exposure and/or vulnerability to climatic risks. Adaptation can consist of *in situ* measures – those that do not require people to move or be moved – and mobility-based options that entail the movement of some or all members of a household or community on a temporary or permanent basis. Here we review examples of each category of adaptations.

4.5.1 In Situ *Adaptation Options for MSLR*

Chapter 2 identified a number of adaptation options commonly used to reduce risks associated with seasonal or occasional flooding in river valleys and with storm surges caused by tropical storms. Such adaptations will continue to be used by coastal populations in the future, with MSLR likely increasing the number and geographic extent of people and settlements obliged to do so. Apart from doing nothing and hoping for the best, *in situ* adaptation options typically involve costly investments in the building and maintenance of engineered structures such as sea walls, groynes, dikes, and other barriers. The exact costs of such structures vary considerably depending on the nature of the construction and the local geography. A study by Jonkman et al. (2013) estimated the costs of raising the height of existing coastal defenses in selected countries and found that, for each meter of additional vertical height, the cost per kilometer would range from EUR 4.5–22.4 million (US\$4.9–24.3 million) in the Netherlands, EUR 2.5–11.8 million (US\$2.7–12.8 million) for New Orleans, and EUR 0.7–1.2 million (US\$0.76–1.3 million) in Vietnam (reported in 2009 currency values). Lower costs in Vietnam likely reflect the lower cost of labor in that country. However, the authors warn that such costs could increase in nonlinear fashion depending on the future rate of MSLR. The costs of building new coastal protection infrastructure can be even more expensive. For example, the Thames Barrier, which protects the city of London and surrounding low-lying lands from floods, cost £535 million (approx. US\$810 million) by the time it became operational in 1983, and costs £8 million pounds (approx. US\$12.4 million) annually to operate and maintain (UK Environment Agency 2021).

The enormous financial costs of building and maintaining protective infrastructure in an era of MSLR will require difficult cost/benefit decisions to be made at national and local levels regarding which areas warrant such protection. Urban centers will almost certainly be prioritized given their high population density. However, even in high-income countries, the financial costs may make it impossible to protect all portions of all urban centers with engineered sea walls and tidal barriers. The rising costs of protecting coastal settlements can be expected to exacerbate the economic gulf between high-income countries that are better able to afford such costs and low-income countries that cannot.

Atoll states are in an especially challenging position when it comes to building protective infrastructure given the very high ratio of coastline to population and the need to protect the entire coastline and not simply a portion of it, as would be done for continental coastal settlements. The atoll nation of the Maldives is pursuing a strategy that includes construction of seawalls and artificial islands adjacent to the capital city of Malé (Brown et al. 2020). Construction of the first large artificial island, known as Hulhumalé, was initiated in the 1990s and is 1.8 m above mean sea level – higher than most of the land on Malé, meaning that floods occur less frequently. Its population already exceeds 65,000, and future expansion is expected to accommodate over 200,000 people (The Maldives Journal 2023). Such investments are possible because of the Maldives' relatively high GDP for a small island state (US$6.2 billion in 2022 [World Bank Open Data Portal 2024]), which is based heavily on high-end resort incomes from international tourists. For atoll states with smaller economies that are too distant for most international tourists, such as Kiribati and Tuvalu (with GDPs of US$0.22 billion and US$0.06 billion, respectively), investments in coastal protection and large-scale land reclamation projects are simply not possible in the absence of outside financial donations.

Besides engineered infrastructure, a second set of options for *in situ* adaptation to MSLR involves making adjustments to the form and structure of settlements and making strategic use of their local natural environment (Mariano and Marino 2022). One way to do so is to construct buildings that are on stilts or floating platforms, allowing the occupied portion of the building to remain above the high water level (Figure 4.2). This is most easily done for new construction, and particularly in large settlements, it may not be practical to replace all existing buildings with floating or elevated ones. Another approach is to employ what are known as *nature-based solutions* – that is, facilitating the maintenance and expansion of coastal marshes, mangroves, beaches, sand dunes, nearshore coral reefs, barrier islands, and other natural features that trap sediments, prevent erosion, and dissipate the energy of waves and storms (Hobbie and Grimm 2020, Toth et al. 2023). A study of 136 coastal cities worldwide found that 75% have opportunities to use nature-based solutions to reduce future flooding risks, with the greatest benefits available

Figure 4.2   School in coastal community of Grand Isle, Louisiana, USA, built on stilts so that storm surges pass underneath. Photo by R. McLeman.

to cities in deltaic locations such as New Orleans, Ho Chi Minh City, and Khulna (Van Coppenolle and Temmerman 2020). Mobilizing nature-based solutions may in some locations be far more cost-effective than building new, engineered infrastructure. A study of San Mateo County, California, found that an MSLR adaptation strategy focused on protecting and improving naturally protective shoreline features could prove to be up to eight times more effective than using engineered infrastructure (Guerry et al. 2022). Such approaches are likely easier to employ in coastal areas with rural/low-density populations. In densely populated cities, the need to protect remaining natural spaces may conflict with pressures for development and growth. There are limits to the amount of adaptation that can be achieved through nature-based solutions alone (Seddon et al. 2020) and, especially in high MSLR scenarios, they may provide insufficient protection for densely populated coastal cities. In many cases, a hybrid approach that combines built infrastructure with protection of natural protective features will likely be needed (Du et al. 2020).

4.5.2 Managed Retreat/Planned Relocations

The sheer number of people globally that will need to be relocated from coastal settlements this century, even under the most conservative estimates, is daunting. Take, for example, the estimate of Kulp and Strauss (2019) that approximately 110 million people currently live in locations that will be below the high tide mark by the year 2100, even under low GHG emissions scenarios, and that this number may rise to 190 million people through population increase. Even with unlimited financial resources, it will not be possible to protect all those people and their communities using engineered infrastructure, much less the hundreds of millions of other people who won't necessarily be living below the high tide mark but whose homes will experience more frequent flooding. Once the financial resources available for investments in protective coastal infrastructure have been spent, what will happen to the people and places that remain? The next set of options involves people moving out of the most highly exposed locations. This might happen autonomously or as a coordinated, *planned relocation* of people, a process also known as *managed retreat* (Box 4.2).

Box 4.2
Relocation of Indigenous Settlements in the Arctic

The Arctic is an especially harsh place to build settlements with modern infrastructure. The extreme temperatures cause manufactured materials to break down much faster than in warmer climates. The permanently frozen subsoil means that infrastructure which in southern communities is buried – sewers, water pipes, gas mains, electrical lines, and telecommunications cables – must be above ground and protected from the elements. Homes and other heated buildings must be placed on stilts or platforms so that heat escaping through the floor does not melt frost in the ground below and cause the building to sink and collapse. Most Arctic settlements are situated on coastal plains and along waterways, exposing them to additional hazards of floods, storms, and erosion. Climate change is exacerbating the hazards faced by Arctic coastal communities, and in Alaska and northwestern Canada, several are now in the process of being relocated or planning to do so in the near future.

Fixed settlements constructed from manufactured materials are a post-World War II phenomenon in the North American Arctic. Indigenous peoples that have lived in and adapted to harsh northern environments for millennia traditionally built homes and shelters from natural materials, and would move between multiple locations throughout the year as they pursued livelihoods based on hunting, gathering, and fishing (Pearce et al. 2015). Beginning in the 1950s, the governments of Canada and the US dispossessed Arctic Indigenous peoples of their lands and forced many to move into fixed settlements, the locations of which were selected by authorities for such reasons as proximity to harbors and to radar installations built to watch for Soviet aircraft during the Cold War (Billson 1990, Mitchell 1997). Coastal settlements were typically assembled of prefabricated buildings shipped in by boat from the south during the brief summer and powered by electricity produced from diesel generators.

Climate change affects the viability of Arctic coastal settlements in multiple ways (Arctic Climate Impact Assessment 2005). Warmer ocean temperatures lead to there being less sea ice and longer ice-free periods. Ice protects shorelines; as the periods of open water along shorelines grow longer, there is more opportunity for erosion of coastal land and infrastructure from waves, tides, and storms. Mean sea level rise hastens the loss of land and allows for greater inland penetration of storms. Milder air temperatures lead to thawing of permafrost along shorelines which lies beneath built infrastructure. For Indigenous people whose livelihoods remain closely tied to hunting and fishing, changes in water, ice, and land conditions affect the distribution of fish and animals they hunt, impede their ability to travel to hunting and fishing areas at critical times, and increase the danger of being caught in storms when out on the land or sea. This in turn leads to greater dependence on food imported from the south and undermines cultural practices and social relations.

In Alaska, over 200 Indigenous coastal villages experience flooding and erosion, and a dozen of these are in the process of relocating or planning their relocations. (Bronen and Chapin 2013). Some villages are planning complete relocation to newly constructed buildings in new sites in nearby areas, such as Shishmaref, Kivalina, and Newtok (Bronen and Chapin 2013), while others such as Teller, Napakiak, and Utqiaġvik (once known as Barrow) are moving individual buildings and infrastructure in a process of gradual retreat from the shore, remaining in roughly the same location (Taylor et al. 2023, Garland et al. 2022). The communities – each of which has locally formed planning groups to lead the relocation process – have encountered many institutional and jurisdictional hurdles that slow the process and make funding difficult to access. For example, US federal government disaster management programs do not provide funding to assist communities experiencing slow-onset climate hazards apart from drought; erosion and sea level rise do not qualify. Further, federal programs discourage wholesale relocation of flood-prone communities, instead favoring a cheaper, less complicated, piecemeal approach of buying out individual properties. Alaska state climate adaptation funding programs exist, but communities must compete with one another for access to such funds. In the case of Newtok, which is one of the more advanced communities in terms of the process of relocation, funding to initiate the construction of a new village had to be cobbled together from a half-dozen different government agencies (Ristroph 2021).

The primarily Inuit hamlet of Tuktoyaktuk in the Mackenzie River delta is one of the first Canadian Arctic communities to face the prospects of climate-related relocation. Erosion, subsidence, and MSLR will make its current location unviable by the 2040s, and storm surges are starting to flood ever larger areas of the settlement on a more frequent basis (Whalen et al. 2022). Successful community-based monitoring programs have been established to document the rate of environmental changes in the area (Mercer et al. 2023). Federal government funding has been received to improve shoreline defenses (Government of Canada 2023), but these will at best delay the need for relocation for a number of years. A potential new townsite in a safer nearby location has been identified, but community consultations are still ongoing, and it is not clear that residents will want to move there. An organization called FutureTuk (futuretuktoyaktuk .org) has been established by the hamlet government, Inuit organizations, and academic researchers to move consultations forward, and their plans include outreach to Alaskan communities to learn from their relocation experience. As more Arctic communities find themselves in similar situations because of climate change and MSLR, community-to-community exchanges of best practices will become an important planning tool.

There is a small but growing inventory of examples where governments have relocated coastal communities away from low-lying locations prone to subsidence, erosion, and frequent flooding (Hino et al. 2017). In some cases, the relocation process is initiated in anticipation of greater risks to come; in others, the relocation has been initiated after an extreme event has occurred, such as a tropical cyclone or a tsunami. Most current examples are small communities with populations numbering in the hundreds or thousands, a disproportionate number of them comprising Indigenous people and/or people pursuing traditional lifestyles. Often, governments have initiated the relocation process, but there are also examples where residents perceive a need to relocate but lack the financial wherewithal to initiate relocation autonomously (see Box 4.2 and case studies in what follows). From existing examples, researchers have been able to identify several categories of challenges for planned relocations:

i Financial Costs

It is very expensive to relocate people in a way that is respectful, humane, and leaves them in economic circumstances that are at very least no worse than those they lived in previously (Hino et al. 2017). There is, however, a surprising lack of economic and financial data with respect to the expected costs of planned relocations for MSLR in the future and how these compare with the financial costs of engineered coastal protection infrastructure (Dedekorkut-Howes et al. 2020). At the moment, the best data for cost estimates for future relocations are pieced together from recent case studies, which come with the caveat that the precise costs for future relocations will inevitably vary from one location to another for a variety of economic, political, social, and cultural factors. As one example, the US government has allocated $48.3 million for the resettlement of approximately 100 residents of Isle de Jean-Charles, a primarily Indigenous community living on a 320-acre island (130 hectares) off the Louisiana coast that is rapidly disappearing due to the combined effects of subsidence, erosion, storm damage, and MSLR (Simms et al. 2021). This is a case of managed retreat being done relatively well: Community members have been actively engaged in most aspects of planning and helped select the location of the new community.

Not all planned relocations in the US have been done so thoughtfully and proactively; more often, relocation takes the form of government-funded buyouts of damaged homes after extreme events occur; approximately 45,000 homes over the past three decades have been bought out in this way (Mach and Siders 2021). The average cost of buyouts varies from one event to another. A well-documented example is that of Hurricane Sandy, a tropical cyclone that made landfall along the coast of New York and New Jersey in 2012, damaging or destroying over 350,000 homes in coastal areas and causing an estimated US$50 billion in property damage

(Koslov et al. 2021). The two states implemented targeted buyouts of homes in the most heavily damaged low-lying areas and less densely urbanized parts of the affected area, such as Staten Island and Long Island. The New York state program cost an estimated US$640 million to buy 1,300 homes from their owners so that they would not be rebuilt (Frank and EE News 2024). Paradoxically, the remaining homes and businesses in nearby areas have increased in value as a result. Studies of the wider areas affected by Sandy show that the average value of properties declined after the storm, but only by single-digit percentages (Cohen et al. 2021). This is consistent with longstanding general evidence that individuals' perceptions of future risks vary. For some people, previous experience with extreme weather and/or a belief in and understanding of climate change risks such as MSLR may influence their perceived value of coastal properties, but this does not apply to all people (Baldauf et al. 2020). The Sandy example is also consistent with studies done elsewhere that show current prices of homes and properties often do not account for future MSLR risks (Filippova et al. 2020, Murfin and Spiegel 2020). As a result, the dollar value of properties in high-risk areas continues to increase over time, thereby increasing the costs of any future relocations that may need to be organized. Although these examples are drawn from the US, similar processes in terms of rising property values in coastal areas are common in many countries.

Recalling from Section 4.4 the low-end estimate by Ohenhen et al. (2024) that 55,000 Americans will be living in locations at direct risk of MSLR by the year 2050, and multiplying that number of people by the $483,000 per person price tag for the Isle de Jean-Charles relocation, the present-day cost of a properly organized relocation of all those Americans would be $26.57 billion. This amount is almost as much as the entire annual current budget for the US Federal Emergency Management Agency (US$29.5 billion). It is within the financial wherewithal of the US government to fund a well-organized, proactive managed retreat program of such a vast scope, although whether it would do so is more a question of politics than it is economics. The cheaper option would be to simply continue with *ad hoc* buyouts after coastal floods and storms as was done in the example of New York state post-Sandy, although it is doubtful whether that program's US$49,230 average per-house buyout price would be satisfactory to most residents of coastal communities today, where average residential property values are typically many times this amount.

A key question in every case of pre-planned or post-event relocations becomes, who pays? The most common options include governments, either through an *ad hoc* allocation of funds or through a public insurance or disaster risk recovery scheme; private insurance, which is typically only available to residents of higher income countries and which typically pays on a household-by-household basis, and only after an extreme event has occurred; or, the residents of an exposed community pay for the relocation themselves (Noy 2020). For low-income countries, international agencies and the humanitarian sector can be a funding option after an extreme event has occurred, but there are few options for raising funds for proactive relocations internationally (Boston et al. 2021). The UNFCCC Loss and Damage fund may provide a new funding stream in the coming years (see Chapter 6), but it is still in its initial stages of establishment. Governments are generally the most common sources of income for relocations, and the cases of Hurricane Sandy and Isle de Jean-Charles illustrate that local governments typically lack the necessary financial means, and need financial assistance from higher level governments to organize relocations from exposed areas (McGinlay et al. 2021; Dundon and Abkowitz 2021). In some cases, local governments dependent on property taxes may be reluctant to support efforts to relocate people away from areas exposed to MSLR risks, given that seaside locations may have high short-term value for development and provide a valuable revenue stream (Shi and Varuzzo 2020). Meanwhile, private insurers are becoming increasingly reluctant to insure properties in areas exposed to floods, storms, and other hazards, refusing to insure many new developments and raising premiums on existing policyholders, and this trend is likely to accelerate as the additional risks associated with climate change grow (Gray 2021, Tesselaar et al. 2020).

ii Legal and Organizational Challenges

In an ideal world, planning for managed retreat would be initiated years in advance, following a systematic process that involves residents, governments, institutions, and other stakeholders working in coordination to establish rules, guidelines, and financial resources to make the eventual implementation go smoothly (Haasnoot et al. 2021) (see Fiji case study in what follows). In reality, such things happen rarely – indeed, if we acted in such a way to tackle GHG emissions, we might be able to avoid the need for planned relocations. A practical challenge is that typically there are multiple levels of government that need to be involved in relocation planning – local, national, and often intermediate-level governments as well – and each of these levels of governments has multiple sub-organizational units and special interests (Mach and Siders 2021). In the US, for example, relocations typically require coordination across local and state governments and multiple federal government departments and agencies, including the Housing and Urban Development, the Federal Emergency Management Agency, and/or the US Army Corps of Engineers, among others. In recent years, the Federal Emergency Management Agency (FEMA) has initiated a program of Hazard Mitigation Assistance grants to be administered by state governments to assist local governments in implementing strategies to reduce disaster-related losses, and these grants include programs to buy out homes in hazardous locations. A variety of implementation challenges have emerged, often associated with a lack of institutional capacity at lower levels of government to implement projects in a timely and cost-effective way (Smith and Vila 2020). In some cases, local governments may not even be aware of the program's existence or believe they are not eligible or lack the ability to participate given local constraints. Although this example is particular to the US, similar challenges can emerge wherever and whenever multiple actors need to be involved in complex decision-making.

An important subsidiary question becomes – and which is inevitably case specific – who ultimately gets to decide when and where to relocate? The answer to this question often depends upon when the decision is made, which can occur:

- after a flood, storm or other climate disaster has occurred, and it is too dangerous and/or too expensive to continue living in that same location;
- before a disaster has yet to happen, but the potential risks to life and/or property are so high it is safer and/or cheaper to relocate people elsewhere (Balachandran et al. 2022).

In both cases, there will be winners and losers and costs to be absorbed, and decisions will inevitably be contentious. In the post-disaster scenario, governments may decide to restrict reconstruction in the hazard area, as has been the case in some coastal areas of Sri Lanka and Indonesia after tsunamis (Gunarathna et al. 2023, Amri and Giyarsih 2022). In other cases, governments may facilitate reconstruction in the hazard area, aiming to 'build back better' (Graveline and Germaine 2022). The latter approach is often the most popular one with residents of the affected area, particularly if they perceive the disaster was a one-off event with little likelihood of happening again. The pre-emptive approach – i.e. deciding to relocate people before the hazard event actually occurs – is more contentious, for residents of the at-risk area may perceive the degree of risk differently than do authorities. People who have not previously experienced a given type of hazard event may be less likely to perceive there being a serious risk as compared with people who have experienced that hazard or a similar one – but even among people who have experienced the hazard before, the perception of future risk depends heavily on the severity of the financial and/or emotional losses they experienced and the amount of time that has elapsed (Bronfman et al. 2020). People may be especially resistant to proactive relocation if they do not trust government authorities or expert opinions (Wachinger et al. 2013).

Planned relocations are made more complicated by private property regimes, where the personal net worth of individuals and households is closely tied to their homes and their homes' contents, and where there are often large financial inequalities between people who own their homes or properties and those that do not. Once a decision is made by authorities that a given area is hazard prone and/or must be abandoned – or in cases where insurance companies decide to no longer provide coverage for homes in that area (insurance being a common prerequisite to obtaining a mortgage) – homes and properties in that area immediately lose monetary value. This in turn means that people who live there and must move no longer have the financial capital necessary to do so. As a result, to facilitate a fair and equitable relocation process, governments must either provide new accommodation for those who must move or provide sufficient financial compensation to help people find their own accommodation. This in turn sets up a cascade of additional challenges. For example, a comparison of hypothetical managed retreat options for Vancouver and Manila found in both cities an overriding problem to be finding places for people to move to. In both metropolitan areas, there is little undeveloped land and affordable housing is already scarce (Doberstein et al. 2020). A decision to begin relocating people from the lowest-lying areas around one of these cities would have the simultaneous effect of reducing the amount of housing available and increasing the number of people in need of it, driving up the cost of accommodation for all. So, although relocation to another neighborhood within the same city would be the fairest and most preferable option should the need arise, it may not be possible for the practical reason of a lack of housing. The question would then become whether and how people might be incentivized or obliged to relocate to more distant locations, far away from existing jobs, family, and friends.

iii Social and Economic Implications

What makes for a "successful" managed retreat/planned relocation? The metrics may vary, but at very least those who are relocated should find themselves no worse off economically and socially than they were before their relocation, and the places they relocate to should be amenable and culturally appropriate. To achieve such outcomes, it is generally believed that the best approach is one where people living in communities faced with the prospects of relocation work collaboratively with governments and other key actors to evaluate options and map out a process with goals and outcomes that are most beneficial to all given the circumstances (Mach and Siders 2021). In practice, this does not always happen, with residents often presented "take-it-or-leave-it" choices that may provide a short-term housing solution but do not address the wider social, economic, and cultural well-being of those being moved (Tubridy et al. 2022). A study of 138 cases of organized relocations from around the world found that slightly less than half might qualify as being successful in terms of the well-being of those who moved, one-quarter were clear failures, and the remainder being too early to tell (Ajibade et al. 2022).

There is a whole raft of additional questions – not easily answered – related to the social, economic, and cultural dimensions of planned relocations. For example, what should be done if people do not want to relocate because their economic, social, and/or cultural ties to their homes and community are too strong? This is a commonly expressed concern of Indigenous and traditional peoples living in areas highly exposed to climatic risks: They will not move unless there is absolutely no other choice, period. Do the authorities still owe them any obligations? What if people say they are willing to consider relocating but want to be able to continue using their former home area for livelihood activities – do they relinquish their land rights and title if they move? These are all legal and jurisdictional gray areas for which there may be few precedents. What about people living in areas to which people are relocated to – do they get any say in the matter? If there are social, economic, and/or cultural implications for them, what are their rights, if any?

4.5.3 Case Study: Carteret Islands

The Carterets form a chain of six small atoll islands approximately 85 km east-northeast of Bougainville Island, which is in turn east of the main island of Papua New Guinea (PNG). The islands, none of which is more than 1.5 m above sea level, sit atop a ring-shaped reef that has built up on the edge of the crater of an extinct undersea volcano (Figure 4.3). The Carterets are home to approximately 1,200 people who call themselves the Tuluun and are related to inhabitants of Buka Island, a much larger, non-atoll island off the northern tip of Bougainville (Connell 2018). In recent decades, there have been more frequent flood events on the Carterets and sea level relative to the islands is rising; whether this is due

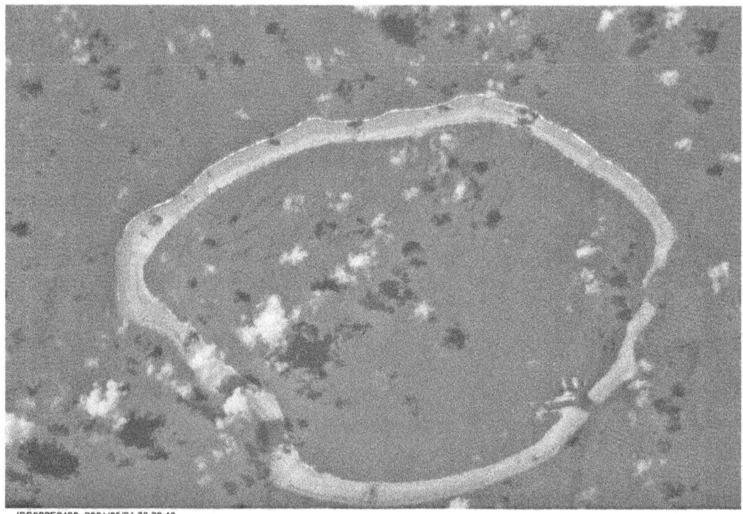

Figure 4.3 Photograph of the Carterets atoll taken from space. Source: NASA. Public domain image: https://eol.jsc.nasa.gov/SearchPhotos/photo.pl?mission=ISS002androll=Eandframe=6439

to subsidence, MSLR, or a combination of both is not clear (Boege and Rakova 2019). Saltwater intrusion into the groundwater and salinization of soils on the islands have made it difficult to grow the islanders' traditional staple crops, and food insecurity has become a chronic challenge for the Tuluun, many of whom wish to relocate to Bougainville. An initial attempt at organized relocation began in 1984 with PNG government assistance, but this plan was abandoned when a civil conflict broke out in 1989 between the government and rebel forces seeking an independent Bougainville (Connell 2018). A peace agreement was reached in 2001 but tensions remain. A 2019 referendum on Bougainville saw 98% of the population vote for independence, and there are ongoing negotiations that could potentially grant Bougainville independence in 2027 (Harding and Pohle 2022).

In 2006, the Tuluun formed a nongovernmental organization to begin planning their own relocation to six parcels of land on Bougainville gifted to them by the Catholic church diocese (Boege and Rakova 2019). New homes and farms are gradually being established, but at 81 hectares in total these parcels are too small to support the entire Tuluun population (Connell 2018), and an estimated US$5.2 million is needed to acquire enough additional land to resettle everyone (Boege and Rakova 2019). Many Tuluun continue to reside on the Carterets despite the growing flood hazard, some having returned from Bougainville because they were homesick. The new settlement sites in Bougainville are not located on the shore, making it difficult to go fishing; are scattered across multiple sites, which weakens social connections; and the people must learn to raise new crops. After generations of living on atolls, Tuluun livelihoods and culture are different from residents of Buka and Bougainville, and they are reluctant to abandon entirely the islands where their ancestors are buried. Not all Bougainville residents are receptive to the Tuluun relocating there, and there have been conflicts between the groups, especially over access to traditional lands of Bougainville residents (Edwards 2013). Most problematically, apart from modest donations from humanitarian organizations, no meaningful financial resources have been made available to the Tuluun from PNG, Bougainville, or international governments, this despite the islanders featuring often in the international media in the early 2000s as being a first example of "climate refugees" (e.g. The Guardian 2005, The New Humanitarian 2008).

If they had the financial resources and were permitted by authorities to do so, it appears most Tuluun would like to be able to move back and forth between their home islands and Bougainville, some people remaining behind as an anchor population so long as the Carterets remain habitable (Connell 2018). Should the islands become uninhabitable on a permanent basis, the Tuluun hope to continue to be able to use the islands as fishing grounds, as part of a flexible, mobility-based livelihood. The future of the Tuluun is very uncertain, particularly should Bougainville gain independence from PNG. Further, the Carterets are not the only islands administered by PNG being threatened by MSLR – the Saposa islands just off the west coast of Bougainville are also gradually disappearing (Fainu 2021) – meaning there will be growing competition for financial resources to assist relocation.

4.5.4 Case Study: Organized Relocations in Fiji

Government-organized relocations of Indigenous and traditional communities have a generally poor track record in most countries. Faced with rising sea levels, frequent extreme weather events and a growing need to relocate small coastal communities on its over 300 islands, the government of Fiji is trying to do better. In 2018, the government published official Planned Relocation Guidelines (PRG) that place an emphasis on collaboration and cooperation between government agencies and residents of communities in need of relocation (Lund 2021). This was followed by the establishment of a trust fund to finance future relocations, a special climate adaptation levy on goods and services, and the development of Standard Operating Procedures in 2020 to guide relocation planning processes under the PRG, done with the support of the German development agency.

In establishing the Standard Operating Procedures, the government drew heavily on the experience of organized relocations initiated in the early 2000s. One of the first villages to be relocated was Vunidogoloa, a remote community of approximately 120 people on the country's second-largest island, Vanua Levu (Bertana 2020). Residents pursued livelihoods based on fishing and farming, but coastal erosion and damage from tropical cyclones had made the location untenable. The relocation process was initiated in 2009 and completed in 2016, at a cost of just under F$1 million (approximately US$450,000). The government led residents through a consultation process that allowed them to consider options such as reconstructing a dilapidated sea wall instead of moving, and to assess various potential relocation sites. Most residents, though not all, favored relocation, particularly younger community members. As is customary in Indigenous Fijian communities, residents had collective title to a large area surrounding the village, which helped facilitate the relocation, since it meant community members could remain together on their ancestral lands. The costs of relocation were offset by allowing commercial logging on the community's lands, and residents provided much of the labor in constructing the new village. Despite being very forward thinking for the time, and though residents were generally happy with the outcome, problems were identified in hindsight. For example, women residents were excluded from aspects of the formal consultation process, something that would be addressed in the new Standard Operating Procedures. Also, the newly constructed village did not have the same layout or housing design as a traditional Fijian community, the houses lacked appropriate kitchen facilities, and it generally felt and looked like an anonymous suburban development.

The importance of cultural considerations in relocation planning was identified by researchers who studied the outcomes of three completed relocations, including Vunidogoloa, and four other villages that were contemplating relocation or where the process was underway (McMichael and Katonivualiku 2020). Indigenous communities have strong emotional and spiritual ties to their ancestral lands and their local environments (referred to as *vanua* [Yee et al. 2022]), and residents are very reluctant to relocate if it means leaving those lands or splitting apart the community. The preference in most cases is for a short-distance relocation of buildings to higher ground on the community's own lands (assuming such high ground exists), which leads to more of a short-distance retreat from the shore than a full-fledged relocation. Another important consideration is the ability of residents to continue pursuing traditional livelihoods that are typically based on fishing and cultivation of coconuts, fruits, and other crops. In some communities, the process of retreat is autonomous and gradual, rather than fully organized and scheduled, with younger residents building new homes on inland, higher ground near the main village, which over time causes the central location of the village to shift (a process referred to as *generational retreat* [Piggott-McKellar et al. 2021]).

Not all villages threatened by MSLR and other coastal hazards elect to relocate, and the government's PRG and Standard Operating Procedures respect this. Residents of villages at risk will often attempt to delay the need to relocate through *in situ* adaptations, such as the village of Tokou on Ovalua island, where residents built a sea wall (which has proven to be ineffective), planted mangroves along the shore (which were washed away by storm surges before they could become established) and moved their wells farther inland (McMichael and Katonivualiku 2020). Residents of the village of Toguru on Viti Levu island have adapted by building their own sea wall, planting more salt-tolerant crops, and raising existing houses vertically (Yee et al. 2022). Some residents migrate temporarily to other places in Fiji in search of wages, but most return or intend to. Government officials have proposed relocating the village on multiple occasions since 2010, but residents have resisted. A key reason is that their ancestral lands do not offer a suitable location for the entire community to move to as one. If they were to be relocated, they would need to split apart to separate locations on lands that belong to families in other communities, contrary to the community's vanua, and their strong local social networks would be disrupted.

Fiji's experience shows that even well-planned, collaborative, and seemingly simple relocations can go awry. For example, the relocation of 100 residents of Narikoso village on Ono island ended up with residents being unsatisfied with the outcome (Anisi 2020). The logistics seemed straightforward: The villagers' ancestral lands offered a suitable location and residents were receptive to the idea of moving. Their location had had erosion problems since the 1960s, and their self-constructed sea wall was ineffective. Residents would be able to continue with their traditional livelihoods, which centered on subsistence fishing, farming, and occasional wage labor employment in nearby tourist resorts. The problems arose from the implementation process itself, which proceeded in stops and starts, ran into problems with finances that were being provided in part by international donors, and was plagued by a lack of ongoing communications between residents and physically distant government officials during intervals between formal meetings. In the end, the government decided to fund relocation of only part of the village. The newly built houses were not of traditional Fijian fashion, and there was no community hall in the new area, which is an important feature of Fijian villages. Even though they could see these problems arising during the process and were becoming increasingly frustrated, residents did not voice their concerns for fear it might cause the government to cancel the project altogether.

Despite the best of intentions, a well-considered approach to planned relocations, and a commitment to continuous collaboration with communities and learning from past experiences, challenges remain for the government and people of Fiji. Over 40 coastal villages have been identified as needing relocation over the next decade (Lyons 2022). Moving entire villages in an intact fashion, even short distances, and doing it well is expensive. Not only are new homes required, but other facilities, including health clinics, schools, and community halls may need to be rebuilt. Ideally, key cultural features and cemeteries are also moved to higher ground. Fiji is not an especially wealthy country; its annual GDP according to World Bank data hovers around US$5 billion – about the equivalent of a small American city – meaning that funds for carrying out the growing number of planned relocations will need to be sourced internationally, from donor agencies that already have any number of countries and communities seeking financial assistance for a wide variety of initiatives. Unless wealthier countries provide a massive influx of new money for climate-related relocations in small island states, the well-thought-out approach being developed by Fiji may never come to full fruition.

4.5.5 Autonomous Relocation or Trapped Populations?

The preceding case studies of Fiji and the Carteret Islands raise an uncomfortable question that applies at a global scale: What will happen to the millions of people who live in countries and locations where governments and institutions are unable to build protective infrastructure for coastal communities and are not able to organize thoughtful, planned relocations where necessary? The most likely outcome will be a disorganized mix of autonomous relocations (i.e. people moving elsewhere of their own volition at their own expense) and trapped, immobile populations that suffer through recurrent, increasingly frequent, and destructive floods and extreme events.

With the exception of small island states, it is reasonable to assume from past experience (and examples of hazard-related migration from previous chapters) that most people who relocate autonomously because of MSLR risks will move to other, safer locations within their home country, ideally those where they can continue to participate in pre-established social and economic networks. Indeed, as summarized in Chapter 1, households that move autonomously are likely to leverage those economic and social connections to facilitate their movement. Conversely, households whose economic, social, and cultural ties are limited and/ or exclusively located within an area exposed to MSLR risks may be unwilling or unable to relocate, and risk becoming trapped. They will be exposed to increasing levels of hardship and suffering, falling into ever greater poverty and declining health. They will be reliant upon the charity of humanitarian organizations and outside institutions, in much the same predicament as people currently living in refugee camps and informal settlements. This represents one of the greatest and most severe long-term risks of climate change, and underlines the importance and urgency of getting GHG emissions under control as quickly as possible (Gilmore et al. 2024).

4.5.6 Case Study: Climate-Related Migration and Displacement in Bangladesh

It seems fitting to conclude this chapter – the third of three chapters assessing the links between specific categories of climatic hazards and migration outcomes – with a brief case study of Bangladesh, a country that has been the subject of research on climate-related migration and displacement for decades, and for good reason. Virtually all the extreme weather events and conditions described in Chapters 2 and 3 happen somewhere in Bangladesh on a periodic basis, and people living along its 1,300+ km of mostly low-lying and subsiding coastline (Snead 2010) are highly exposed to the long-term risks of MSLR. Much of the country straddles the world's largest river delta, formed by the Ganges, Brahmaputra, Padma, and Meghna rivers, and Bangladeshis' lives and livelihoods are strongly shaped by the continuous physical processes of monsoonal precipitation, floods, sediment deposition, subsidence, coastal erosion, and accretion (Raff et al. 2023, Islam et al. 2021). Coastal settlements experience tropical storms and cyclones that move inland from the Bay of Bengal more years than not, roughly 70 million people live within 2 km of rivers that swell during the summer monsoon season (including

6 million residents living in flood-prone locations of the capital city of Dhaka), and residents in the north of the country often grapple with severe droughts (Mojid 2020, Masrur et al. 2022).

The country's population of 171 million is growing at an annual rate of 1.1% (World Bank 2024). Roughly 60% of Bangladeshis live in rural areas, but that percentage is shrinking (World Bank 2024). Bangladeshis are an increasingly mobile people, using temporary and permanent migration within the country, to India, and to more distant destinations to diversify livelihoods and access economic opportunities. Rural out-migration rates are high in northern and coastal areas of the country, with the largest share of internal migrants destined for Dhaka, now one of the world's most populous cities at 21 million residents and counting (Alam and Al Mamun 2022). Remittances from Bangladeshis living abroad, which total over US$20 billion annually, play an important role in supporting many households in a country where GDP per capita is roughly $2,700 (IOM 2023, World Bank 2024). At the same time, Bangladesh hosts over 975,000 Rohingya refugees from neighboring Myanmar (Burma), over 200,000 of whom live in the world's largest refugee camps at Cox's Bazaar and Bhasan Char (IOM 2024).

Most migration within Bangladesh occurs for reasons related to marriage, family commitments, education, employment, and economic opportunity seeking, but on average 2.1% of all migration is related to environmental hazards (Alam and Al Mamun 2022). The most commonly occurring of these is flooding, displacing on average a million people each year (IDMC 2022) (see also Section 2.14). Tropical cyclones are the next most important environmental driver of displacement in Bangladesh, averaging 100,000 people per year, but the actual numbers vary considerably from one year to the next depending on storm activity (IDMC 2022). For example, in 2019 over 3.6 million people in Bangladesh and West Bengal state in neighboring India were displaced by two cyclones, in 2020 another 2.5 million were displaced by a single cyclone, but in 2021 only 18,000 Bangladeshis were displaced by cyclones (IDMC 2022). Drought is a less common hazard and is less likely to suddenly displace people from their homes, but it is nonetheless an important factor motivating migration out of rural areas of northern Bangladesh, as was reviewed previously in Section 3.2.6.

Mean sea level rise can be expected to exacerbate existing climatic risks in Bangladesh, especially in the lowest lying southern areas of the country. The extent to which it does, and the number of people to be affected, is uncertain, and will depend not only on the rate of MSLR but also on future population and migration trends within the country. For example, one study estimated that up to 900,000 people will live in areas likely to be inundated by MSLR by 2050 and 2.1 million by the year 2100, depending on the RCP used, and that most of those displaced will move to Dhaka or surrounding cities (Davis et al. 2018). A different study by De Lellis et al. (2021) arrived at somewhat higher estimates of displacement by MSLR by 2050 (1.35 million) but found that Dhaka would not necessarily grow, and might actually lose a small number of people. The authors suggest those forced to move would likely spread across much wider areas of the country, seeking out economic opportunities and avoiding coastal areas with growing environmental risks. In yet another study, models created by Bell et al. (2021) suggested that, notwithstanding increased rates of displacement due to flooding and inundation associated with MSLR, economic motivations will continue to be the primary influence on migration patterns within Bangladesh in future decades. As a result – and perhaps counter-intuitively if one considers only the climatic hazard – migration into cities and coastal areas highly exposed to floods and cyclones will continue to grow in spite of MSLR, as it is in these locations where economic opportunities are most greatly concentrated. An additional way in which MSLR might potentially affect future migration patterns in Bangladesh is through inland penetration of saltwater into groundwater, rendering it undrinkable and less usable for irrigation. A study of the effects of saline groundwater encroachment on current migration patterns found that in some rural areas, its effects on out-migration were greater than those of rapid-onset flooding (Chen and Mueller 2018).

Given the wide range of overlapping and cascading climatic risks facing Bangladeshis today and in coming decades, it is unsurprising that Bangladesh was among the more vocal proponents for the Loss and Damage fund that was established under the UNFCCC in 2022 (ICCAD 2024) (see Chapter 6). Envisaged as a mechanism for transferring large amounts of financial assistance from high-income countries that are disproportionately responsible for GHG emissions to low-income countries that are experiencing and will continue to experience the greater harm, the Loss and Damage fund – if fully implemented as intended – will become a key resource for communities and countries threatened with displacement by the impacts of MSLR.

5

Data and Methods for Modeling Climate-Related Migration

Climate-related migration is a complex phenomenon that is influenced by social, economic, political, and environmental factors across multiple spatial and temporal scales (see Chapter 1 for more detail). This complexity can present challenges for researchers who aim to model climate-related migration. For one, accessing appropriate data related to both climate change and migration can be difficult, and the results of modeling efforts can vary greatly based on the data used. Even with appropriate data, it can be very difficult to disentangle climatic drivers of migration from other factors. Despite the challenges, considerable progress has been made in recent years in modeling climate-related migration. Researchers employ a range of data sources and modeling techniques to try to better understand how climate change interacts with migration. These studies span a wide range of study areas, scales of analysis (from local to country level), rapid- and slow-onset climate hazards, and conceptualizations.

In this chapter, we review approaches to model climate-related migration including the multiple goals of modeling efforts and why modeling climate-related migration is of interest to researchers, commonly used sources of climate and migration data and data-related challenges, and various modeling methods used. The chapter is not meant to be an exhaustive inventory of approaches to modeling climate-related migration, but rather is intended to present the reader with an overview of the most common approaches and possible pitfalls associated with those approaches. We end the chapter with a discussion of some of the future directions and opportunities for data and modeling of climate-related migration.

We do not in this chapter provide an in-depth review of qualitative approaches to conducting climate-related migration research. This is because firstly, the methods most often used – such as literature reviews, interviews, surveys, focus groups, and other ethnographic techniques – tend to follow well-established methodological practices that are common across the social sciences, that are regularly taught in mid-level undergraduate courses, and for which there already exist many useful books and references. There are far fewer introductory sources available for those thinking about modeling climate-related migration, and so we see an opportunity to add value here. Secondly, as observed in a review of qualitative research methods done by Gemenne (2018), there are any number of qualitative methods that researchers might use depending on the circumstances, and the decision of which one(s) to use is downstream of a more important decision – What question is it the researcher wants to answer? We do describe toward the end of this chapter the opportunities that exist for researchers to use participatory and mixed methods – the latter a blend of whatever quantitative and qualitative approaches might best uncover the answers to questions at hand.

5.1 Goals of Modeling Climate-Related Migration

5.1.1 Understanding Relationships between Climate and Migration

In the most basic sense, the goal of modeling climate-related migration is often to understand the relationship between various hazards associated with climate change and changes in human migration patterns and/or behavior. As described in preceding chapters, relevant effects of climate change might be slow in their onset (meaning that they emerge gradually over extended periods of time such as droughts and rising sea levels) or rapid onset (such as hurricanes and wildfires). Depending on the kind of climate hazard being investigated, the impact on migration might vary greatly. For example, the occurrence of a hurricane might increase internal migration while a drought might decrease internal migration in the same study area.

The relationship between climate hazards and migration also varies significantly by location and context. For example, research based on modeling has suggested that decreases in rainfall (or drought) significantly correlate with increases in human migration away from agricultural communities in Bangladesh (Gray and Mueller 2012) but a similar study involving many of the same authors using similar techniques found that in rural Pakistan there was no significant relationship between changes in rainfall but a significant effect between temperature increases and migration (Mueller et al. 2014). Because connections between climate hazards and migration outcomes are so complex (recall Figure 1.13), it is often difficult to say definitively that climate event X *caused* migration outcome Y. Rather, we are often at best able to state only that event X is associated with outcome Y, or that there is a statistically significant correlation between X and Y. In statistics, "correlation" simply refers to a relationship between two variables. If two variables are highly correlated, then, that means that a change in one variable is strongly associated with a change in the other variable. If two variables are weakly correlated or not correlated at all, then a change in one variable does not result in a significant change in the other (Box 5.1).

Box 5.1
Statistical Significance

In quantitative research (and in this chapter), the term "statistical significance" is commonly used. For example, a researcher modeling climate-related migration might describe their findings by saying that "the relationship between temperature change and rates of human migration is statistically significant" or, more simply, "there was a significant relationship between temperature change and migration." Conversely, a researcher might say that their results were "not statistically significant" or just "not significant." When researchers use the term "significant," this is not a judgment but refers to a very specific statistical concept.

Statistical significance allows researchers to quantify the likelihood that their findings capture a "real" relationship between variables rather than just by chance. A researcher begins by setting a significance level, specifically a probability that the found results would occur if there was actually no true effect. In other words, a significance level quantifies the degree to which a researcher is willing to accept that their results might be a false-positive (known as Type 1 error). Commonly, this significance level is 0.05, but it can be higher or lower. After defining a significance threshold, researchers can calculate p-values corresponding to their results. If a p-value is less than the significance threshold, then a researcher will accept those results as significant. A very, very small p-value means that there is a very small probability that the results are occurring randomly. So, when we refer to results as being significant, we mean that there is a very small chance that those results occur randomly. Instead, we can be confident that the results are capturing a true underlying relationship in the data. If results are not significant, then we cannot put much stock in them because there is a high probability that they are occurring randomly and are probably not representing a true relationship.

Studies investigating the relationship between climate change and migration can also vary greatly by the scale of analysis. When we think about scale, we consider spatial and temporal dimensions. Spatially, studies can range from very fine scales (e.g. a neighborhood within a city, or a grid scale on a map of a hundred meters square) to a global scale. Temporal scales can also vary from daily or seasonal (less than a year) to multiple years, decades, or longer. Studies investigating the relationships between climate and migration also reflect choices in scales, from micro-level studies that focus on the behavior of individuals and households, or macro-level studies that focus on large, aggregated migration flows within or between regions or countries (Hoffmann et al. 2021). The scale of analysis that researchers select may be influenced by their specific research interests and questions they are aiming to answer, the availability of data, methodological limitations, and many other factors.

Because modeling efforts vary so much depending on context, scale, data, etc., every modeling effort will generate unique results. One review of 127 studies of climate-related migration found 4,962 separate relationship coefficients between some type of climatic factor and migration (Hoffmann et al. 2021). In another study of 30 papers focusing on both internal and international climate-related migration studies at the country level (i.e. a macro-level scale), researchers found another 1,803 different coefficients estimating the relationship between environmental factors (including, but not limited to climate) and migration (Hoffmann et al. 2020). This huge number of combinations and permutations of climate variables and migration outcomes should be viewed the way one views a jigsaw puzzle: Each additional one helps us better see the larger picture of how climate-related migration functions across contexts and scales.

5.1.2 Prediction and Projection

In addition to understanding and quantifying existing relationships between climate change and migration, some models may attempt to go one step further by attempting to predict or project future climate-related migration. The ability to predict with some degree of accuracy future climate-related migration is of significant importance to inform policy, as we discuss in the next chapter. Like modeling for other purposes, models with the specific goal of making future projections can vary in their scale both spatially and temporally. Early efforts in this area focused on trying to identify geographic "hot spots" – areas that might experience especially high levels of climate-related migration in the future (Box 5.2).

Box 5.2
Hot Spot Analysis

Hot spot analysis, as the name suggests, aims to identify geographic areas that might be especially likely to experience climate-related migration in the future (Piguet 2010, Piguet 2022). Studies that use hot spot analysis often try to generate maps that combine multiple layers of data for such things as projections of the future spatial frequency and/or severity of climate hazards, data for expected population change in the area of study, and sometimes (though not always) data for factors that might be indicative of vulnerability to such hazards, such as poverty, housing types, or elevation. Because they lend themselves to easily visualized outputs, hot spot analyses can be appealing to policymakers and planners who must decide how to identify and direct resources to communities where future risks may be greatest. Examples for multiple countries and regions are found in the World Bank's *Groundswell Report: Preparing for Internal Climate Migration* (Rigaud et al. 2018).

Models attempting to project future climate-related migration will often use a range of scenarios. This is because there is a wide range of uncertainty related to future climate change, population dynamics, and possible policy responses. Rather than presenting a single result, such studies present a range of possible futures. For example, modelers may use data related to future climate change based on representative concentration pathways (RCPs) that capture a range of future emissions and warming scenarios. For scenarios of population and economic growth, researchers will commonly draw on established shared socioeconomic pathways (SSPs), which include scenarios for population, economic development, urbanization, and so on (see Chapter 2, Box 2.1 for a description of these).

Developing models for future projections or forecasting comes with numerous challenges. Rather than relying on existing data, developing forecasting models requires researchers to make assumptions about both future climate patterns and future population dynamics (including overall population changes, distributions of populations spatially, and mortality and birth rates). In addition, researchers must also make assumptions about how migration behavior will respond to particular climate hazards, which as has been shown in preceding chapters can play out in a variety of ways. Often, modelers use data on past or present climate–migration interactions to inform assumptions about how future migration will unfold, even though it is not certain these relationships will hold in the future. Uncertainty is therefore unavoidable in any model projecting future scenarios. Models for predicting future climate-related migration have come a long way in the past decade, despite the large degree of uncertainty and complexity associated with climate-related migration processes, and are of increasing interest to policy and decision-makers – but they still have a long way to go before their accuracy can be truly relied upon (Schewel et al. 2024). In the meantime, they provide useful learning tools for visioning what migration could look like in a climate disrupted future.

5.1.3 Understanding Underlying Dynamics

One of the biggest challenges in modeling and understanding climate-related migration is the question of *causality*. Causality refers to the idea of cause and effect that one event leads to the occurrence (or causes) another. Causality is so very difficult to pinpoint in climate-related migration because of the sheer number of potential climatic and non-climatic factors that might influence migration outcomes and the complex nature of their potential interactions (as described in Chapter 1). It is exceedingly difficult to say whether any one factor or group of factors has had a clear causal effect on a particular migration outcome, and even more difficult to estimate the relative strength of influence of one causal factor relative to others in a multi-causal relationship. There may also be factors that on their own have little or no causal influence, but which do have an effect when they occur simultaneously with others (for example, when a drought that coincides with high interest rates and an economic downturn, as happened in rural eastern South Dakota in the 1970s [McLeman et al. 2022]). When describing research results, it is important to remember an old saying that "correlation does not equal causation." This is especially important when describing results from modeling research because, by definition, models are simplified representations of real-world phenomena and should not be expected to capture the full complexity of reality.

That said, there are modeling methods that have been used to move beyond correlation and help researchers better understand causality in climate–migration processes. One example is agent-based modeling (ABM) (discussed in more detail later in this chapter), a simulation-based modeling approach that can help researchers explore different ways that various climate and non-climate factors influence the decisions of individuals, which in turn shape migration outcomes.

In theory, one might assume that by increasing the number of potential causal variables captured in a model and increasing the number of potential interactions between variables within the model – i.e. by increasing the complexity of the model itself – it might lead to more realistic outputs from the model. But in practice, there are limits to what we can do with models. One limit is that, as models become complex in the simulations they run, they become more computationally expensive – that is, they require ever larger amounts of data which in turn requires more computational power, more storage capacity, and more time to run a single simulation. We are also limited by the types and quality of data available to us, which in turn shapes what factors can (and should and should not) be considered in a model, and this in turn influences the results and outputs of the modeling exercise.

5.2 Data Sources

The results of any modeling activity are only as good as the data used (an old saying used by people who do modeling is "garbage in = garbage out"). If messy, inaccurate, or incomplete data are used in models, the results will be just as messy, inaccurate, and incomplete. It goes without saying, then, that data sources are extremely important influences on studies of climate-related migration that use modeling methods.

The specific data a researcher will want to use in a particular study will depend on the nature of the research question(s) being asked, the scale of the study, the study location, the characteristics of the population in question, and similar factors. However, the data the researcher might like to have are often not well aligned with the data that are actually available. It may be that the necessary data exist only for a specific time period, may be only partially complete, or may not exist at all. In the case of climate-related migration research, the modeler typically needs to combine data about particular climatic events or conditions with data about migration or population change at particular spatial scales, and may also need to include data about other aspects of human systems other than migration, such as cultural, economic, political, and social conditions. It is very rare to find the exact data one needs in each of these areas, and even rarer to find datasets that align well with one another in terms of spatial and temporal scales. In this section, we describe key categories of data and the opportunities and challenges associated with each.

5.2.1 Migration and Population Data

On the "migration" side of the climate-related migration interaction, researchers often find that even basic data about migration itself can be difficult to find. National censuses are the go-to source for information about populations and their characteristics (Box 5.3), but information specifically about migration is often not collected in censuses. In such cases researchers will instead use changes in population counts between censuses to estimate migration patterns (Fussell et al. 2014). This does not provide an exact measure of migration, since the observed change over time in the number of people living in a given census unit is the sum of new people moving in, prior residents moving out, deaths, and births. If it can be assumed that the birth rate and death rate have not changed significantly from one census period to the next, any change in population is assumed to be reflective of a change in net migration for that spatial unit. Census data can also offer useful information about population characteristics that are relevant to migration such as age, sex, and household characteristics. Depending on the nature of the research being done, census data may or may not be the best fit for purpose. The census may simply not contain the sort of data being sought. For example, a researcher might want to know if the effects of a particular climate hazard on migration vary depending on a person's occupation or livelihood; not all censuses contain such information. Even where the desired types of data have been collected, the spatial scale at which census data are reported may be too coarse for making meaningful analyses, or the intervals between census counts (typically ten years for most countries) are too far apart to make useful comparisons over time.

Box 5.3
What Is a Census?

A census is a process of systematically collecting information about a population and is carried out on a regular basis by most national governments. The Department of Economic and Social Affairs of the United Nations (UN DESA) publishes guidelines for countries on how to conduct a population census, the latest version of which ran to roughly 400 pages (UN DESA 2021). At a minimum, a census seeks to identify and record the number of people living within a country or jurisdiction at a given point in time and specifically where they live and the type of housing they live in. In many countries, censuses have evolved to collect a much wider array of data, and may ask people questions about their age, marital status, occupation, and any number of other things. The principle behind a census is that every household is counted separately, and each individual within the household is also recorded separately. All the individuals and households within the country or jurisdiction should be counted at the same point in time, and this count should be repeated on a regular basis at a minimum of every ten years.

A census is logistically a very difficult thing to do. A variety of ways might be used to do a census count, such as sending enumerators door to door to do a manual count to having people complete a hard-copy form that is mailed to the census agency to having people complete online forms. Unforeseen events can affect the process of carrying out reliable censuses at regular intervals, such as conflicts, social disruptions, or lack of financial resources. The COVID-19 pandemic created logistical challenges for censuses in many countries (Wilson 2023), particularly because governments are encouraged to conduct censuses in years ending in zero (or as close as possible) to allow for inter-country comparisons of results. The spatial scale at which census results are reported varies from one country to another. Most will disaggregate results according to one or more sub-national jurisdictional boundaries such as states or provinces, counties, cities, or townships, and they may also create spatial units specifically for reporting census data, especially for rural areas that have no large population centers. These latter units can create headaches for researchers comparing changes in data from one census period to another should the boundaries of census data reporting units change over time (McLeman et al. 2010).

For the researcher seeking migration and/or population data, there can be other sources besides national censuses, such as the International Organization for Migration (IOM), the UN High Commissioner for Refugees (UNHCR), the World Bank Global Bilateral Migration Database, the OECD Migration Database, and annual displacement reports from the Internal Displacement Monitoring Centre (IDMC) (although reliable data from the latter only go back to 2010). The UNHCR Refugee Data Finder provides information about displacement by sex, age, and country and, depending on the country, may also have anonymized displacement data collected via household surveys. Of these various data sources, IOM has been particularly active in developing and publishing new data collection tools for human mobility in the context of environmental and climatic change, as well as slow- and rapid-onset climate hazards (Nicoletti et al. 2023). A useful tool is IOM's Displacement Tracking Matrix (DTM), which collects data about multiple factors that can influence mobility in various disaster and conflict contexts. IOM also has projects to collect comparative data across countries with common survey instruments and methodologies for both quantitative and qualitative data collection. However, where and when IOM collects data for the DTM can sometimes be inconsistent or not the best fit for analyzing long-term trends. For example, after hurricanes Eta and Iota, IOM collected data on internal displacement only up to 6 months after the disasters occurred, and only in shelters or other official environments. This meant that data collection was *ad hoc*, did not necessarily capture people sheltering in hard-to-reach places, and was ultimately hard to reconcile at later dates (Schmidtke and Ober, 2023).

Datasets for migration and displacement at granular levels (i.e. fine spatial scales) can be generally difficult to source, especially for countries with limited financial resources for data collection and data management. This is often one of the bigger challenges for people conducting research on climate-related migration. Researchers can however sometimes find roughly what they are looking for depending on the country and the purpose. For example, a recent study compared the availability of data for four countries where flood- and/or drought-related migration is known to occur: Somalia, Bangladesh, Afghanistan, and the Marshall Islands (Thalheimer and Sok Oh 2023). In Somalia, where displacement due to conflict and repeated droughts has occurred, the researchers could obtain through UNHCR estimates of internal displacements at the regional and district levels on a weekly and monthly basis, and the dataset includes information about the reason(s) for displacement, including drought. For Afghanistan, where displacements occur due to conflict, floods, and droughts, the UN Office for the Coordination of Humanitarian Affairs and IOM DTM both collect regularly updated data on internal displacements. In Bangladesh, data on climate-related displacement is available at the household level via the Bangladesh Integrated Household Survey (BIHS), a national survey conducted by a nongovernmental organization called the International Food Policy Research Institute (IFPRI). The data include climate- and disaster-related information, and household demographic information including age, sex, and education level. Although the displacement-causing climate hazards are similar across all three countries, the spatial and temporal scales differ, meaning that the researcher can reach conclusions about the climate-migration dynamics within each country, but cannot easily compare findings across the three countries.

Box 5.4

Looking for Data on Population and Migration? Try IPUMS

The largest online repository for censuses, demographic surveys, and other population datasets from countries around the world that may be useful for climate-related migration is IPUMS (ipums.org; the acronym has no current meaning). It was first created to serve as a consolidated depository for US Census and American Community Survey reports, dating back to the year 1790; IPUMS now also contains over a billion records from the international censuses of over 100 countries.

IPUMS data and services are available free of charge for researchers.

An alternative to using pre-existing data is for researchers to collect their own data via individual and/or household surveys, an approach that is most practical and financially viable when doing research at national or subnational scales and which has been increasing noticeably since 2010 (Piguet 2022). Examples of projects that have produced large household survey-based datasets that directly focus on migration include the Mexican Migration Project (MMP) (Durand and Massey 2004) and the Bangladesh Environment and Migration Survey (Carrico and Donato 2019). This approach allows researchers to develop survey questions and collect information specifically focused on topics of interest and can generate rich detail that is often missing from official datasets of migration information, such as how many times members of a household have moved and their motivations for doing so, their aspirations for the future, and how climate factors into their decision-making process. Although the data can be incredibly rich, they require considerable time and effort on the part of researchers and participants. To collect representative samples, the surveys may need to be administered to a large number of households (often hundreds or thousands), which may require teams of survey administrators and a large amount of funding for support.

5.2.2 Use of Climate and Environmental Data

Unlike migration and population data, off-the-shelf climate datasets are much more widely available at spatial and temporal scales conducive to modeling, including data for average weather conditions, extreme weather events, and medium-term climate trends for recent decades (Hoffmann et al. 2021, Hoffmann et al. 2020). Daily precipitation and temperature data (including high, low, and average daily temperatures) are available for most land areas for recent decades, as are data for droughts (including SPEI and PDSI estimates [see Chapter 3]) and hydrological information for many watersheds, including flood records. The data are usually of good quality and typically georeferenced, meaning they can be easily mapped using standard geographical information software. The spatial resolution can vary; in some cases, the data are point-source (i.e. specific to one particular spot); in other cases they may be gridded, meaning that the value is estimated for an area, which can vary depending on the nature of the dataset from a few hundred meters squared to a commonly used size of 10 km^2. As one goes back in time, the quality and spatial coverage of climate and environmental datasets erodes, but unless a researcher is interested in studying climate-related migration patterns that occurred more than fifty years ago, the research project is more likely to be constrained by limited population data than a lack of useful climate data.

A greater challenge than the availability of climate data is the question of how to use it in conjunction with migration and population data. Let's assume, for example, that we want to model if and how migration patterns in location Y respond to a particular type of climate hazard. A number of logistical questions then arise. A first one is how best to define the climatic attributes of the hazard in question. This is usually more straightforward for sudden onset events like cyclones than for slower onset events like droughts. As an example of the latter, an immediate question becomes how to define a "drought" climatically. Is it going to be derived from an SPEI index, a PDSI index, or some customized combination of temperature and precipitation data of the researcher's choosing? (There are pros and cons to each). A next set of logistical questions concern the temporal connections between the independent variable(s) (i.e. the measures of drought) and the dependent variable (migration). In the case of sudden-onset events like cyclones, where the risk of displacement occurs almost immediately, with slower onset events like droughts there is often a lag time before migration patterns start to show a response (see Chapter 3). Another temporal consideration for droughts and other slow-onset events is, how long must the climatic conditions persist before they start to influence migration outcomes (for example, short-duration droughts may have little or no influence on migration outcomes, but longer ones do – so what time period distinguishes short from long?). And yet another temporal question is, how long after the occurrence (or conclusion) of the climate event can we expect the influence on migration to persist? The effects typically decrease over time, but for how long after the event should we continue to look for an influence in our model? Once we sort out the temporal considerations, there is next the question of direct versus indirect causal pathways between climate event and migration outcome. In the case of a cyclone, the impacts on short-term displacement and long-term migration in and out of the affected area depend heavily on damage to housing stocks and infrastructure – are we able to get data on these? (Probably.) The migration impacts of droughts depend on their impacts on livelihoods and food security, so again, climate data alone may not be sufficient to detect migration responses without adding additional data to the model such as soil conditions (clay soils retain water better than sandy soils) or irrigation rates (incomes on farms with irrigated land are much less sensitive to droughts than farms where production is entirely rainfed) (McLeman and Ploeger 2012, McLeman et al. 2022).

What we tend to see in existing studies is that those using climate data for periods of longer than a year – i.e. where the climate data capture longer term changes in climate – tend to find less of an influence on migration outcomes than those that use climate data for shorter duration changes or anomalies in climate (Hoffmann et al. 2020). In other words, gradual changes in climate tend not to show a response in migration patterns. We also see that, the longer the amount of time that passes after a climate hazard event occurs, the less evidence there is of an influence on migration (Sedova et al. 2021).

Some studies model connections between climate events and migration outcomes in binary terms across areas or across time periods (e.g. did a wildfire occur in this community versus that one, and did migration patterns change in the one where it did? [Winkler and Rouleau 2020]), while others model and compare migration patterns before and after an event (e.g. Lu et al. 2016, Alexander et al. 2019; see other examples in Chapter 2). Still other studies analyze the effects on migration of repeated exposures to the same type of event (e.g. impacts of repeated tropical storms on temporary migration in Vietnam [Berlemann and Tran 2021]) or to see whether migration outcomes depend upon the characteristics of a type of event, such as the maximum sustained wind speed of a storm, total precipitation over a specified period of time, or the severity of a drought (Berlemann and Tran 2021, Hoffmann et al. 2020). In the case of precipitation and temperature, researchers looking for possible impacts on migration typically look for anomalies in the climate data, identified as deviations from a mean for a given period of time. A heat wave, for example, might be identified as periods when temperatures exceed 95th percentiles from the base period for a given location.

Sometimes absolute measures of climate variables alone do not provide enough data to correctly identify a hazard. Earlier, we noted that climate-based data for droughts might need to be supported by additional data for irrigation or soil in order to build an accurate model for detecting a migration response. But even that might not be enough. Sometimes the impacts of a drought or a wildfire or other climate event may ripple outward through social or economic processes to affect people in places not directly exposed. So, for example, we might see that when people choose to relocate from a drought-stricken farming area, farmers in neighboring areas might also leave, not because they were directly hit by the drought but because the departure of others had adverse impacts on the economy of the wider area (McLeman and Ploeger 2012). Similarly, counties that are adjacent to counties where a wildfire occurs in the US experience population change attributable to the fire (Winkler and Rouleau 2020). Or, it may be something even more personal to individuals that determines whether they decide to move or not. In such cases, the researcher may want to draw upon *self-reported* climate data, where they survey people in the area to understand their perceived experience of risks associated with climate events or how they experienced harm, and build these into their models (the researcher may even be able to correlate this information with participants' migration histories). Studies that use self-reported climate data may yield very different results than those that utilize other data (Hoffmann et al. 2021).

The examples given previously in this section consider the use of observed climate data for modeling past or current migration responses. To model projections of future climate-related migration, researchers typically use gridded data from general circulation models (GCMs; sometimes also referred to as global climate models). These are complex numerical models of the major components of the climate system including the atmosphere, ocean, land, and cryosphere that use mathematical equations to simulate physical processes in each component (such as fluxes of heat, air, and water) and interactions or feedbacks between the elements of the system (NOAA & GDFL n.d.). GCMs break the Earth down into a three-dimensional grid, with a horizontal resolution of between 250 and 600 km in each direction across the surface, with anywhere between 10 and 20 stacked vertical layers up into the atmosphere and a number of stacked layers into the ocean (sometimes as many as 30) (IPCC n.d.). Some GCMs may be downscaled to higher spatial resolutions to project local or regional-level changes in climate (NOAA & GFDL n.d.).

The accuracy of a GCM is validated by seeing how well it can reproduce historical climate data. For example, we might set the GCM start date at the year 1975 and see how well it can "predict" air and ocean surface temperatures and precipitation patterns recorded between that date and the present. Once a model has been validated, it is then used to simulate future temperature and precipitation patterns under different concentrations of GHGs in the atmosphere. GCMs require considerable computational power and human resources to operate and are continually being refined. Agencies and researchers that operate GCMs tend to work together in collaborative fashion so as to improve the quality and reliability of their models' projections. Reports of the Intergovernmental Panel on Climate Change (IPCC – see description in Chapter 1) make use of GCM projections generated through the Coupled Model Intercomparison Project (CMIP6), a consortium of approximately 100 different GCMs. Every GCM will inevitably project slightly different outcomes for a given grid cell at the same level of assumed warming. These variations can be smoothed out by averaging the outputs from multiple GCMs (or ensembles). This technique was used, for example, by Smirnov et al. (2023), who used projections of future drought conditions derived from 16 different GCMs to project that the potential for drought-induced migration globally doubles by the end of this century at current rates of greenhouse gas (GHG) emissions.

5.2.3 Additional Challenges with Data

In each of the preceding sections, we have identified a number of caveats or challenges with respect to data collection and interpretation; there are a few more worth noting.

One is that standard, off-the-shelf data from government agencies and institutions for migration and population change may not allow the researcher to differentiate between voluntary and forced movements, much less place individual decisions along a spectrum of voluntarity or agency (Nicoletti et al. 2023) as was described in Chapter 1. Further, the data available may not allow the researcher to distinguish among non-migrants those who stay because they do not wish to migrate or have not even contemplated moving (i.e. voluntary immobility) from people who might want to move but by cannot (i.e. involuntary immobility). Unless one is using survey data that have explicitly asked as questions intended to capture such distinctions, these challenges are difficult to overcome. Even then, a question becomes where the survey data are collected. Have they been collected in the sending area, the destination area, or both? The researcher will have access to incomplete information about motivations and agency unless similar surveys have been done in both locales, as has been done in the MMP, for example. The MMP (mmp.opr.princeton.edu) is notable in that it has collected data longitudinally, attempting to follow the migration behavior of the same individuals, households, and communities over time. It is consequently not surprising that the MMP has been the data source for a large number of studies about the effects of drought on internal and international migration cited in Chapter 3 of this book.

Even large, well-resourced survey initiatives can produce data with biases or errors. A survey that is conducted only once provides information from a single snapshot in time. When participants are asked to recount past events, such as past migration experiences or motivations, it can sometimes be difficult for them to remember specific details. Surveys can also by nature of their design can be prone to sampling biases, such as having too small sample size, the timing of the survey (e.g. seasonal workers may be away from home at certain times of the year), and the inability to reach a representative sample of people. Certain types of migrants and community members may be especially difficult to reach through surveys, such as undocumented migrants, low-income people, people living in temporary accommodations, people with disabilities, or people who are distrustful of authorities (Borderon et al. 2021).

A particular bias in research on climate-related migration is geographical: Far more studies have been done on selected countries in low- and middle-income countries of Asia, Africa, and Latin America than elsewhere (Piguet et al. 2018). Better geographical coverage of research would help generate results that may be more generalizable across countries. At the same time, much of this research and the funding for it has been produced by universities and research institutes based in high-income countries. What types of research questions might be asked, what methods might be used and, ultimately, what more might we know if there were greater diversity in the people doing the research (Hoffmann et al. 2023)?

One final notable challenge relates to potential spatial mismatches between climate datasets and migration datasets. The spatial distribution of human settlements

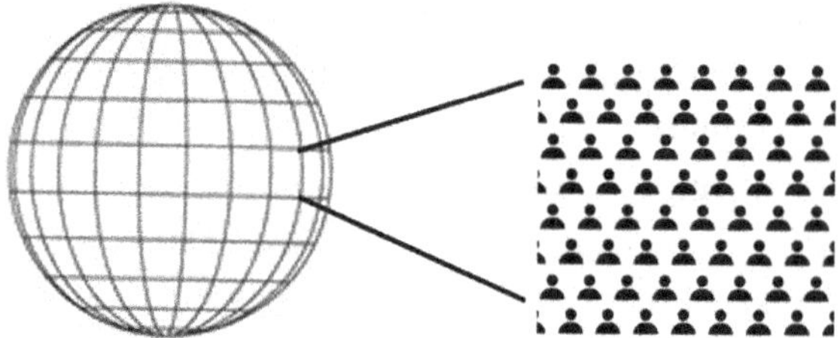

Figure 5.1   Differences in the spatial scale of different datasets (for example, gridded climate data versus individual or household-level social data) can pose a challenge to climate-related migration research.

is neither random nor even. There are areas with few homes or settlements, and other areas that are densely packed urban settlements. When population or migration datasets are gridded, people are assigned to a single grid cell. But unlike as is shown in Figure 5.1, people are not going to be uniformly distributed within that cell. The cell might include a single town or village of 5,000 people surrounded by virtually unpopulated space, or it might contain two towns of roughly 2,000 people each plus a rural area of another 1,000 people in scattered homes, or it might capture 5,000 people living in a single neighborhood of a larger urban center. The actual distribution of people will be concealed, and the grid cell in question will have a value of 5,000.

5.3 Modeling Methods

A variety of modeling methods have been used by researchers to try to capture climate–migration dynamics, and each approach has its own set of strengths and weaknesses. The selection of a given methodology reflects the specific research questions, study context, and data available. Researchers modeling climate-related migration and environmental migration more broadly come from a range of disciplinary backgrounds including economics, sociology, geography, and environmental science, among others, and the methods they favor may reflect disciplinary norms (Hoffmann et al. 2021, de Sherbinin et al. 2022, Fussell et al. 2014). Methods and results also reflect differences in the scale of analysis (individual to regional) and in the kind of climatic conditions being considered. Following is a summary of some of the more common methods used.

5.3.1 Regression-Based and Statistical Methods

One common approach to modeling climate-related migration involves what is called regression analysis (Box 5.5), which allows researchers to model mathematically the relationship between climatic variables (such as temperature or precipitation) and migration data. Regression analysis also allows researchers to "control" for other potential causal factors at macro levels (such as economy, political events) and/or micro-levels (such as household size, age, income, etc.) and attempt to isolate the effects of the climate variable(s) of interest on migration.

If a migration outcome is treated as a binary variable (i.e. assigned the value of 1 if migration occurred and 0 if migration did not), researchers will use what is known as logistic regression. If the migration outcome variable is a continuous variable (e.g. the researcher is counting the number of migration trips a household or individual has taken (0,1,2,3…etc.), then researchers may use an ordinary least squares (OLS) regression model. When the researcher is looking at factored outcomes (e.g. where migration might occur to one of multiple possible destinations), they may use a multinomial model. We will not go into the technical dimensions and limitations of these various regression methods, but introduce them simply to alert readers that regression models are commonly used in this area of research, that there are different types that are used for different purposes, and that it is important to select the appropriate one based on the nature of the migration outcome being looked for.

Box 5.5

Regression Analysis

Regression analysis is a statistical method that quantifies the relationship between two or more variables. In a regression analysis, there is always at least one dependent variable (or outcome variable) and at least one or more independent variables. A regression analysis allows researchers to investigate how a change in an independent variable (e.g. rainfall or temperature, or some other factor that is not influenced by the dependent variable) influences the dependent variable. The simplest version of a regression is a linear regression, which, as the name suggests, identifies a line of best fit to quantify how a change in each independent variable linearly affects the dependent variable. The simple equation for a linear regression is

$$Y = A + BX,$$

where Y is the dependent variable, X is the independent variable, B is the fitted coefficient that defines the relationship between X and Y, and A is a fitted constant. In the case of climate-related migration studies, the outcome variable Y might be related to migration (such as the number of migration trips that a household reports), and X could be a climate-related variable such as a change in temperature. B would then provide quantitative insight to the researcher about how a change in temperature corresponds to a change in migration outcomes.

Regression analyses do not have to be linear. They can range in complexity from linear relationships to highly nonlinear (see Figure 5.2). How "well" a regression fits the data can be evaluated by a range of approaches, including by evaluating error metrics such as root-mean-square deviation (RMSE) or by assessing the fit based on a coefficient of determination, also known as R^2.

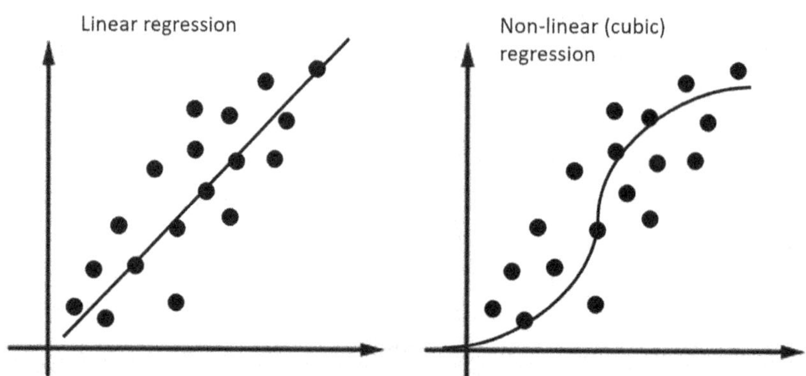

Figure 5.2   Regression-based methods are used to quantify relationships between data. They can range in degree of complexity and can be linear (left) or nonlinear (right).

5.3.2 Machine Learning

Machine learning is a term that refers to a range of quantitative methods in which a computer (the "machine") is used to recognize existing patterns in data and then use those patterns to construct a model of that data. It is an especially useful approach when one is using very large sets of data with multiple potential causal variables and the patterns or relationships within the data are highly nonlinear or not easily recognizable. Machine learning methods do not require the researcher to explicitly program a model based on theory or underlying assumptions (e.g. the researcher decides, "I will build an OLS model to look at the effects of these two particular factors on the number of migration trips made"); instead, the machine learns directly from the data the model to build. Machine learning methods fall generally into one of two types, supervised and unsupervised. Supervised machine learning algorithms are used to predict one or more dependent variables based on relationships between independent variables in the dataset (similar to regression analysis). Unsupervised models do not predict any dependent or outcome variable but are used to identify patterns within the data, such as by identifying clusters of similar data points.

Machine learning uses algorithms that range in complexity from simple linear regression to complex neural networks. As models become more complex, there may be a tradeoff between that complexity and the ease with which one can interpret the results (Best et al. 2022) (Figure 5.3). For example, a simple linear regression model does not allow for nonlinear interactions between variables but is straightforward to interpret. By comparison, advanced machine learning algorithms such as random forests or neural networks can capture a large degree of complexity and nonlinear relationships within the data but are often considered

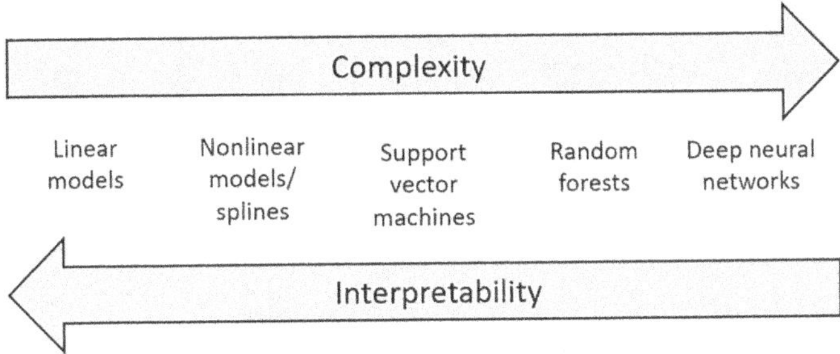

Figure 5.3   There can be a tradeoff between complexity and interpretability of machine learning algorithms. Researchers should take care to select the appropriate model, based on research goals. Diagram adapted from Best, K., Gilligan, J., Baroud, H. et al. (2022), Springer, Open Access, Copyright © 2022, The Author(s).

"black box" – i.e. it is difficult to interpret how the outcomes were arrived at. Machine learning methods are being increasingly used in climate-related migration studies; the specific type of machine learning approach and the algorithms the researcher will select and develop depend on the data and the goals of the research. For example, if prediction is the goal, then a more complex but less interpretable algorithm may improve predictive performance, but if the goal is to understand relationships between variables in a conceptual sense, then a simpler but more interpretable approach may be desirable. An attractive feature for climate-related migration research is that machine learning algorithms help the researcher explore and model complex, nonlinear relationships between variables – which is often the nature of how climatic factors and non-climatic factors interact to influence migration – and can analyze large and complicated datasets such as household surveys that might take a prohibitively long time to explore using other methods.

As examples of how this approach has been used in climate-related migration research, Best et al. (2021) used a type of machine learning algorithm known as a random forest to analyze two household surveys to identify key social and economic factors that predicted migration patterns in southwestern Bangladesh. Although it could identify the important factors, the model could not explain precisely how they affected the migration decision or the dynamics of how the variables interact.

In a different study, McLeman et al. (2022a) combined random forest models with regression tree analysis – the latter a simple machine learning method where the goal is to create a model that predicts the value of a target variable (e.g. % population change over given areas over given time periods) based on several input variables (in this case, variables such as growing season precipitation, daily high temperature, presence of irrigation) – to explore possible linkages between rural population change and droughts on the North American Great plains between 1970 and 2010. This approach proved to be very useful in identifying where and when to look more closely for possible sites of local population losses through out-migration caused by droughts, but the researchers caution that "ground truth-ing" through other mixed-method approaches, including site visits, is essential to confirming the existence and nature of the drought–migration relationship. In subsequent field investigations, the researchers found that drought-related migra-tion identified by the models as having occurred in rural eastern South Dakota in the 1970s was more complex in nature than suggested by the models, with rural out-migration being concentrated among specific age groups, with different par-ticipation rates between men and women (McLeman et al. 2022b). In other words, machine learning approaches are especially valuable in climate-migration research when used in combination with other, more interpretable methods (see section on "Mixed methods").

5.3.3 *Gravity Models*

Gravity models are a commonly used modeling approach to explain migration between two areas. Inspired by Newton's law of gravity – the concept that the force of attraction between two objects is proportional to the product of their masses and inversely proportional to the square of their distance – these models assume that the strength of migration between two areas is proportional to the product of their populations and inversely proportional to the square of the distance between them.

Newton's law of gravity:

$$F_{\text{gravity}} = \frac{\text{Mass}_1 \times \text{Mass}_2}{\text{distance}^2} \quad F_{\text{gravity}} = \frac{\text{Mass}_1 \times \text{Mass}_2}{\text{distance}^2}$$

Gravity models of migration:

$$F_{\text{migration}} = \frac{\text{Population}_1 \times \text{Population}_2}{\text{distance}^2}$$

The implication of gravity models is that migration is stronger between areas of higher populations and areas that are closer to one another. Other assumptions, such as the effects of climate change or environmental conditions, can also be incorporated.

Because of the very simple assumptions underlying gravity models, they are most commonly used to model migration across large geographic areas such as between countries or regions. But, their simplicity also makes them very useful tools, as they require fewer underlying assumptions than more complex types of models. They have been used successfully to reproduce past shifts in populations, and have poten-tial for forecasting future migration at large scales (de Sherbinin 2022).

Gravity models were used in the World Bank's *Groundswell* reports to estimate future internal migration due to slow-onset effects of climate change (including water stress, crop failure, and sea level rise) across six regions, beginning first with sub-Saharan Africa, South Asia, and Latin America (Rigaud et al. 2018) and then expanding to include East Asia and the Pacific, North Africa, and Eastern Europe and Central Asia (Clement et al. 2021). The reports used a scenario-based gravity model that combines gridded population and climate data to predict population shifts under three combined climate and development scenarios. The scenarios include a "pessimistic" reference scenario of high GHG emissions and high levels of inequality in development, a "more inclusive development" pathway of high emissions but more equitable development, and a "more-climate friendly" scenario of lower GHG emissions but inequitable development. Under pessimistic scenarios, the *Groundswell* reports suggested as many as 216 million people might migrate internally by 2050 as a result of slow-onset impacts of climate change.

The *Groundswell* reports highlight the ability of gravity models to project climate-related migration across wide geographic areas. An analysis of similarly large geographical scale would be much more difficult with a more complex, computationally expensive modeling approach. Scenario-based model projections of this type are useful to policymakers looking to understand what the future might look like under different emissions and development scenarios. A similar modeling initiative inspired by Groundswell was used to project future migration scenarios for Africa for the year 2050 (Africa Climate Mobility Initiative 2022).

5.3.4 Agent-Based Modeling

Agent-based modeling (ABM), sometimes also called "individual-based" modeling, is another method that has been used to study climate-related migration. ABMs are models that simulate the behavior and decisions of individuals (described as agents) allowing the researcher to explore the outcome their interactions might collectively produce (Bonabeau 2002, Railsback and Grimm 2012). To study climate-related migration, a modeler might simulate how agents representing individual households make decisions about migration under various climate hazards or scenarios and then observe from the simulation what patterns emerge from their aggregated decisions (Figure 5.4). Unlike gravity models, which focus on larger-scale population movements, ABMs place more emphasis on how migration evolves from the collective behavior of individuals.

A strength of ABMs is their ability to explicitly model in considerable detail the linkages and feedback effects between human systems (from individual decision-making about migration to community-level dynamics) and changes in the environment (DeAngelis and Diaz 2019). In ABMs, agent decision-making rules need to be set by the modeler, and the choices are critically important to the overall model behavior. In ABMs of migration broadly, decision-making rules

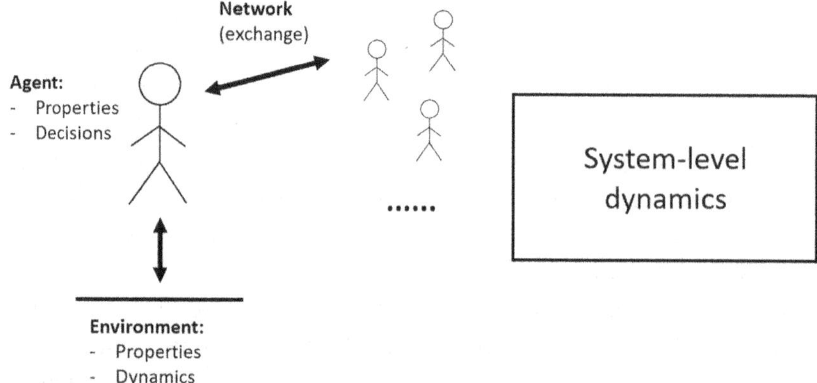

Figure 5.4   Agent-based models (ABMs) represent the properties and decision-making of individual agents. These agents can interact with other agents and exchange information or resources. They can also interact with their environment, which may change dynamically. The goal of ABMs is to identify what emergent, system-level dynamics arise from the behavior of many individuals.

vary from simple numerical models to heuristics (i.e. a small set of pragmatically selected decision-making criteria) to more complex behavioral theory (Klabunde and Willekens 2016). With ABMs, a modeler can code for as much detail in the way that agents make decisions as desired and for how the behavior of one agent may influence others, such as by assuming agents will share information as it is learned or may exchange resources with one another.

An example ABM being applied to studying climate-related migration is a model used by Hassani-Mahmooei and Parris (2012) to simulate climate-related migration between districts in Bangladesh using 10 heuristics or migration "rules" related to socio-economic factors, livelihood prospects, house ownership, employment that might moderate migration decisions made in response to a selected set of climate shocks. The model also considered population growth and mortality. The researchers predicted using the model that between 3 and 10 million people in Bangladesh would migrate internally over the following 40 years (2012). In a more recent study of climate-related migration in Bangladesh, Bell et al. (2019) developed an ABM with more complex decision-making rules that included individual perceptions and place attachment. Their results found that sea level rise would not likely lead to migration away from coasts, the opposite of what was projected in the Hassani-Mahmooei and Parris study (Bell et al. 2021). This highlights how differences in the decision-making rules and factors selected by the modeler generate different outcomes.

Another notable ABM of climate-related migration is one developed by Kniveton et al. (2011) to model migration responses to precipitation in Burkina Faso. This model used a decision-making method based on a theory from psychology known as the Theory of Planned Behavior (TPB), suggesting it provided a more realistic representation of decision-making than simpler, heuristic-based models. In Kniveton et al.'s model, as agents move around they are assumed to exchange information about their migration experiences with other agents that they encounter, thus potentially influencing the decisions of others. The model assumes that each agent is part of a peer network connected to 50 other agents. The agents' migration patterns are then filtered through a range of future scenarios of demographic, economic, and climate conditions in the region. Their model results suggested that a drier environment would produce higher amounts of internal and international migration under inclusive social and political conditions, and that international migration would be low in a wetter climate and less inclusive political and social conditions.

ABMs can also be used to test theories about climate-related migration (which might in turn refine future modeling) in addition to trying to project how many people might migrate, and where, under scenarios (Grimm et al. 2005). For example, Best et al. (2021) used a pattern-oriented ABM of climate-related migration in a simulated rural community in Bangladesh in order to compare the ability of multiple decision-making approaches to reproduce known migration patterns.

Despite their potential to advance our understanding of climate-related migration, only a few studies have made use of them (Thober et al. 2018). A strength of ABMs is that they allow researchers to more explicitly capture known complexities and multiple influences and factors in climate-related migration, but they also depend upon a very subjective set of assumptions on the part of the modeler about how individuals make decisions and interact with one another, and how much detail to include or leave out.

5.4 Future Directions in Data and Modeling

New datasets and new methods for data collection, analysis modeling efforts are continually emerging, and these can be expected to lead to promising future directions for climate and migration research. We touch now on several of these.

5.4.1 Data Collection Using Mobile Devices and Social Media

There are a variety of ways that the globally expanding use of mobile phones may be harnessed to facilitate research on climate-related migration. As one example, mobile phones are continuously collecting data on the location of their users, and under certain circumstances, researchers may be able to access pooled, anonymous user locational records from service providers for mobility-related research. One such study analyzed mobile phone records from 6 million users in Bangladesh and allowed researchers to identify where people moved to in the immediate aftermath of Cyclone Mahasen, which made landfall in May 2013 (Lu et al. 2016).

Mobile phones may also help reduce the time requirements and costs associated with data collection, such as for collecting survey data in hard-to-reach locations. For example, Coupland (2020) used WhatsApp messages to collect information from research participants living in temporary housing on small islands in the Bahamas in the months following Hurricane Dorian and the subsequent COVID-19 pandemic, when researchers were required to leave the islands. Another benefit of using mobile phones for surveys is they allow researchers to engage with participants for longer periods of time and through short, regular mobile check-ins (Bell et al. 2016, Bell 2017). Such "high frequency" survey interactions may also reduce recall bias when participants are asked to recount migration histories or remember previous migration decisions, and facilitate longitudinal studies (Bell et al. 2019).

Data collected from social media posts and digital transactions (what Bell calls "big secondary data") can be used for a variety of research purposes. For example, analysis of Twitter/X and Facebook posts can help researchers monitor and map large, complex social networks that might be otherwise difficult to quantify (Boas 2020). Crowd-sourced data and "citizen science" approaches can facilitate collection of high-resolution information about climate conditions such as temperature, precipitation, and air quality data, which can in turn be incorporated into a variety of different modeling approaches (Bell 2017).

Despite the potential benefits, the use of mobile phone data is not without its challenges and limitations. One challenge related to using high-resolution data gathered from mobile devices, social media, and other online sources is that it could potentially be used to identify the movements of individuals and households, and researchers should keep privacy concerns in mind when using it (de Montjoye et al. 2018). Also, as mobile phones and communication technologies make it easier to identify and collect information from "hard to reach" groups of people, it is important to recognize that some groups, especially those that are persecuted, susceptible to exploitation, or otherwise especially vulnerable, may not wish to participate or be identified, and so their safety and wellbeing may depend on the maintenance of their anonymity (Borderon et al. 2021).

5.4.2 Participatory Methods

The concept of participatory methods refers to the idea that the people who are the subjects of research should also be actively involved in its design, implementation, and dissemination as much as possible. It is closely related to the concept of "co-production of knowledge," which is the idea that researchers work closely with communities and community members to conduct the research together, rather than researchers working without the input of local stakeholders. By working closely with participants, researchers are not the removed creators and keepers of knowledge, but rather co-learn and co-create a deeper understanding together.

Although participatory methods are often associated with qualitative research, they are equally as important to quantitative studies, geospatial/mapping projects and mixed-methods research. An important strength of participatory methods is that it facilitates the integration of local knowledge into research that the outside researcher or modeler may not possess. As noted in previous sections, assumptions about interactions between climate, non-climatic factors, and individuals' migration decisions can make or break a model or study. Bad assumptions in model and study outputs can in turn lead to bad decisions by decision-makers who rely upon it, which can in turn have adverse consequences for people living in affected areas. Having local participation and including local knowledge from the very beginning of a research project can significantly enrich research while also helping to ensure that results are relevant to the context.

Depending on the context, researchers may want to incorporate and weave the knowledge of Indigenous and traditional peoples into their study (a process sometimes described as "braiding" [McGregor et al. 2018]) and participatory methods offer an avenue for doing so. Indigenous peoples often possess deep knowledge, built up across generations, about the local environment and the effects of a changing climate, and a growing number of studies have engaged with Indigenous communities to improve our understanding of climate change impacts and adaptation options (Pearce 2018). Although Indigenous communities displaced by climatic and environmental factors have been studied by outside researchers using data collection methods that include surveys and interviews, very few studies have attempted to engage with Indigenous communities in participatory research and co-production of knowledge about or experience with climate-related migration (Manrique et al. 2018). This represents an opportunity for both doing research and developing new methodologies. However, engagement with Indigenous peoples and integration of Indigenous knowledge requires in-depth considerations of ethical responsibilities and cultural context. It is important that researchers avoid being "extractive" in their engagement with all local communities, especially Indigenous communities. Researchers must never "take" Indigenous knowledge to improve their research, but should deeply consider how the research or activities surrounding the research could be designed to benefit these communities, meet their needs, and center their voices.

Participatory methods can also help researchers better understand the ways that climate-related migration differentially affects populations within a community, especially vulnerable populations such as women, people with disabilities, undocumented people, and historically excluded minorities, whose experiences may not be captured or reflected in more traditional sources of data such as censuses (Borderon et al. 2021). Participatory modeling, in which the people and communities being studied are actively involved in designing and developing models, is a promising way of doing this. The researcher provides participants the opportunity at multiple points in the development process to provide feedback on the model and as results emerge, they are provided first to participants to validate the reliability of the outputs, with participants also then helping to decide how those results are communicated more broadly.

One example of participatory modeling for migration that included climate considerations was a participatory ABM of rice production and labor migration in Northeast Thailand (Naivinit et al. 2010). In this work, the modelers used an iterative modeling approach called Companion Modeling (Com Mod) in which local farmers' and academic experts' knowledge is integrated into the ABM. The model structure itself was co-developed with local farmers to ensure that the most important aspects of the system were captured. This provided researchers with rich insights into how farmers make decisions about rice production and migration that could then be incorporated into the model, including information about rainfall, water storage, and rice yields. Farmers who participated in this work reported that it was also meaningful for them to more clearly understand their different options when adapting to changes in their environment.

5.4.3 Artificial Intelligence (AI) and Big Data

As *artificial intelligence (AI)* technologies become more powerful and accessible, and massive amounts of new data are continuously generated, there is considerable potential for researchers to utilize AI in climate-related migration studies. It is a fast-moving field of technology, and by the time the present book is published, there will no doubt have been any number of new developments and publications about the possibilities, challenges, and ethical concerns about using AI in research.

The term AI refers to a range of technologies that simulate human intelligence and decision-making, and the potential applications are broad. Many of us use some kind of predictive text AI almost daily when we send text messages or emails (when your email software suggests words to complete a phrase or sentence you are typing, that's AI). More advanced kinds of AI include chatbots, virtual assistants, and advanced search tools like ChatGPT. Image recognition tools can use AI to identify and classify elements of interest within photos and imagery.

AI can be used at almost any stage of the research process. AI tools can be used to identify data sources, harvest data from those sources, help researchers generate code to analyze the data and build models, conduct the analysis and identify patterns of interest, and even test the reliability of models. AI can also be used to generate images and visualizations of data, facilitate systematic literature reviews, scan non-scholarly reporting of information on particular topics of interest, write up results, and much more. All of these uses of AI could be applied to specific research questions on climate-related migration.

As an example of the potential utility of AI in research, in July 2023, one of the present authors tested the ability of ChatGPT to reproduce research done by the author and colleagues over a decade earlier when they assessed drought-related rural population change in western Canada in the 1930s (McLeman et al. 2010). The first stage of the original research required months of preliminary work, with the team manually combing through hard-copy Canadian census volumes to identify and enter into a GIS model population counts for hundreds of small rural census divisions across three census periods (the census boundaries for those three censuses, which changed multiple times, also had to be manually digitized). With only a few short queries, ChatGPT was able to correctly identify and rank all the rural census divisions in Saskatchewan, Canada that lost population during the 1930s. It was not immediately clear if this was the authors' own work being fed back to them (probably), if the AI was actually accessing original census records (possible but unlikely), or if other sources were being accessed (ChatGPT could not or would not identify its source(s)). This opaqueness of the data source represents one of the challenges of using AI for research at this stage of its development. Nonetheless, this simple exercise showed how AI has the potential to reduce much of the drudgery of data collection and preparation. Interestingly, ChatGPT was unable to do the same for ranking for US states, suggesting that the AI had not by mid-2023 been trained to directly harvest and analyze online US Census data – perhaps because of something as simple as census data portals requiring the creation of a username and password.

While AI presents a range of opportunities to advance research in climate-related migration, it also poses challenges. For example, a user of AI might not necessarily know the reliability of the underlying data, and, if not careful, could overlook limitations or errors in the data that the AI algorithm utilizes. In some cases, AI may be able to point to useful datasets but may not have the permissions to access the data itself. In any application of AI, users may be tempted to use it to take shortcuts in research rather than using more rigorous or time-consuming methods. Ultimately, the onus is on the researcher to ensure the validity of their work, whether AI tools are used or not.

AI can also pose ethical challenges. AI is being increasingly used at borders and to monitor migration, with massive amounts of data being collected every day, raising many legal concerns (Beduschi 2021). If data collected from migrants without their knowledge or consent are used to conduct migration research directly or are used to train AI models that are in turn used by researchers, it raises serious ethical concerns. In such cases, researchers may want to reflect upon whether they wish to be complicit in the use of algorithms and data sources that might conceivably be weaponized against migrants.

5.4.4 Mixed Methods

Every methodological approach has its strengths and limitations. Researchers studying climate-related migration are often working to find ways to use multiple methods in combination, where possible. A common approach is for researchers to collect both qualitative and quantitative data in and for their research location in order to gain insights from rigorous, numerical data, and the richness of qualitative data related to lived experience. This is referred to as a *mixed methods approach* to research. While it may sound like a straightforward idea to use multiple methods in combination, it can be very difficult, time consuming, and resource intensive to implement. But when done well, mixed methods approaches can lead to synergistic findings about the question at hand.

A common mixed-methods approach to studying climate-related migration has been to use qualitative interviews in combination with regression analysis (Ty Miller and Thai Vu 2021). For example, in one multi-year, multi-institutional project a large interdisciplinary research team combined data from household surveys, participatory methods (including focus group discussions, transect walks of agricultural land, mobility mapping, and impact diagrams), semi-structured expert interviews, and publicly available data to investigate the ways that households use migration to manage risks of rainfall variability and food security in eight countries: Guatemala, Peru, Ghana, Tanzania, Bangladesh, India, Thailand, and Vietnam (Warner and Afifi 2014). In this way, they were able to collect robust quantitative and qualitative data across a broad geographic scale and identify commonalities and differences between migration strategies in the different country contexts. The researchers also used insights from both their quantitative and qualitative analyses to inform an ABM of rainfall variability and migration in Tanzania (Smith 2014). This study is an excellent example of combining multiple methods across countries and contexts to investigate climate-related migration.

Personal narratives can be powerful tools in advancing both research and policy for climate-related migration, communicating research findings, and giving voice to individuals and communities most affected by climate change. Stories have power in their ability to influence and evoke emotion. Integrating narrative and qualitative information into any of the modeling approaches described previously in this chapter through a mixed methods approach should be widely considered by aspiring and experienced researchers working in this field.

6

Policy Considerations

Governments, institutions, communities, and individual members of the public hold a wide range of views with respect to how to respond to climate change and its underlying causes, and this extends to climate-related migration and displacement as well. Preceding chapters have illustrated the complexity of the processes that give rise to climate-related migration and displacement, which further complicates the development of appropriate policy responses and programs. Effective policy response requires the engagement of governments and institutions at all scales, from the local to the international, as well as a wide range of nongovernmental and humanitarian organizations, other non-institutional actors, and representatives from affected communities. Despite this, policymaking and institutional decision-making in general tend to be done in a siloed approach: that is, decisions about financial policies are made by one set of actors, decisions about health policies by another, decisions about environmental policies made by still another, migration policy by another, and so on, often without much consultation with decision-makers across silos. This is further reflected in the internal structures of governments, which have dedicated ministries and units to deal with specific issues, and in the network of international agencies that work on specific issues of global importance. This siloed approach does not work well when attempting to respond to a challenge like climate-related migration and displacement that demands the coordinated attention of decision-makers in multiple silos – but a move away from silos is starting to emerge.

Relative to other aspects of climate and environmental policy, climate-related migration and displacement have only recently attracted concerted attention from policymakers. To date, policy discussions have tended to center around questions of how to prevent or minimize involuntary migration due to climate hazards and whether to create legal and institutional protections for climate-displaced people (and if so, what would be the best mechanism(s) for doing so) (Refugees International 2021). As policymaking for climate-related migration evolves,

"

increasing attention is and will be given to possible points of intervention across stages of the migration process, including prior to, during, and after migration or displacement occurs (Rigaud et al. 2018, McInerney et al. 2022). Some of the specific interventions being considered can be developed within the context of existing national and international policies, laws, agreements, and institutional arrangements, such as those pertaining to climate change mitigation and adaptation, disaster risk reduction, development, migration, displacement, and refugee policy, while other interventions being discussed would require the establishment of new laws, policies, and agreements (McAdam 2011 Schraven 2011, Hall 2015, Ober and Sakdapolrak 2017, Rosenow-Williams and Gemenne 2016). The types of policies that we agree upon today may well end up shaping the rates, patterns, and types of climate-related migration we see in the future (Clement et al. 2021, Geddes 2012, McLeman 2020a, Rigaud et al. 2018).

This chapter provides an overview of policymaking that is being done at international, regional, and national levels, highlighting some of the key processes and frameworks relevant to climate-related migration and displacement. We discuss the extent to which these policies respond adequately (or not) to the needs and complexity of the challenge. We also describe a range of actors that have been actively engaged in these policymaking processes and the nature of their influence on these processes. We assess the different levels of actors in descending order of scale, starting with an overview of international policy frameworks and processes and then moving through regional and national level processes and approaches. Although we assess each level separately, they should be viewed as a network or web of interconnected actors and processes that influence one another, even as they evolve. Because policymaking relevant to migration and displacement in a changing climate is rapidly evolving, it is inevitable that new developments will occur frequently, and so readers are encouraged to view this chapter as a primer that provides the background information needed to begin their own research and exploration of the policymaking landscape.

6.1 Key International Policies on Climate, Disasters, and Development

High-level international agreements on climate, disasters (Box 6.1), and development have been a natural home for developing policies and actions for climate-related migration and displacement for several reasons. The first and most obvious is that they deal with the underlying causes by addressing the need to reduce greenhouse gas (GHG) emissions and responding to poverty and socio-economic inequalities that make people vulnerable to the effects of climate change (Rigaud et al. 2018). If we want to avoid involuntary displacement of people and involuntary immobility in the face of climate hazards, the most effective

Box 6.1
What Is a Disaster?

A *disaster* is defined under the *Sendai Framework on Disaster Risk Reduction* (see below) as being, "A serious disruption of the functioning of a community or a society at any scale due to hazardous events interacting with conditions of exposure, vulnerability and capacity, leading to one or more of the following: human, material, economic and environmental losses and impacts." This definition overlaps considerably with the description of climate risk given in Chapter 1, the main distinction being that a risk is something that may happen, but a disaster is an event that has happened. Climate hazards described in Chapters 2 and 3 fall within the description of hazardous events used in the Sendai Framework.

way is to take steps to reduce the frequency and severity of those hazards in the first place. Second, climate-related migration and displacement may potentially be minimized by ensuring people and communities have the resources, capacity and overall financial means to adapt successfully to climate hazards, and sustainable development is a key aspect of this. And third, when people become displaced by climate hazards and/or are trapped in locations of high risk, high-level international agreements and funding mechanisms in these areas provide potential mechanisms for providing financial assistance to people and communities in vulnerable countries for the losses and damages they experience. An overview of the key agreements and institutions involved now follows.

6.1.1 International Climate Policy under the UN Framework Convention on Climate Change (UNFCCC)

The key international agreement pertaining to climate change is the United Nations Framework Convention on Climate Change (UNFCCC). There are three aspects of it that are particularly relevant to climate-related migration and displacement, and will be discussed in detail in what follows:

- Actions to reduce GHGs that would reduce the future frequency and/or severity of climate hazards.
- Adaptation initiatives that include climate-related displacement as an area of concern.
- Newly emerging initiatives to provide financial assistance to low-income countries for loss and damage due to the impacts of climate change, potentially including involuntary displacement and planned relocations.

The UNFCCC was a product of an international conference held in 1992 in Rio de Janeiro known as the Earth Summit, at which conventions on biodiversity protection and combating desertification were also developed. The UNFCCC came into effect in 1994 and has been signed and ratified by almost all UN member states. Article 2 of the Convention states that the ultimate objective of the UNFCCC is the

… stabilization of greenhouse gas concentrations in the atmosphere at a level that would prevent dangerous anthropogenic interference with the climate system. Such a level should be achieved within a time frame sufficient to allow ecosystems to adapt naturally to climate change, to ensure that food production is not threatened and to enable economic development to proceed in a sustainable manner.

Signatories to the UNFCCC commit to six key actions:

- Providing regular reports on their country's GHG emissions
- Participating in coordinated actions to reduce GHG emissions
- Promoting the sustainable management of natural systems such as forests that absorb GHGs from the atmosphere
- Promoting technical, scientific, and educational cooperation in achieving GHG emissions reductions
- Cooperating in preparing for adaptation to the impacts of climate change
- Meeting annually to advance action in response to the preceding five commitments

All of these commitments are directly or indirectly relevant to responding to climate risks that stimulate migration and displacement. The first four commitments listed earlier are aimed at limiting increases in atmospheric concentrations of GHGs (generally referred to as *mitigation*), and if effective would reduce future climate hazards and risks associated with slow- and sudden-onset hazard events and sea level rise. The fifth commitment – to facilitate adaptation to the impacts of climate change – has led in recent years to formal discussions under the UNFCCC on how to reduce the risks of climate-related displacement. The sixth commitment, to meet annually, is done late each calendar year by holding a Conferences of the Parties (or COPs for short) in a different city, and it is at these meetings that key decisions and actions to be taken under the UNFCCC are negotiated (Figure 6.1). An important guiding principle of the UNFCCC is "common but differentiated responsibilities," meaning that all countries share the responsibility of addressing climate change, but not all countries have the same resources and capabilities to do so, and some countries have much higher historical carbon emissions than others and thus have a greater responsibility to take action.

All decisions reached under the UNFCCC and at COPs must be approved through consensus, that is, all participating countries must agree. This can be very challenging and is a reason why international climate negotiations and actions proceed at a slow pace. The 190+ signatories to the UNFCCC include countries that

Figure 6.1 Bicycles parked outside the venue for COP23 in 2017. Although the government of Fiji was responsible for organizing COP23, the country lacked sufficient hotel space to accommodate the thousands of attendees, so meetings were held in Bonn, Germany, where the UNFCCC secretariat is based. Photo by R. McLeman.

have high GHG emissions (e.g. China, USA), wealthy countries with relatively small populations but high *per capita* GHG emissions (e.g. Canada, Australia), oil exporting countries that would suffer economic losses if fossil fuel use were to be rapidly reduced (e.g. Qatar, United Arab Emirates, Saudi Arabia), and low-income countries and small island states that have comparatively low emissions but are highly vulnerable and highly exposed to climate hazards (many of which have been described in earlier chapters, including Bangladesh, Pakistan, and Pacific atoll nations). For obvious reasons, low-income countries and small island states continually press other nations at COPs for more urgent action; oil producers and high emitters prefer a more slow-going approach; and high-income countries (which tend to be both high emitters and countries with high historical emissions) worry about how much money they will be asked to contribute to fund adaptation assistance, technology transfers, and, lately, compensation for climate-related loss and damage.

A first coordinated attempt at reducing GHG emissions through the UNFCCC emerged at the third COP held in Kyoto, Japan in 1997. The Kyoto Protocol, as it became known, established GHG emission reduction targets that industrialized countries were expected to meet starting in 2012. Low- and middle-income countries, including growing emitters such as India and China, were not assigned targets under Kyoto. The US ended up not ratifying Kyoto (ratifying means having the national government formally adopt and implement it) and many countries that ratified Kyoto never came close to meeting their targets, such as Canada and Japan. Despite this, the Kyoto Protocol did have measurable success by reducing the total emissions of countries that ratified it by an estimated 7% as compared with what their emissions would have been in its absence (Maamoun 2019).

In addition to setting emissions reduction targets, the Kyoto Protocol also required signatories to create national and regional programs that would facilitate adaptation to the impacts of climate change and to establish "clean development mechanisms" that facilitate GHG emission reductions, any proceeds from which would be used to assist the most vulnerable countries in meeting the costs of adaptation. What actually constitutes "adaptation" was not specified in the Kyoto Protocol or in the main text of the UNFCCC itself. What Kyoto did do was establish a dynamic whereby adaptation became something for which industrialized states are expected to provide financial support for vulnerable ones. Subsequent COPs since 1997 have increasingly sought to identify which states are most vulnerable (and therefore entitled to financial assistance), and to develop programs and mechanisms to fund adaptation efforts. At the 2010 COP in Cancun, UNFCCC signatories stated that adaptation should receive the same priority as mitigation efforts to reduce atmospheric GHG emissions, and that adaptation required additional funding. Among other achievements, a Cancun Adaptation Framework was created to assist countries in developing formal national adaptation plans (NAPs) and a Green Climate Fund was established to help countries organize financial support to facilitate emissions reduction and adaptation strategies. Notably, Section II of the Cancun Adaptation Framework (UN Climate Change n.d.) explicitly mentions migration and displacement in the context of adaptation, stating that the Conference of the Parties

14. Invites all Parties to enhance action on adaptation under the Cancun Adaptation Framework, taking into account their common but differentiated responsibilities and respective capabilities, and specific national and regional development priorities, objectives and circumstances, by undertaking, inter alia, the following:

…

(f) Measures to enhance understanding, coordination and cooperation with regard to climate change induced displacement, migration and planned relocation, where appropriate, at the national, regional and international levels;

Through this statement, the 190+ member countries of the UNFCCC for the first time explicitly identified a need to address climate-related migration and displacement within the broader context of adaptation policymaking and adaptive capacity building (Warner 2012). The decision to include migration and displacement within the Cancun Adaptation Framework was a product of extensive debate and recommendations from people and countries actively engaged in the UNFCCC process and outside organizations working adjacent to the process (Serdeczny 2017, Warner 2018). A practical outcome was that it encouraged a wide range of development- and migration-focused organizations and institutions that had not previously engaged with the UNFCCC process to start doing so, and to frame their work as "migration as adaptation" (Nash 2015, Ober and Sakdapolrak 2017).

A small but important step for the future of migration and displacement policy within the UNFCCC process took place at the Warsaw COP in 2013. Although not remembered for any sweeping announcements or ambitious new initiatives, this technically oriented COP established the Warsaw International Mechanism for Loss and Damage associated with Climate Change Impacts. The executive committee of this "mechanism" was tasked with identifying ways of assessing potential losses and damages associated with the adverse impacts of climate change, coordinating discussions between parties about loss and damage, and making recommendations to the COP on how to move forward. As part of the Paris Agreement signed at COP21, 2 years later, the executive committee of the Warsaw International Mechanism would additionally be tasked with developing recommendations on how best to respond through the UNFCCC process to climatic risks that could lead to involuntary displacement.

The Paris Agreement that came out of the 2015 COP represented a significant step up from Kyoto and Cancun in terms of ambition and responsibility in multiple ways. It has an explicitly stated aim of limiting global warming to no more than 2°C over pre-industrial temperatures, and to pursue action that could ideally limit the increase to no more than 1.5°C which, as was reviewed in Chapter 1, is the level that scientists have long warned that the consequences of climate change become "dangerous" (Hoegh-Guldberg et al. 2019). Unlike Kyoto, all countries are required to reduce GHG emissions. The specific amount of reduction is to be determined by each country itself based on its capabilities, and is referred to as a "nationally determined contribution" (NDC). Each country is required to update its NDC plan every five years, and there is an expectation that countries will increase their emission reduction efforts in each new 5-year cycle (a process referred to as "ratcheting up" (Iyer et al. 2022). The first 5-year renewal occurred in 2021 and over 150 countries provided new, updated plans of action at that year's COP in Glasgow. If all of these NDCs are fully implemented and followed up with more aggressive "ratcheting" over the long term, it may be possible to limit warming by the end of this century to 2°C (van de Ven et al. 2023). However, given actual implementation to date and the need for very aggressive reductions to start occurring by the year 2030, a more realistic outcome is end-of-century warming closer to 2.5°C.

As part of the Paris Agreement, the executive committee of the Warsaw International Mechanism was asked by the COP to establish a task force that would develop recommendations for integrated approaches to avert, minimize, and address displacement related to the adverse impacts of climate change. Interestingly, the request asked that in addition to existing groups working within the UNFCCC process such as the Adaptation Committee and the Least Developed, the new Task Force on Displacement should include relevant organizations and expert bodies from outside the Convention. The creation of the Task Force was arrived at as a compromise during Paris Agreement negotiations, with high-income countries wanting alternatives to proposals from low-income countries asking for creation of a UNFCCC facility to provide coordination assistance and funding for climate-related migration, displacement, and planned relocation (Serdeczny 2017, Calliari et al. 2020).

Since its creation, the Task Force on Displacement has included members from a variety of multilateral organizations, some of which will be discussed further, including the International Organization for Migration (IOM) (Box 6.2), the UN High Commissioner for Refugees (UNHCR), the UN Development Programme (UNDP), the International Federation of Red Cross and Red Crescent Societies, and the Platform on Disaster Displacement, a selection of UNFCCC member states, and civil society members drawn from the Children and Youth Constituency of the UNFCCC (YOUNGO) and the Advisory Group on Climate Change and Human Mobility. In its first few years of existence, the Task Force assessed existing policies and practices at national and international levels; analyzed the effects of slow-onset events on displacement; and identified relevant mandates within the UN system. The Task Force's work plan for 2022–2024 (current at the time of writing this book) included developing technical guides and training for UNFCCC members to incorporate migration and displacement planning in their NAPs, building bridges to the UN's Global Compact on Migration and the Global Compact on Refugees (more on these below) and raising awareness of its existence and its role (UN Climate Change n.d.).

Box 6.2

What Is IOM?

The International Organization for Migration (IOM) is a UN-affiliated agency established in 1951 that has become the leading intergovernmental organization in the field of migration (International Organization for Migration 2024). It has 175 member states and works with governmental, intergovernmental, and nongovernmental partners to improve the lives of people on the move and to assist in the management of all forms of mobility and their impacts. Its three key objectives are: to protect the safety of migrants, especially people in crisis situations, and to reduce the risks and impacts of climate change, environmental degradation, conflict, and instability that can give rise to involuntary displacement, and, facilitating pathways for regular migration. It is a key go-to source for policy information and data about climate-related migration.

A third set of actions being taken under the UNFCCC that are directly relevant to migration and displacement is to provide funds to low-income countries for climate-related loss and damage. This is distinctive from other UNFCCC initiatives and funding mechanisms in that the goal here is to help countries recover from the aftermath of extreme events, rather than help countries take proactive adaptation steps and reduce their GHG emissions (Naylor and Ford 2023). After a decade of negotiation, a decision to establish a formal Loss and Damage Fund was reached in 2022 at the COP 28 meeting in Sharm el-Sheikh. This agreement, which had been pushed for by small island states and low-income countries, particularly Bangladesh, was reached only after all parties agreed that high-income countries with historically high emissions would not be held legally liable for impacts of climate change that cause loss or damage, and that the fund would not be seen as providing reparations or simply compensation (Tietjen and Gopalakrishnan 2023). The stated purpose of the new Fund for responding to Loss and Damage is to provide direct financial assistance to low-income countries that are particularly vulnerable to the adverse effects of climate change in responding to economic and non-economic loss and damage associated with the adverse effects of climate change, including extreme weather events and slow onset events (UNFCCC n.d.). A transitional committee was established to work out specific details with respect to what things would be eligible for funding, which of these would be a priority, which countries would be eligible to receive funds, how money should be allocated, and other logistical questions. In the end, the committee decided not to identify the specific countries eligible to receive funds but instead recommended that the fund's administrative board devise a resource allocation system that would guarantee a minimum percentage be available to the least developed countries and small island states. It also mandated that the fund's board devise a "resource allocation system" based on available evidence, with a guaranteed minimum percentage earmarked for Least Developed Countries (LDCs) and small island states. The final text also does not oblige specific countries to pay into the fund; rather it encourages developed countries to provide financing. Wealthy nations promised approximately US$660 million to establish the fund, and it was agreed that the fund would in its first four years be administered by the World Bank (Box 6.3).

At the time of writing in May 2024, the board responsible for overseeing the Fund for Responding to Loss and Damage is holding its first meetings, and so readers of this book will need to track its future progress through the UNFCCC website (unfccc.int). It will be very interesting to see how much money is made available in coming years for this fund, as there is no binding requirement or set amount that wealthy countries contribute to it. There will also be considerable debate as to what sorts of non-monetary losses should receive funding. Nonetheless, this Fund is a very important development, as the COP explicitly intended that it would

Box 6.3
What Is the World Bank?

The World Bank is a group of five international financial institutions that provides funding and knowledge to reduce poverty, increase shared prosperity, and promote sustainable development in low- and middle-income countries (worldbank.org). Headquartered in Washington, DC and with over 180 member countries, the World Bank group has since 1947 funded over 12,000 development projects through loans, interest-free credits, and grants. The World Bank receives its funds through donations from wealthy member countries and through issuing its own bonds and financial securities. In addition to being actively involved in the UNFCCC Loss and Damage Fund and supporting projects aimed at meeting the Sustainable Development Goals (SDGs) (see below), the World Bank also commissioned a detailed study of climate-related migration and displacement risks in low-income countries known as the Groundswell Report, which has been cited on multiple occasions in the present book (Rigaud et al. 2018).

bridge gaps in existing climate financing such as, "climate-related emergencies, sea level rise, displacement, relocation, migration, insufficient climate information and data, or the need for climate-resilient reconstruction and recovery" (UNFCCC 2023). Looking to the future, it is almost certain to be a key international mechanism for responding to involuntary displacements and involuntary immobility of the types described in Chapters 2–4 of this book.

6.1.2 Disaster Risk Reduction Policy

Running parallel to the UNFCCC is an ongoing international initiative known as Disaster Risk Reduction (DRR) aimed at reducing loss and harm caused by a wide range of disasters that include, but are not limited to climate hazards. Other common types of disasters are geotechnical hazards (e.g. earthquakes, tsunamis, and volcanoes), pandemics, and human-caused disasters (e.g. chemical spills and nuclear accidents). Recognizing that the impacts of disasters are often felt across national borders, UN member states have in recent decades been working toward greater international cooperation and coordination in planning for and responding to disasters. International DRR policy aims to clarify roles and responsibilities, facilitate mutual assistance and support, and promote cooperation in disaster response and recovery efforts. It also seeks to promote cross-sector collaboration in addressing factors that amplify disaster risk, including poverty, socio-economic inequality, climate change, environmental degradation, urbanization, and social vulnerability.

A key international DRR initiative is the *Sendai Framework for Disaster Risk Reduction*, which is a voluntary international agreement adopted by UN member states in 2015 (UNDRR n.d.). The Sendai Framework identifies seven global targets to be achieved by 2030, including reducing disaster mortality, reducing the number of people affected by disasters, reducing economic losses, and reducing damage to critical infrastructure and disruptions to basic services. The Sendai Framework provides indicators to measure progress in disaster risk reduction and a basis for monitoring and accountability.

While displacement is not addressed as a standalone issue, it is indirectly referenced in several parts of the Sendai Framework with respect to the impacts of disasters and the importance of addressing the needs of affected people. The Framework emphasizes the importance of understanding disaster risk in all its dimensions, including the vulnerabilities and needs of different population groups, along similar lines to how vulnerability to climate hazards was described in Chapter 1 of this book. It recognizes that disasters can result in displacement or migration of affected populations, particularly those living in hazard-prone areas or in vulnerable conditions. It also encourages investments in measures to reduce disaster risk and enhance resilience at all levels, such as providing access to safe shelter, livelihood opportunities, and social protection mechanisms to support their recovery. The Sendai Framework also underscores the importance of taking into account the needs of displaced populations in disaster and emergency planning, including ensuring that these plans are designed to support their safety, well-being, and dignity during and after disasters. And finally, the Framework includes provisions for monitoring and reporting on progress in disaster risk reduction efforts, which includes collecting data on displacement and migration related to disasters (IDMC 2017).

A midterm report on progress toward meeting the Sendai Framework goals was released to the UN member states in January 2023 (UN General Assembly 2023). It found that low-income countries, especially landlocked ones and small island states, have made little progress in disaster risk reduction. For them, access to disaster early warning systems is poor, they have little access to financing for disaster risk reduction, and they were especially hard hit by impacts of the COVID-19 pandemic. Globally, the number of people dying from and affected by disasters increased since the Sendai Framework was agreed to in 2015, the pandemic being an important contributor to this. Economic losses from extreme weather events and other disasters also increased over the first half of the Sendai implementation period. Overall, while the Sendai Framework provides an admirable set of goals and ambitions, it has to date proven to have limited practical effect in reducing the underlying causes of disasters in general, and of extreme events that lead to involuntary displacement and involuntary immobility specifically. A key limitation of it is the fact that it is not a binding international agreement, and that it lacks meaningful mechanisms to generate international financial support for countries that lack the capacity to reduce disaster vulnerability on their own (see Box 6.4 for more on important terminology).

Box 6.4

Conventions, Protocols, Frameworks, Goals, and Compacts: What Are the Differences?

International agreements go by a variety of titles, terms, and descriptions, as is seen throughout this chapter. The most important distinction is that when countries sign a *convention*, they are agreeing to be legally bound by it. This has implications under international law, and it may also create new rights under the national laws of countries that sign on. For example, countries that have signed the UN Refugee Convention described below are expected to establish procedures to adjudicate refugee claims made by people arriving at their borders and to make their determinations according to principles contained in the Convention. This in turn may create legal rights for the refugee claimant under the receiving country's own legal system. Unless explicitly stated to be optional, *protocols* are legally binding agreements on additional actions to be taken under an existing Convention. Conventions and protocols agreed to during international negotiations must then be ratified (i.e. officially approved) by each government. It may happen that what negotiators agree to is not ultimately approved by their own government when they return home and seek ratification, as was the case with the US and the Kyoto Protocol.

International agreements with other titles, such as frameworks, goals, compacts, and principles are not legally binding on signatory states unless otherwise stated, although an individual country or its legal system may consider itself bound if it so decides. These types of agreements are best seen as statements of intentions and/or vehicles for cooperation between states on subjects of shared concern. As a result, they are typically less successful than conventions in achieving progress and compliance. There is no way of forcing countries to participate in international agreements if they do not wish. Regardless of whether an agreement is binding or non-binding, in reality there are very few ways of forcing states to comply with agreements they have signed, as was seen with the Kyoto Protocol's emission reductions' targets. The most common approach to getting compliance is for states to use diplomacy, exert peer pressure, and/or offer financial incentives that prompt action on the part of their non-compliant peers.

6.1.3 Sustainable Development Goals (SDGs)

Sustainable development has been a priority for the UN community since the 1980s when it was first defined in the 1987 report of the *World Commission on Environment and Development* as being, "development that meets the needs of the present without compromising the ability of future generations to meet their own needs." (World Commission on Environment and Development 1987). While the term "development" when applied to low-income countries is often used in purely economic terms, *sustainable development* is achieved through increased levels of economic growth, social inclusion, and environmental protection (United Nations 2024). In 2015, UN member states agreed to an ambitious set of 17 SDGs to be achieved by 2030 that address a wide range of interconnected challenges including poverty, inequality, climate change, environmental degradation, peace and justice, health, education, and economic growth (Figure 6.2). Each of the 17 goals has a set of targets and an annual report is released that tracks progress toward meeting them. A midterm progress report was released in 2023, halfway to the 2030 target date, which reported mixed results (United Nations 2023). More rapid action was identified as being necessary for all 17 goals, with the greatest early progress being made toward goals with respect to SDGs 9 (innovation), 12 (responsible production and consumption), 14 (aquatic biodiversity), and 15 (terrestrial biodiversity). By contrast, weak progress had been made toward achieving goals with respect to SDG 1 (eliminating poverty), 4 (quality education for all), 6 (clean water and sanitation), 8 (economic growth), 13 (climate action), and 16 (peace, justice, and strong institutions).

There are multiple ways in which achieving the goals of the SDGs would help reduce involuntary climate-related displacement and involuntary immobility, especially in low- and middle-income countries and communities. At the broadest level, achieving the SDGs, and especially goals 1–6, would reduce many of the root causes of involuntary displacement and involuntary immobility, and provide households greater agency when it comes to making migration decisions (USAID 2020). In a specific sense, many of the goals contain targets that are directly relevant to climate-related migration and displacement. The most obvious of these is SDG 13, which aligns directly with the UNFCCC in requiring more urgent action to reduce GHG emissions and to strengthen adaptive capacity to climate-related hazards in all countries. SDG 8 on decent work and economic growth has a specific target of protecting labor rights and promoting safe and secure working environments for all workers, including migrant workers. It also has a target to eliminate human trafficking and another target to improve access to financial services. This latter target, if achieved, would reduce the transaction costs for migrants when they send remittances, which in turn help recipient households get out of poverty, pursue new opportunities, and help them recover from disasters (Le De et al. 2013, Yang and Choi 2007). SDG 10, which focuses on reducing societal inequalities, explicitly calls upon countries to facilitate orderly, safe, regular, and responsible migration and mobility of people, and to implement planned and well-managed migration policies. SDG 17 encourages greater international cooperation and coordination in achieving the SDGs, and specifically calls upon all countries to assist in building adaptive capacity in low-income countries and small island developing states. This overlaps closely with adaptation objectives of the UNFCCC, and is

Figure 6.2   Areas of focus of the 17 Sustainable Development Goals set by the UN in 2015. www.un.org/sustainabledevelopment/.
The content of this publication has not been approved by the United Nations and does not reflect the views of the United Nations or its officials or Member States

a reminder that the UNFCCC and the SDGs should not be seen as alternatives or as discrete choices, but as a pair of initiatives that, if fulfilled, would dramatically reduce and potentially eliminate involuntary displacement and immobility (climate-related and otherwise) in most regions of the world. It cannot be understated how important it is that all governments work actively and cooperatively in this direction.

6.2 International Policies with Respect to Migration and Refugees

A number of international policy initiatives exist that deal explicitly with migration, displacement, and refugees, most of which are directly relevant for considerations with respect to climate-related migration and displacement. The following is an overview.

6.2.1 UN Convention Relating to the Status of Refugees

The term "climate refugee" is often used by the media, advocacy organizations, and the general public to describe people who are involuntarily displaced by climate-related hazards. This is incorrect. There is an explicit, legally binding, internationally recognized definition of the term refugee and it does <u>not</u> include people who move for reasons related to the climate or any other environmental factors. Under the United Nations 1951 Refugee Convention, along with its 1967 Protocol, a *refugee* is defined as someone who

owing to well-founded fear of being persecuted for reasons of race, religion, nationality, membership of a particular social group or political opinion, is outside the country of [their] nationality and is unable or, owing to such fear, is unwilling to avail [themself] of the protection of that country; or who, not having a nationality and being outside the country of [their] former habitual residence, is unable or, owing to such fear, is unwilling to return to it.

Nowhere is there any mention in this definition of disaster, climate change or any other environmental consideration. It is explicitly reserved for people fleeing their country of residence for fear of persecution. The 149 countries that have signed the Refugee Convention are legally obligated to protect refugees on their territory and must not punish refugees that arrive without proper documentation or through irregular means. The term "climate refugee" simply does not exist under international law.

There have been efforts at times by researchers, legal scholars, and organizations to seek expansion of the definition of a refugee to include people on the move because of climate change, or to at least understand refugees *in the context of* climate change, environment, and disasters (Cremins et al. 2023, Ober 2021, PDD/

UNHCR 2023, Weerasinghe 2020). For example, in 2020 and 2023 the UN agency responsible for refugee protection – the UN High Commissioner for Refugees (UNHCR) – released studies that summarized legal considerations with respect to how climate change and factors that give rise to refugees interact. The 2020 study suggested that the adverse effects of climate change are often worsened by factors such as poor governance, scarce natural resources, socioeconomic inequality, political and religious tensions, and xenophobia. The interaction of these factors could give rise to violence and conflict that in turn could lead to persecution. The 2023 study provided specific examples of how these contextual interactions may work, including:

- Situations where a state is unable or unwilling to protect people fleeing conflict or violence exacerbated or caused by a climate disaster may leave those people at risk of persecution
- Environmental defenders, activists, or journalists could be targeted and persecuted for criticizing government responses to climate change
- People belonging to particular social groups may be denied access to resources or assistance for disaster risk reduction or post-disaster recovery, which may take the form of persecution under certain circumstances

In offering these examples, the UNHCR emphasizes that it does not seek to redefine "refugee" or change international refugee law in any way. This in turn raises an important point: Countries that are signatories to the Refugee Convention are free to interpret it more widely should they choose (we provide below examples of regional agreements that do so). The country receiving an asylum claimant is expected to provide a fair process to determine if the arrival does indeed merit legal protection as a refugee, and if that country wishes to look favorably upon climate-related factors when making that determination, there is nothing to stop it from doing so. The reality, however, is that few (if any) countries that receive asylum claimants on a regular basis express any interest in expanding formally or informally the legal definition of a refugee (Box 6.5). There is consequently little chance of there being in the foreseeable future international efforts to rewrite the Refugee Convention to include people fleeing the impacts of climate change.

There are other reasons why seeking to change the Refugee Convention would be impractical. It is difficult and expensive to seek refugee status in another country, only a small percentage are ever permanently resettled, and that number is declining (Van Hear 2004, MPI 2019, Radford and Connor 2019). The international community already does a poor job of protecting and caring for people who meet the existing definition of a refugee. Today, a growing number of countries that are signatories to the Refugee Convention are making it more difficult for

---

Box 6.5
What Is the Difference between a Refugee and
an Asylum Claimant or Seeker?

Under the UN Refugee Convention, a person who has a well-founded fear of persecution in their country of habitual residence has a legal right to seek protection in another country. The receiving country makes a determination whether the person does indeed meet the definition of a refugee under the Convention and, if so, that person is entitled to protection and has a variety of other legal rights. In the period between the person's arrival and the date their refugee claim is decided, the person is referred to as an asylum seeker or *asylum claimant*. In principle, the refugee claim should be adjudicated promptly, but in practice people may wait years to receive a hearing. Countries that receive asylum seekers treat them in a variety of ways. In Canada, for example, people that arrive and request refugee protection are given temporary permits to remain in the country and are eligible to work and go to school while they wait for their refugee claim to be adjudicated (Government of Canada 2024). Temporary accommodation will also be provided if needed. By contrast, asylum seekers entering Australia are subject to detention until an initial assessment can be made of their case, and this can take weeks or longer. Asylum seekers arriving in Australian waters by boat may be detained in detention centers in Nauru or Papua New Guinea and not on Australian territory (Refugee Council of Australia 2020). At time of writing, the UK is preparing to launch a program whereby some asylum claimants will be flown to centers in the central African country of Rwanda (even if they have not arrived from Africa) where their refugee application will be processed.

---

people to approach their borders to seek refugee protection (McLeman 2019), detaining refugee claimants in prison-like conditions and deporting unsuccessful claimants to other countries. In addition, leaders and residents of small island states threatened by rising sea levels have explicitly stated they do <u>not</u> want to be considered refugees. They have seen how the world treats refugees and they want no part of it (Randall 2014).

A final reality is that, as was highlighted in multiple chapters of this book, most people that move because of climate-related hazards remain within their home countries and therefore by definition are not refugees. They are considered by law to be internal migrants or internally displaced people (IDPs), and countries have different rights and obligations to them.

6.2.2 Global Compact for Safe, Orderly, and Regular Migration

The *Global Compact for Safe, Orderly, and Regular Migration (GCM)* is an international agreement adopted by the United Nations in 2018 that aims to

---

Box 6.6
Human Rights Law, Climate Change, and the
Principle of Non-refoulement

The principle of *non-refoulement*, rooted in international human rights law, prohibits governments from deporting individuals to locations where they face a genuine risk of arbitrary deprivation of life or exposure to inhuman or degrading treatment. Many national governments have interpreted this principle as not including deportations to locations experiencing climate change risks, but recent decisions in international courts are clarifying this principle. A notable example is the case of Ioane Teitiota, who is originally from the atoll island of Funafuti in the Pacific nation of Kiribati. He and his family traveled to New Zealand in 2007 and overstayed the length of their permitted temporary entry. In 2011 he was apprehended, at which time he made an asylum claim arguing that his atoll was becoming unlivable due to rising sea levels, storms and related impacts of climate change. His claim was rejected by the court, and a deportation order was issued in 2015 (BBC News 2015). Teitiota next brought his case to the UN Human Rights Committee, alleging that New Zealand had violated his right to life under the *International Covenant on Social and Political Rights*. The Committee ruled in 2020 that although his claims were true, the risk to his life if he returned to Kiribati was not imminent, and that there was time for the government of Kiribati to implement protections for its citizens. The Committee did, however, state in its decisions that, "given that the risk of an entire country becoming submerged under water is such an extreme risk, the conditions of life in such a country may become incompatible with the right to life with dignity before the risk is realized" (UN Office of the High Commissioner for Human Rights 2020). The legal implication for future cases is that, should the impacts of climate change become so severe it creates an imminent risk to habitability in a given country, people who flee to another country would have a legal right not to be forced to return.

---

address policy issues related to international migration in a comprehensive manner (other than movements of refugees, which is covered under the Global Compact on Refugees). The GCM consists of 23 objectives covering various aspects of migration, including governance, protection of migrants' human rights (Box 6.6), addressing the drivers of migration, enhancing cooperation between countries, and promoting sustainable development. Unlike the UNFCCC it is a non-binding agreement. Like the *Sendai Agreement on Disaster Risk Reduction* discussed above, it provides a framework for countries to collaborate on managing migration more effectively, while respecting the sovereignty of nations and the rights of migrants.

The GCM was the outcome of a multilateral effort that involved extensive consultations, negotiations, and diplomatic discussions between UN member states, civil society organizations, and other stakeholders. The process leading to the

GCM began with the adoption of the *New York Declaration for Refugees and Migrants* in September 2016 by the UN General Assembly. This declaration recognized the need for a comprehensive approach to human mobility and initiated the development of two Global Compacts: one on migration and another on refugees (the latter is not discussed here). After extensive negotiations, the final text of the GCM was agreed upon by member states in July 2018 and formally adopted by the UN General Assembly in December of the same year. The Czech Republic, Hungary, Israel, Poland and the US voted against adoption of the GCM and a dozen other countries abstained from voting.

Three of the GCM's 23 objectives make explicit mention of climate change. Objective 2 aims to, "minimize the adverse drivers and structural factors that compel people to leave their country of origin." It acknowledges that environmental factors, including climate change, can contribute to displacement and migration by destabilizing livelihoods, exacerbating resource scarcity, and increasing vulnerability to natural disasters. It also includes an explicit theme on "Natural disasters, the adverse effects of climate change, and environmental degradation," which calls for better data and analysis to understand climate-related migration, develop and strengthen adaptation and resilience strategies, integrate displacement into disaster planning, and make coherent policies and approaches at different governance levels. Objective 5 asks signatories to, "Enhance availability and flexibility of pathways for regular migration." This objective focuses on expanding legal channels for migration, and explicitly calls for cooperation in finding solutions for people compelled to leave their country of origin for climate-related reasons. Objective 21, which requests that signatories "Cooperate in facilitating safe and dignified return and readmission, as well as sustainable reintegration" also refers to climate change, stating that some migrants may face challenges in returning to their countries of origin due to ongoing environmental degradation and climate-related hazards.

Two other GCM objectives are directly pertinent to climate-related migration even though they do not explicitly mention climate change. Objective 13 asks signatories to, "Provide access to basic services for migrants," including humanitarian assistance to migrants affected by environment-related disasters and other emergencies. Objective 15 seeks to, "Empower migrants and societies to realize full inclusion and social cohesion." This objective highlights the importance of promoting social inclusion and cohesion in communities, including those affected by climate-related migration, to foster resilience and support sustainable integration.

The GCM is a truly remarkable achievement. The advice contained within the GCM is expertly conceived and would, if fully implemented, help ensure that climate-related migration occurs within a context of high migrant agency, and it would greatly increase the prospects for climate-related migration to generate benefits for sending areas, receiving areas, and migrants themselves. However, its prospects for success are diminished by the reality that it is a non-binding agreement. Countries are free to embrace or reject its recommendations. As of 2022, implementation of GCM recommendations related to climate change has been slow, with most attention being given to how to address climate as a driver of migration, rather than offering new or expanded pathways for migration (Mokhnacheva 2022).

6.2.3 Guiding Principles on Internally Displaced People

Internally Displaced Persons (IDPs) are individuals or groups of people who have been forced to flee their homes or places of habitual residence due to armed conflict, violence, disasters (including climate-related ones), human rights abuses, and/or other reasons that have compromised their safety, livelihoods, and well-being. Although they often find themselves living in refugee-like conditions, IDPs remain within their country of habitual residence and are therefore not legally protected under the UN Refugee Convention (although the UNHCR does keep track of their numbers and advocates for greater protections on their behalf). Generally, IDPs have the same legal protections as all people under the UN's *Universal Declaration of Human Rights*, a variety of other international human rights agreements (Box 6.7) and the laws of their nation of residence. There is no binding international legal agreement aimed directly at IDPs, but a document known as the *Guiding Principles on Internal Displacement* was introduced to the UN in 1988 by the Representative of the UN Secretary-General on Internally Displaced Persons. Although they are non-binding, they provide a useful framework for the international community on how to provide protection and assistance to IDPs.

The *Guiding Principles* state that IDPs have the right to be shielded from arbitrary displacement and cannot be forcibly removed from their homes or habitual places of residence without lawful justification. Discrimination based on displacement status is prohibited, and IDPs are entitled to equal treatment and access to the same assistance and protection as are other people. During their period of displacement, IDPs are entitled to protection from violence, harassment, and exploitation, and particular attention should be given to vulnerable groups including children, women, and the elderly. IDPs have the right to receive humanitarian assistance, and states must facilitate the passage of such assistance without hindrance and ensure that humanitarian assistance organizations have access to IDPs. IDPs have the right to durable solutions to their displacement, which may include returning to their homes, local integration,

Box 6.7
Key Agreements Under International Human Rights Law

International human rights laws flow from the UN's 1948 *Universal Declaration of Human Rights* (UDHR), which at its essence is described at Article 3 which states, "Everyone has the right to life, liberty and security of person." Of the more than two dozen other articles of the UDHR, two are directly applicable to migration and mobility, including Article 13, which states

1. Everyone has the right to freedom of movement and residence within the borders of each state.
2. Everyone has the right to leave any country, including his own, and to return to his country

   and Article 14, which states

1. Everyone has the right to seek and to enjoy in other countries asylum from persecution

In the decades since the adoption of the UDHR, an additional ten human rights-related agreements and mechanisms have been reached by UN member countries, some binding, some voluntary, and some of which have direct or indirect relevance for the protection of the rights of migrants, IDPs, and/or refugees. The most obvious of these is the *International Convention on the Protection of the Rights of All Migrant Workers and Members of Their Families*, signed in 2003. This convention has been signed by only 56 nations as of 2024, none of which are high-income countries. The largest geographical concentrations of signatories are countries in Central and South America (notable exceptions being Brazil and Panama), North and West Africa (notable exceptions being Tunisia and Cote D'Ivoire), and Asian nations that have large numbers of migrant workers living abroad, including Bangladesh, Indonesia, the Philippines, and Sri Lanka.

Following is a list of other potentially relevant international human rights agreements; further details about each can be obtained through the website of the Office of the High Commissioner for Human Rights (ohchr.org):

- International Convention on the Elimination of All Forms of Racial Discrimination
- International Covenant on Economic, Social and Cultural Rights
- International Covenant on Civil and Political Rights
- Convention on the Elimination of All Forms of Discrimination against Women
- Convention against Torture and Other Cruel, Inhuman or Degrading Treatment
- Convention on the Rights of the Child
- International Convention on the Rights of Persons with Disabilities
- International Convention for the Protection of All Persons from Enforced Disappearance

or resettlement depending on the circumstances, with states responsible for creating conducive conditions for these solutions. IDPs also have the right to participate in decisions affecting their lives, including those related to their return, resettlement, or integration. IDPs returning home or resettling elsewhere should receive assistance to rebuild their lives, including support for housing, land, and property restitution. Lastly, states bear the responsibility to prevent arbitrary displacement by addressing root causes and finding durable solutions to prevent recurrence.

In many cases, those displaced internally due to climate-related hazards can seek out and be covered by the same protection and assistance offered to other IDPs, such as those displaced by conflict. Protection of IDPs is complicated by the reality that each individual state (and not an international organization) bears responsibility for protecting and assisting all IDPs within its territory. The governments of some states may be predisposed toward protecting and assisting people within their borders, but in other states the government may be an active participant in conflicts or violence and itself be causing people to be displaced in the first place. Conceivably, in circumstances where climate-related hazards are the cause of displacement, governments may be more inclined to assist IDPs and be more willing to ask for international support (PDD 2020). However, this logic may not hold in situations where civil conflict and climate-related disasters coincide, or in the case of authoritarian governments that fear asking for assistance would undermine their legitimacy. For example, in May 2008, Cyclone Nargis caused widespread destruction in Myanmar (Burma), killing at least 85,000 people and displacing millions more (Özerdem 2010). Despite being almost completely incapable of providing assistance to its own citizens, the country's repressive military rulers refused any outside humanitarian assistance for 3 weeks, until they finally gave way to international pressure and allowed assistance to be delivered via a working group consisting primarily of neighboring Southeast Asian countries. As a result, much additional harm and loss of life occurred that could have been avoided. This is an extreme example, but it highlights the significant limits of a non-binding international agreement such as the Guiding Principles (for another example, see Box 6.8).

Box 6.8
Nansen Initiative/Platform on Disaster Displacement: Advocating for the Rights of Climate-Displaced People

The *Nansen Initiative* was launched in 2012 to identify and advocate for actions that respond to gaps in protection for people displaced across borders by disasters, including climate-related hazards. It was primarily funded by the governments of Norway and Switzerland, with additional support from Germany, the European Commission, and the nongovernmental MacArthur Foundation. The initiative was named after Fridtjof Nansen, a Norwegian polar explorer and humanitarian who was the first UN High Commissioner for Refugees, and began with consultations of governments, international organizations, civil society, and experts. The initiative did not aim to establish new legal protection mechanisms, but to build consensus that this category of displaced people warrants protection, to map out existing laws and protections that may be relevant and to strengthen implementation of these. In 2015 the Initiative released a "Protection Agenda" consisting of principles, guidelines, and recommendations that were endorsed by governments of 109 states (Nansen Initiative 2015). Having achieved this, a successor organization known as the *Platform for Disaster Displacement* was established with funding from the government of Germany to continue advocating for implementation of the protection agenda, to facilitate information sharing, and to foster cooperation between governments, humanitarian organizations, and other relevant actors. More information about the Platform is available at its website disasterdisplacement.org.

6.3 Regional Policy Solutions

In addition to global agreements, countries, or groups of countries may enter into formal agreements on a bilateral (i.e. one-to-one) or regional basis. In this section, we identify a number of such existing agreements that contain useful elements and can serve as models for future agreements at bilateral, regional, or international scales.

Beginning in the 1960s, African countries have participated in an organization originally known as the Organisation of African Unity, now known simply as the African Union, with goals of improving international cooperation, security, and social and economic development across the continent (African Union 2024). An important early agreement reached by organization members was the *Convention Governing the Specific Aspects of Refugee Problems in Africa* that came into effect in 1974. A key element of this agreement was that it expanded the definition of a refugee used in the UN Refugee Convention to also include a person who, "owing to external aggression, occupation, foreign domination or events seriously disturbing public order in either part or the whole of his country of origin or nationality, is compelled to leave his place of habitual residence in order to seek refuge in another place outside his country of origin or nationality" (African Union, 1974). It also called upon signatories to cooperate with one another in refugee protection, stating that when a signatory country is struggling with an influx of asylum seekers, it may, "appeal directly to other Member States and through the OAU, and such other Member States shall in the spirit of African solidarity and international cooperation take appropriate measures to lighten the burden of the Member State granting asylum." Many humanitarian organizations and refugee advocates would argue that a similar expansion of the refugee definition and a call for greater global solidarity is needed. This African Union convention highlights how countries need not feel limited by the language of existing UN agreements but should feel encouraged to build upon them to address national and regional priorities, one of which being climate change and its impacts.

African Union members have also moved much farther ahead than countries elsewhere in the world when it comes to developing policies for IDPs. In 2012, the African Union Convention for the Protection and Assistance of Internally Displaced Persons – informally known as the *Kampala Convention* – came into effect, and was eventually ratified by 33 of 55 African Union member states. The Kampala Convention is aimed at promoting cooperation between countries on eliminating the root causes of internal displacement, establishing legal frameworks to protect IDPs, and finding durable solutions (African Union 2012). Particularly notable in this convention is the definition used for IDPs, describing them as:

persons or groups of persons who have been forced or obliged to flee or to leave their homes or places of habitual residence, in particular as a result of or in order to avoid the effects of armed conflict, situations of generalized violence, violations of human rights or natural or human-made disasters, and who have not crossed an internationally recognized State border.

This definition clearly includes people displaced by climate-related hazards as people warranting assistance and having rights under the Convention. Signatories to the Convention pledge to, "devise early warning systems, in the context of the continental early warning system, in areas of potential displacement, establish and implement disaster risk reduction strategies, emergency and disaster preparedness and management measures and, where necessary, provide immediate protection and assistance to internally displaced persons." Most importantly for climate-related displacement, in Article V section 4 signatories agree to, "take measures to protect and assist persons who have been internally displaced due to natural or human made disasters, including climate change." Also notably, in Article XII, which deals with compensation rights, section 3 says that signatories, "shall be liable to make reparation to internally displaced persons for damage when such a State Party refrains from protecting and assisting internally displaced persons in the event of natural disasters." As with all international agreements, compliance with the Kampala Convention varies across signatory countries, and IDPs across the continent often experience a lack of practical protection, are socio-economically marginalized, and may suffer higher rates of mortality and poor health than other people (Cantor et al. 2021). Nonetheless, the *Kampala Convention* provides a clear and useful model that the international community would do well to emulate as it moves ahead through the UNFCCC process to create formal mechanisms to protect people involuntarily displaced by the impacts of climate change.

Somewhat similar to the African Union's refugee convention (Arboleda 1995), in 1984 a group of 10 Central America countries adopted a non-binding agreement known as the *Cartagena Declaration on Refugees*. It sought to improve regional cooperation in refugee protection and expanded the definition of a refugee to include, "persons who have fled their country because their lives, security or freedom have been threatened by generalized violence, foreign aggression, internal conflicts, massive violation of human rights or other circumstances which have seriously disturbed public order." The Cartagena Declaration has been updated each decade to include measures such as improving labor mobility between signatories, and additional countries have signed onto its successor declarations (Castillo 2015). Important refugee destination countries including Mexico and Brazil have included language from the Cartagena Declaration in their national refugee laws, and have applied these to, for example, Venezuelan asylum seekers fleeing the political and economic chaos of that country (Freier et al. 2022). Although the *Cartagena Declaration* definition of a refugee does not explicitly include people displaced by the effects of climate change, it has been argued that there could be occasions when they fall within the phrase "other circumstances which have seriously disturbed public order" (Pajuelo 2020). As with other such declarations, its non-binding nature limits its implementation across signatory countries. What most warrants emulation is that the *Cartagena Declaration* is regularly updated each decade by signatories and is therefore a living, evolving document, and participants in the process speak of a spirit of cooperation when describing it (de Castro Simão 2024).

Another regional policy approach to mobility and migration that would be of great benefit to managing climate-related migration of all types is the creation of *freedom of movement* agreements that establish legal provisions for people to travel, reside, and work in other countries regardless of their underlying motivation for seeking to move. The European Union's *Schengen Agreement* is one such agreement. These agreements can be especially helpful when extreme climate events occur that necessitate temporary movement between countries. For example, freedom of movement agreements involving the Caribbean Community (CARICOM) and the Organisation of Eastern Caribbean States (OECS) allowed people displaced by 2017's Hurricane Sandy to move to other Caribbean islands, in some cases for indefinite periods of time and often with the right to work (Francis 2019). A free movement agreement has also been established in East Africa by the Intergovernmental Authority on Development (IGAD), a trade bloc consisting of Eritrea, Ethiopia, Kenya, South Sudan, Sudan, Somalia, Djibouti, and Uganda. This agreement is notable in that it requires that members allow the entry of citizens of other member countries who are moving in anticipation of, during or in the aftermath of disaster, and that such people should be allowed to stay if returning to their state of origin is not possible or reasonable (Intergovernmental Authority on Development 2020).

6.4 National-Level Policy Considerations

Climate-related migration and displacement will require new ways of planning and programming at the national level. Options for countries to do so include dealing with new challenges on an *ad hoc* basis as they occur (an unfortunately common approach), deliberately creating new policies and programs specifically aimed at migration and displacement associated with climate change (examples of which are yet to emerge), or to gradually incorporate consideration of climate-related migration and displacement into existing policies and programs, an approach referred to as *mainstreaming*. Mainstreaming can be done in any number of areas within government that touch upon such things as climate policy, adaptation, social protection, economic development, and/or migration/mobility policy. Mainstreaming of climate and migration considerations could, for example, form part of expanding policies and programs focused on green growth and urban development or rethinking public health programs and social programs so that they consider climate risks that may lead to migration or displacement (Aleksandrova 2020, Friend et al. 2014). As noted earlier, countries are encouraged to include climate-related migration and displacement risks as part of their wider NAPs under the UNFCCC, and lessons on how to do so may be drawn from the significant experience many countries have in mainstreaming climate adaptation more generally into existing policies and programs. Limits on the success of mainstreaming include the fact that mainstreaming itself is not always done in a systematic way and that is often subject to political commitment to the process, which can ebb and flow (Braunschweiger and Pütz 2021).

An obvious area where policy innovation and evolution will be needed at national levels to account for climate-related migration and displacement is in the area of immigration and settlement policies (Cantor 2021). As an example, in the US initiatives known as Temporary Protected Status (TPS) and Deferred Enforced Departure have been developed to manage cases where migrants from other countries are unable to return home because of environmental disasters in their country of origin (Ober 2021). Although the programs do not explicitly reference climate change, their past application indicates they can be used for climate-related displacements in the future. Although it has not to date led to any specific decisions with respect to US immigration policy, the White House commissioned a detailed study on the implications of climate change for migration, identifying options for responding to displacement abroad through development assistance and recommending the creation of a cross-government process for developing policy in this area (White House 2021).

Other national governments have used existing regular immigration channels flexibly to admit and accommodate people displaced by disasters, such as through streamlined procedures, waived or relaxed requirements, and discretion in issuing visas (Scissa 2022). Another option is to use special humanitarian visas and immigration programs to assist people in neighboring countries or who have transnational family ties to immigrate temporarily or permanently after an environmental disaster, such as those offered by Brazil to Haitian nationals after the 2010 earthquake (Jubilut et al. 2016). In 2022, Argentina established a new special humanitarian visa aimed at providing protection and relocation support for nationals and residents of Mexico, Central America, and the Caribbean displaced by "socio-natural" disasters (Ministerio del Interior de Argentina/Migraciones and PDD 2022). The government of New Zealand has developed a variety of immigration programs aimed at citizens of Pacific small island states that include preferential labor migration agreements with some (e.g. Vanuatu, Solomon Islands, and Papua New Guinea), others receiving automatic dual citizenship (e.g. Cook Islands and Niue), and still others being eligible for a special permanent resident category called the Pacific Access visa that operates on a lottery basis (Kiribati, Tuvalu, Tonga and Fiji) (Neef and Benge 2022). The possibility of creating targeted programs for climate-displaced Pacific islanders has been a topic of study and consideration by the New Zealand government in recent years (New Zealand Ministry of Foreign Affairs and Trade 2018).

Finally, it is also worth noting that planning and policy for climate-related migration and displacement can also be undertaken by subnational levels of governments including cities. As has been shown in many examples in Chapters 2–4, cities often end up being the destination for migrants moving away from climate hazards in other areas of the same country, and this dynamic is likely to intensify in coming decades in many parts of the world (Adger et al. 2020, Rigaud et al. 2018, Rosengaertner et al. 2023). One recent example of coordinated action on climate and migration at the city level is the Mayors Migration Council (mayorsmigrationcouncil.org), a group consisting of municipal government leaders from around the world working collaboratively to exercise influence on migration and climate policy, to promote implementation of the UN Global Compact for Migration, and to more broadly improve the quality of urban life. In 2021 the Mayors Migration Council issued a report outlining their priorities and recommendations for increasing the resilience of urban residents in the face of climate hazards and climate displacement, ensuring the protection and inclusion of people who move into cities due to climate hazards, and including migrants and IDPs in processes that make cities more climate resilient (C40 Cities Leadership Group Mayors Migration Council 2021).

To summarize this chapter, the need to develop policy and programming to respond to climate-related migration, displacement, and immobility is growing. Such actions can and should take place at all scales of government, from the local to the global, and to be most effective should seek to draw upon the expertise of multilateral organizations, humanitarian groups, representatives of migrant communities, researchers, and the wider public. Governments themselves should seek to avoid policymaking siloes and instead work across government to ensure policy decisions and outcomes reflect an assessment of the best possible options. It is likely that such decisions will emerge as a mix of planned, *ad hoc*, and mainstreamed policies and programs at national levels, with political will being an important determinant of the speed and efficiency with which they are devised and implemented.

7

Emerging Issues and Future Directions

The previous six chapters have provided a detailed review of the current state of scientific knowledge, methods, and policies as they relate to climate-related migration, displacement, and immobility. Scholarly understanding of these topics has evolved rapidly over recent decades, yet there is still much we don't know about the complex ways that climate change interacts with migration decisions. In this final chapter, we discuss a number of emerging issues and future research needs including:

- Gendered dimensions of migration in the context of future climate change
- How climate-related migration affects Indigenous populations and cultural heritage
- The interplay between climate-related migration and human health
- The impacts of climate-related migration on receiving communities
- Identification of critical thresholds in climate-migration connections
- Unforeseeable climate-migration outcomes

7.1 Gendered Dimensions of Climate-Related Migration

Much of our existing knowledge of climate-related migration does not consider the role that gender and gender identity play throughout the migration processes (see Box 7.1 for an important explanation of terminology). Research and scholarship on migration broadly, and climate-related migration specifically, has often focused disproportionately on the experiences of male migrants and tacitly implied that the experiences of migration are "gender neutral." There has also often been an implied assumption that men are more likely to migrate than women. While this may be true in some contexts, it is not in others, and globally women's migration has increased in recent decades. This includes a growing "feminization" of international migration being driven by a number of factors including male unemployment

Box 7.1
Differences in the Terms *Sex* and *Gender*
While they are often conflated, the terms sex and gender are not synonymous. *Sex* refers to biological differences between males and females, while *gender* refers to ways in which expectations, differences, and norms are constructed by societies and cultures and imbued on an individual (Pessar and Mahler 2003). In this book, we have used the words "women" and "men" to describe biological differences and gender identities, but we do so recognizing that gender is not a simple male/female binary. In this chapter, we use the terms *masculinity* and *femininity* to describe society's gendered expectations of men and women, again recognizing that these terms are not mutually exclusive and that their meaning evolves over time.

and underemployment in many countries and labor markets structured such that women workers earn lower wages, prompting women to increasingly seek work opportunities abroad (Dannecker 2005; Yeoh 2016). These are just small examples of the importance of understanding better how gender impacts migration and migration experiences (Choi, Hwang, and Parreñas 2018; Donato, Enriquez, and Llewellyn 2017).

As was detailed in Chapter 1, migration is a complex process and there are many aspects of the process that may be influenced by gender, from the underlying motivation to migrate, the decision to move (or not), the agency of the would-be migrant, the migration experience, and short- and long-term outcomes of it. We can see the influence of gender in the macro-level political, cultural, economic, and social factors that shape household adaptation and migration decisions. Sometimes this is obvious and overt. For example, there have been extreme cases where governments have forbidden women to migrate internationally, and have required women to obtain the permission of their husband or an adult male "guardian" before being allowed to move (Shivakoti et al. 2021). As an example, the government of Bangladesh placed restrictions on the international migration of women from 1981 to 1998, meaning that up until 2003 less than 1% of international labor migrants from that country were women (International Labour Organization 2014). Governments in Cambodia, Sri Lanka, Nepal, Indonesia, and the Philippines have all implemented bans of varying length on the migration of domestic workers (primarily women) to particular countries because of incidents of violent acts and exploitation. The justification for such bans has often been a paternalistic desire to protect women migrants. The exploitation of migrants is obviously a serious concern in every society, but it is notable that governments only feel a need to restrict the migration of women and not men in such circumstances.

Cultural norms surrounding femininity and masculinity can be powerful forces that influence an individual's decision to migrate. In many communities and societies, women may be faced with strong expectations to perform a particular kind of femininity. This may include the expectation that women are nurturers, mothers, and wives, and should therefore stay at home rather than move (Fernández-Sánchez et al. 2020). Other cultural norms that shape women's decisions may include norms that suggest that women who migrate alone may be perceived as sexually impure and that women should only leave their parents' household to marry. In contrast, norms surrounding masculinity may place pressure on men to be the primary financial providers and decision-makers for their household, which can in turn lead to them to make decisions and/or undertake migration that may be ultimately harmful for them and/or their families (Broughton 2008). For both men and women, challenging deeply embedded gender norms can come with significant consequences including stigmatization and violence.

Gendered cultural, societal, and economic influences on migrant agency are often entangled. On average, women around the world are less likely to have control over their personal and/or household's financial resources, which may limit their ability to make migration decisions – which we described in Chapter 1 as migrant agency (Kartiki 2011). Women may also have less access to particular types of employment opportunities, social networks, and other resources in potential migration destinations, placing further restrictions on their decision-making options (IOM 2024). At the very least, the kinds of opportunities and resources available to migrants are generally different for men and women with respect to both internal and international migration. This is shown, for example, in how Bangladeshi women are more likely to migrate for reasons related to marriage, while men are more likely to migrate for economic reasons or labor opportunities (Carrico and Donato 2019).

The impacts of climate change are heavily gendered (Pearse 2017), and recognition of this is critical to understanding how migration and displacement will play out in a changing climate (Hunter and David 2009). There is mounting evidence that gender can influence a person's exposure, vulnerability, and ability to adapt to climate hazards (Alston 2014, Enarson et al. 2018). The many examples of this in recent research show a wide range of ways in which this plays out, from the fact that Bangladeshi women experience higher rates of mortality in cyclones in part because they fear evacuating to cyclone shelters due to the threat of gender-based violence in those shelters (Rahman 2013) to research showing that Australian men who migrate from rural areas because of drought have higher rates of suicide (Alston and Whittenbury 2012). The causal pathways that give rise to gendered impacts of climate hazards and the outcomes that ensue can be very complex. For example, research by Carrico et al. (2020) found that heat waves in Bangladesh led to women and girls marrying at younger ages. Those who did get married during heat waves were also more likely to marry into poorer households and face higher risks of intimate partner violence. Additional research that has a clearer and more deliberate focus on the experience of women and girls in climate-related migration research would be welcome given their high levels of previous exclusion and underrepresentation in past studies (Hondagneu-Sotelo and Cranford 2006).

Although gender does not exist as a binary, research on the impacts of climate change and on climate-related migration in particular has been slow to move beyond focusing on differences between men and women, and has generally to date failed to move beyond assumptions of *cisgenderedness* of people (*cisgender* meaning the male or female biological sex assigned to a person at birth). It is starting to be recognized that LGBTQIA+ (lesbian, gay, bisexual, transgender, queer, intersex, asexual, and other emerging gender identities) communities face unique vulnerabilities to climate change and natural disasters (Goldsmith et al. 2022), but there has yet to be significant research into the experiences of gender nonconforming people (including transgender and nonbinary people) and how their gender identities influence climate-related migration decisions and outcomes. This presents an important opportunity for future research to broaden the inclusion and diversity of experiences considered in climate-related migration scholarship.

Considering gender and evolving understanding of gender in climate-related migration research is not straightforward, and it should not be seen as simply being a need to include gender as simply one more unit of analysis in methodologies that have been used before. This has been referred to as an "add and stir" approach to incorporating gender into migration scholarship, and it fails to adequately capture the complexity of gender (Hondagneu-Sotelo and Cranford 2006). There is a need for more research to help understand how gender and gender identity interact with climate-related migration across all stages of a migration decision and across different contexts. This also means that more data collection is needed. Future work should also consider how race, class, religion, and nationality add additional layers and complexity to how gender influences climate-related migration (Box 7.2). Analysis that does not consider these complex interactions between gender, climate change, and migration ignores and potentially erases the experiences of women and gender-nonconforming persons. Policies that result from such analyses could exacerbate differences in vulnerability to climate change across gendered lines.

Box 7.2
Intersectionality

The concept of *intersectionality* refers to the ways that multiple aspects of a person's identity (i.e. gender, race, sexual orientation, class, disability, etc.) intersect and overlap (Crenshaw 2017). Inequality is often embedded within these various aspects of identity and contributes to unique dynamics and experiences (Figure 7.1). For example, Black women in the United States often experience a combination of misogyny and racism due to their identity (often referred to as *misogynoir* [Kwarteng et al. 2022]), and both of these aspects of their identity can contribute to exclusion or discrimination. Because people's experiences and the systems of oppression that impact them typically operate beyond a single identity, it is important to analyze and consider how identities overlap and how different systems of oppression (such as patriarchy, colonialism, and white supremacy) reinforce one another. The many overlapping identities that people possess also influence their vulnerability to climate change (Kaijser and Kronsell 2014). In the case of climate-related migration, it is important, then, for research going forward to take an intersectional approach when considering gender. For example, the experience of a wealthy, cisgender woman in the Global North choosing to migrate as a result of increased wildfire risk is likely very different from the migration experience of a poor, transgender man in a fire-prone area in the Global South, and all aspects of identity, as well as their intersections, are salient.

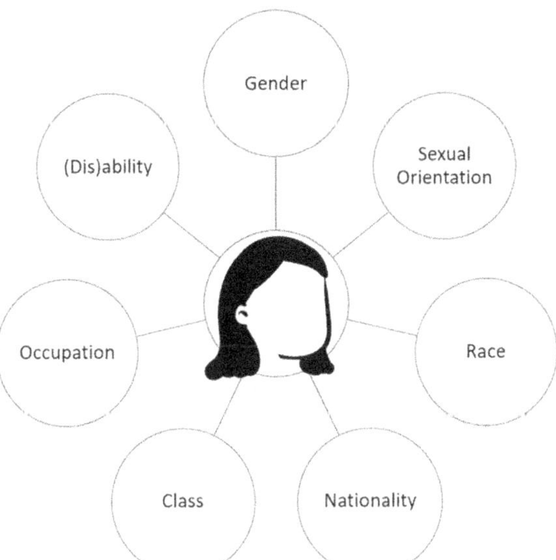

Figure 7.1　Intersectionality recognizes the ways that multiple aspects of an individual's identity interact and overlap. The seven bubbles are common examples of the many more potential aspects of an individual's identity.

7.2 Indigenous Peoples and Cultural Heritage

7.2.1 Climate-Related Migration and Indigenous Peoples

Climate change threatens the traditional livelihoods, health, and cultural heritage of Indigenous and traditional communities around the world (Cissé et al. 2022). The potential disruptions range from tangible cultural heritage, such as culturally or religiously significant sites, monuments, and buildings, and intangible cultural heritage such as livelihood practices, cultural identity, and local knowledge, traditions, and folklore (Sesana et al. 2021; Orr et al. 2021). Cultural heritage may be especially threatened when communities are forced or elect to leave their homes due to climate change, as was illustrated in the case study of relocations of coastal villages in Fiji described in Chapter 4.

Many Indigenous communities around the world have experienced significant and traumatic experiences of forced displacement at the hands of colonial settlers who dispossessed them of their original lands, and this has often left them and their descendants today occupying locations at much higher levels of exposure to extreme heat, wildfires, droughts, floods, and other hazards (Farrell et al. 2021). It is not coincidental that many of the cases and examples presented in Chapters 2–4 describe the impacts of hazards on Indigenous communities. The history of forced displacement into areas of higher climate risks makes it especially important for future work to understand specifically how Indigenous communities perceive, experience, and make decisions surrounding climate-related migration. Some have argued that climate change may even pose a "double dispossession" of Indigenous communities from their lands and their cultures (Derickson 2022). More research on experiences of climate-related migration within Indigenous communities would be most welcome.

At the same time, Indigenous communities are at the forefront of climate knowledge and adaptation. Each Indigenous community and group is unique in its culture, experience, and local knowledge of the environment, providing them with a strong ability to adapt to climate change in place (Schlingmann et al. 2021). However, climate change threatens traditional livelihoods and ways of life and is in some areas generating physical changes to local environments for which past experience no longer fully prepares them, placing new strains on the ability to adapt in place (Ford et al. 2020). Still, the deep, multigenerational knowledge of the environment possessed by Indigenous and traditional communities has considerable potential for informing climate science and adaptation planning for all communities. The process of *co-production* of knowledge (i.e. bringing together traditional knowledge with scientific knowledge to generate improved understanding) works most effectively when Indigenous communities maintain an ongoing connection to their traditional territories to maintain their Indigenous knowledge, and when partnerships between Indigenous and non-Indigenous participants are based on trust, respect, and reciprocity (Hill et al. 2020). This dynamic is particularly important in the context of planning for Indigenous communities faced with climate risks that may potentially necessitate movement or relocation. When forced to move, Indigenous populations often experience increased risk of adverse mental health due to loss of home, social networks, and cultural heritage (Torres and Casey 2017, Cunsolo and Ellis 2018). For such reasons, Indigenous communities often prefer to move together when necessary, in order to attempt to preserve their community and culture (Maldonado et al. 2013). Ensuring that Indigenous communities have the ability for self-determination and the highest possible level of agency when confronting existential climate hazards is an area where more research of a very practical nature is warranted.

Readers will recall the concept of *place attachment* was first introduced in Chapter 2, and refers to the relationships between people and their surroundings and the close bonds and emotional connections that people develop with the areas where they live (Devine-Wright 2013). It is not surprising that for many Indigenous communities, attachment to place is often especially strong, going back many generations and deeply connected to cultural identity. For some groups, the land itself may be even considered a relative or family member. Place attachment can make households and communities more likely to resist moving in the face of climate hazards, preferring *in situ* adaptation and voluntary immobility instead (Adams 2016; Swapan and Sadeque 2021). It is also dynamic. Historic and present changes (including colonization, political influences, changes in the environment, and resource degradation) can shape an individual's sense of place attachment (Barnwell et al. 2021). As climate change disrupts a place itself through changes to the environment, it has the potential to disrupt the sense of attachment for people living there. More research would be very welcome on the complex connections between cultural identity and heritage, how place attachment is expressed in different Indigenous communities, and how decisions to move or stay will be affected as the impacts of climate change increase. If required to relocate beyond their traditional lands, how does (or will) the severing of place attachment affect them? How can they be supported in that process and how can both tangible and intangible cultural heritage be preserved when a place is left behind?

7.3 Climate-Related Migration and Human Health

There is considerable evidence that climate change will pose serious challenges to human health, including increased exposure to extreme heat, higher rates of injury due to intensifying extreme events, changes in the range of vector-borne and water-borne diseases, strains on food security, and increased risks to mental health and well-being (Romanello et al. 2023, Cissé et al. 2022) (e.g. Box 7.3). Lately, there has been growing attention to the connections between mobility, climate change, and human health (McMichael et al. 2023, Schwerdtle et al. 2020), and this is another area where much more research is warranted. The connections between climate, migration, and health are complex and multidirectional, as health influences the decision or ability to migrate, and migration itself can influence health outcomes (Cisse et al. 2022). There are many ways by which these interactions may play out, and they vary by context. For example, the health effects and needs for migrants vary based on the kind of migration. Was the migration voluntary or involuntary? Was it within country borders or across borders? People displaced by extreme weather events often have lower access to basic health care, sanitation, food, and adequate housing and may be obliged to endure such conditions for months or years (McMichael et al. 2023, Cissé et al. 2022). Displaced populations also often experience increased mental health challenges including anxiety, social isolation, and trauma (Shultz et al. 2019, Schwerdtle et al. 2018).

People who move voluntarily may have different health outcomes. There is long-standing evidence of a "healthy migrant effect" by which migrants on average have better health than people who do not migrate, but it is not clear whether this will persist in a changing climate. There is some evidence that people are equally likely to migrate regardless of health during severe droughts, especially where social networks can facilitate migration (Hunter and Simon 2017). Moving away from hazardous locations may in turn have beneficial impacts; for example, households in New Jersey and New York, USA that participated in home buyout programs after Hurricane Sandy reported lower levels of stress than those households that stayed and rebuilt in place (Koslov et al. 2021). There may also be better access to health services, depending on the migration destination. However, movement from one's home can also impact sense of place and contribute to feelings of loss and bereavement (Lopez et al. 2022).

The interactions between climate-related migration and health outcomes will vary by the individual. Very young and very old individuals or individuals with preexisting health conditions, for example, are more vulnerable to certain health-related challenges such as heat-related illness and morbidity. Women also face specific health needs related to sexual and reproductive health and access to such care (Baada et al. 2021, van Daalen et al. 2021). Displacement may also place women and gender minorities at higher risk of gender-related and sexual violence, as noted in the earlier example of Bangladeshi women who risk being victims of violence in cyclone shelters (Ahmed et al. 2019). The elderly may also experience health challenges during the migration process and in destinations due to existing health issues, decreased mobility, and reduced access to care and medication (Cissé et al. 2022).

An influx of migrants can pose challenges and place strain on healthcare systems and resources in destinations, and this can be expected to be especially pronounced in areas where climate change places additional strain on health services. Migrants may have less access to healthcare in their destinations, be more likely to live in poorer-quality housing, may lack reliable access to water and sanitation, and have fewer financial resources to pay for healthcare needs (Ebi and McLeman 2022). Arriving migrants may also have lower immunity and be exposed to diseases in destinations to which they have not previously been exposed, and there may be instances where migrants introduce diseases to destination locations (McMichael 2015). Much more research is needed to understand how climate-related migration may place additional strain on health systems around the world, and how these systems can respond in order to meet the needs of migrants and non-migrants alike. This includes a need to consider how health systems can respond to short-term increases in needs after extreme weather events, and also a need to consider how to meet increased demand for care over the longer term. There would be considerable benefit in having more longitudinal studies to track individuals and groups over time to understand the longer-term effects of climate change and migration on health according to types of migration and climatic drivers. More work on how health needs, both physical and mental, differ across stages of the migration process from origin to transit to destination would also be beneficial. Additional research on climate-related migration and health considerations for socially disadvantaged groups such as racial and ethnic minorities, women and gender minorities, the elderly, and the very young would also be exceedingly useful for health policy and health care delivery in a changing climate.

Box 7.3

Health Challenges for Migrants in Korail, Dhaka

Many people in Bangladesh who migrate for climate-related reasons end up living in informal settlements in and around the capital city, Dhaka. These settlements are not formally recognized by the government and residents therefore do not receive formal access to services and infrastructure. With more than 50,000 residents, the Korail community is one of the largest such informal communities in Dhaka. Residents of Korail, many of whom are migrants from other parts of the country, face challenges in their day-to-day lives that impact their health. People who migrated for climate or environmental reasons to Korail describe food and water insecurity as pressing challenges. One migrant, Rahima, explained to author Kelsea Best, "We don't get water. If water comes one day then it stops for 2 to 3 days." Households often have to pay exorbitant amounts of money for clean water, which leaves even fewer resources for food and other necessities. Without sufficient sanitation infrastructure, and exacerbated by crowded living conditions, the Korail community is highly susceptible to the spread of infectious diseases, especially water-borne diseases. In 2019, Dhaka experienced a serious outbreak of dengue fever, a mosquito-borne virus, and the Korail community was especially hard hit. The lived experiences of environmental migrants living in the Korail informal community in Dhaka are just one example that highlights the many health-related challenges that migrants can face, especially when they do not have sufficient access to infrastructure and services in destination locations.

7.4 Research on Receiving Communities and Climate Destinations

Much of the current research on climate-related migration focuses on why people choose to leave or stay in a specific place in the face of a particular climate hazard. Much less attention has been given to important questions of where people will go in a changing climate, and how they make such choices. Understanding where people are likely to move to has serious implications for security, equity, infrastructure needs, and resources in potential destinations (e.g. Box 7.4). Understanding where people are likely to move will enable proactive (rather than reactive) planning for that movement and possible population changes. Yet, the question of where people will go can be difficult to predict, especially when that migration is not planned. We know in general that most current examples of movement due to climate change are internal and across short distances, and we can assume from this that most future climate-related migration will take place similarly, within borders (Findlay 2011, Rigaud et al. 2018). We also know that people tend to go to places where they have social networks and support (Sue et al. 2019, Torres and Casey 2017). We also have a very good idea of which populations are most highly exposed to particular climate hazards, as was described in Chapters 2 through 4, and are therefore most likely to become sources of out-migration in coming decades. But beyond this, more progress can be made toward identifying with precision the most likely future destinations and transit routes of migration in a changing climate.

Some work has begun to attempt to define different types of possible climate-migrant receiving areas. One existing typology for the US has differentiated between three types of cities: vulnerable cities, recipient cities, and climate destinations (Marandi and Main, 2021). Vulnerable cities are those cities that will experience out-migration and population loss due to climate change. These areas will lose tax revenue as population declines and will face growing challenges in meeting the needs of those residents left behind who may be unwilling or unable to move. Recipient cities are defined as those areas that will unexpectedly receive migrants following a sudden onset hazard – often for which they have not planned or prepared – which will also create significant challenges. The third group, climate destinations, refers to areas that have explicitly planned to receive climate-related migrants, and may even deliberately attempt to attract these migrants to spur population and tax revenue growth (see Box 7.5). While no part of the world will be unaffected by climate change, some US cities in the Midwest, the Northeast, the Great Plains, and the Great Lakes region have already been identified by popular media as possible "climate havens" given their relatively temperate climates and relatively lower exposure to sudden-onset climate hazards (Lopez 2023). Whether they will indeed experience future influxes of migrants remains to be seen.

Box 7.4

**How Many More People Will Move to Seattle
and Portland in a Changing Climate?**

Internal migration from other parts of the US to the Pacific Northwest, and especially
the metropolitan areas of Portland and Seattle, has grown steadily in recent decades,
with Californians and Arizonans being among the largest sources. In 2016, a
workshop was held at Portland State University that brought together researchers and
city planning officials with the aim of improving projections of future climate-related
migration to the region and identifying priorities for future planning (Whitley Binder
and Jurjevich 2016). As one Seattle city planner explained to author Robert McLeman,
it would be very beneficial to know how many additional migrants are likely to move
from the Southwest to the Northwest because of fires, heat, water shortages, etc.
in addition to the people that are already expected to move for other reasons. Will
it be an extra 2% per year? 5% per year? Or more? The reason is that new urban
infrastructure – housing, roads, sewerage, water treatment facilities, electrical utilities,
etc. – must be planned for two or three decades in advance so as to allow sufficient
time for the raising of funds, allocation of lands, environmental impact assessments,
and all the other planning procedures that need to be implemented in advance. To
be managed effectively, an influx of people that may be expected in the year 2050
must be planned for starting in 2025 or 2030. This particular workshop is one of a
surprisingly small number of examples worldwide where urban planning has begun
the first tentative steps toward planning for receiving future climate-related migrants.

Box 7.5

Climate-Resilient and Migrant-Friendly Towns in Bangladesh

In Bangladesh, more than 50% of urban residents live in informal communities
without adequate access to safe housing, water and sanitation, electricity, and other
essential infrastructure (BRAC 2023). At the same time, rural-to-urban migration is
very common, and some studies suggest it is likely to increase with climate change.
The capital city of Dhaka is already under considerable strain, with a population
of more than 23 million in the metro area placing mounting pressure on the city's
infrastructure and resources. To address this challenge and plan for possible increases
in internal climate-related migration, Bangladesh has developed a plan to develop and
promote "climate-resilient and migrant-friendly" cities (BRAC 2023). The idea is that
these cities will attract rural migrants and provide them with access to infrastructure,
education, and employment opportunities. Mongla – Bangladesh's second busiest
port, situated in the southwest of the country – is one such city that is eager to grow
and working to position itself as a destination for climate migrants. Will the idea
of proactively designing cities to attract climate migrants and draw them to places
that want more population growth become a model for other countries? If so, more
planning-oriented research will be needed on the identification and implementation of
critical resources and infrastructure that will be needed to support newcomers while
simultaneously maintaining and improving living standards for existing residents.

Destination communities for climate-related migrants will likely find that the incorporation process occurs in stages, with specific needs evolving over time. In the case of people displaced by extreme events, their arrival in receiving communities will initially place additional demand on emergency services, health care, and temporary shelter, and will then transition quickly to additional demands for permanent housing, employment, and social services that facilitate integration (Junod et al. 2023, Teicher and Marchman 2024). Although such needs are already understood and expected through experience with past disaster experiences, researchers, and planners should expect there to be possible unintended consequences and outcomes, especially for historically marginalized and disadvantaged populations such as low-income groups, racial and ethnic minorities, the very old and very young, and people with health conditions. This includes both climate migrants and prior residents of receiving communities. Indeed, this latter group has often been overlooked in the literature, and researchers have flagged the possibility that influxes of climate-related migrants could increase social and economic inequities that adversely affect prior residents (Box 7.6). Achieving healthy, just, and equitable cities in a climate-disrupted future will require coordinated efforts on the part of researchers, policymakers, and planners.

Box 7.6
Climate Gentrification

One way that climate change may influence where people chose to move or stay is via direct and indirect effects on housing prices. The emerging field of climate gentrification suggests that climate change could contribute to making certain areas and certain neighborhoods within cities more or less desirable compared with others, and therefore more or less affordable (Best and Jouzi 2022, Keenan et al. 2018). Climate-driven changes in property values could result in the displacement of low-income, socially disadvantaged groups to areas of higher exposure, and there is evidence of this occurring in cities such as Miami and New Orleans. In Miami, property values are appreciating faster in higher elevation areas of the city that are less susceptible to flooding and the effects of sea level rise (Keenan et al. 2018). In New Orleans, evidence identified that historically Black parts of the city experienced more severe flooding and were more likely to gentrify following Hurricane Katrina (Aune et al. 2020).

In Little Haiti, a neighborhood in Miami that is predominantly Black and low-income, the effects of climate gentrification are being felt by residents (Figure 7.2). Little Haiti happens to be higher elevation than surrounding areas, a fact that is becoming noticed by developers. New, high-end developments are underway in Little Haiti, and residents are concerned that they will be pushed out of their neighborhood and displaced by gentrification. If they are forced to move due, the residents of Little Haiti may be obliged to move to lower-price areas that are less safe

Figure 7.2   A community mural in Little Haiti, Miami, a location where climate gentrification is a concern and may contribute to displacement of local residents. Photo by K. Best.

and more susceptible to flooding. Despite the multiple pressures of climate change and gentrification, the residents of Little Haiti and local activists are resilient and steadfast in their commitment to their culture and community. Climate gentrification is a complex phenomenon that highlights how climate change and migration could contribute to inequities in nonlinear, indirect ways. Future research into how best to provide safe, affordable housing for all people, especially as the effects of climate change worsen, is needed.

7.5 Thresholds in Climate–Migration Relationships

Complex systems are continually changing. These changes can occur within the functioning of the system and to the system as a whole. They can occur as gradual, barely observable changes as well as wholesale, radical changes that occur suddenly. Consider a forest in northern Canada that passes through seasonal changes from winter to summer and back again each year, over the course of centuries, seemingly stable over the long run until one year a large wildfire sweeps through, changing the structure and species composition within a matter of days. Both types of change – gradual and imperceptible, sudden and dramatic – are perfectly natural.

When we talk about climate-related migration and displacement, we are in practice describing the outcomes of interactions between complex physical systems (atmosphere, oceans, and biological systems) and human systems (agricultural,

cultural, economic, political, social, and urban systems). Changes within any of these interacting systems have potential consequences for migration outcomes.

It is human nature when thinking about the future to assume that any changes that occur are likely to be familiar to us, to be similar to changes we have experienced in recent memory. It's why financial advisors and companies that sell investment products will boldly advertise that they have earned their clients X% profits over the last 1-year, 5-year, or 10-year period, but the fine print contains an explicit statement that past performance is not a guarantee of future performance. It is no exception when people think about climate change: They often assume that future changes will unfold gradually and incrementally, as they have over recent decades, allowing us the opportunity to adjust and adapt. This is reflected in the slow pace of international negotiations to control greenhouse gas (GHG) emissions and the even more lethargic pace at which governments are implementing practical measures to meet emission-reduction commitments and keep promises they have already made. But the reality is that there is no reason to assume that future changes in the climate will be gradual or within our capacity to adapt. The atmosphere has never been so heavily loaded with carbon dioxide or methane as it is today (at least, never within the period that humans have been on Earth), and so it is very rational to assume the climate will start behaving in unfamiliar ways, and that this will have a vast array of unforeseen consequences for physical and human systems, which will in turn affect human migration, population, and displacement patterns.

The massive expansion in research on climate-related migration and displacement over the last twenty years allows us to identify the key linkages between climate, risk, adaptation, and migration as was done in Chapter 1. But, although we are able to sketch out these interactions, we do not have a complete picture of how the systems interact and function. An important aspect that we presently understand only a little is the presence of critical thresholds or tipping points that, once passed, may cause radical changes in outcomes. These thresholds can exist within the functioning of physical and human systems. Some we know exist and can take steps to avoid; others we may not know exist until it is too late and we have already passed them. An example of a known threshold is 0°C (32°F), the temperature at which water changes state from liquid to frozen. Average global temperatures have risen since the pre-industrial period by approximately 1.2°C. This may not sound like much, but it has a huge impact in higher latitudes, which are actually warming at faster rates than the global average. The difference between -1°C and +1°C is the difference between ice forming on water or melting, and precipitation falling as rain or snow. It means that across Canada, Scandinavia, and Russia, winters are getting shorter and milder with less snow and ice, spring temperatures arrive earlier in the calendar year, and summers are warmer and drier. This in turn means longer periods of fire weather each summer, and wildfires displace people, as was reviewed in detail in Chapter 3. In western Canada, the frequency of large wildfires had been declining over the course of the twentieth century, but suddenly wildfires larger and more severe than any previously experienced in living memory occurred in 2017, 2018, 2021, and 2023 (Parisien et al. 2023). Is this evidence that a threshold has been crossed? Should people living in western Canada now expect large wildfires and the resulting evacuations and health risks from smoke to become a regular phenomenon – the "new normal" as people are fond of saying? The next few years will tell us for sure. Scientists had been warning for decades that wildfires would likely increase in a changing climate, but until 15 years ago, that seemed like a distant, hypothetical risk with an undetermined onset date. Research using global scale data suggests that for many regions around the world, the climatic threshold between a low-fire frequency regime and a more frequent fire regime was crossed in the late 1990s or early 2000s (Shi and Touge 2023).

The preceding example is just one of many critical thresholds in climate systems science has identified that, once crossed, create hazards that can potentially generate migration and displacement. Others that have already been referred to in preceding chapters include ocean surface temperatures that create conditions favorable for producing tropical cyclones (26.5°C and warmer) and wet bulb air temperatures warmer than 35°C that make human survival difficult. But human systems and their outcomes, including migration, also have critical thresholds and tipping points, and these are less well known and studied, particularly as they pertain to climate-related migration.

Migration scholars have long referred to the existence of a *threshold of dissatisfaction* – a point at which an individual or household that has an established place of permanent residence is no longer satisfied with living there, perhaps because they experience adverse events, or perhaps new opportunities arise somewhere else that are too tempting to pass up (Wolpert 1966, Speare 1974). Once this threshold is crossed, migration is likely. Each individual or household will have their own particular threshold, and any number of combinations of events might push them across it. In the case of climate-related migration, examples from past research suggest that this threshold is reached when *in situ* forms of adaptation to a particular hazard are no longer effective or beneficial (Adger et al. 2009, Bardsley and Hugo 2010, Meze-Hausken 2008). When thinking toward understanding the future of climate-related migration, greater research that helps communities identify the cost and effectiveness of various *in situ* adaptation options for the climate hazards they are most likely to experience would be very useful.

Another important threshold in migration dynamics relates to what other people do. Although migration is ultimately a decision to be made by an individual or household, humans by nature are influenced by what other people do. A household that sees its neighbors moving away might begin to ask whether it should also leave – in other words, the migration decision of others can push us across the threshold of dissatisfaction. Research that has looked at the social and economic dynamics of US cities has shown, for example, that small changes in the number of people from a particular social or economic group moving into or out of a neighborhood can quickly lead to rapid changes in the number of people moving in and out (McLeman 2018). Such changes are driven by residents' perceptions of the future demographic and socio-economic makeup of the community, as opposed to any specific material change. In the case of smaller communities, out-migration of even a few families can have very practical effects that undermine the well-being of the community as a whole and therefore stimulate rapid additional outmigration. For example, in many jurisdictions, the provision of schools, health clinics, retail banking outlets, and other important services in rural areas is determined by the number of residents. The departure of a small number of residents can push the overall population below the threshold needed to maintain those services, and once they are no longer provided, many of those remaining will need to relocate. In situations where climate-related migration has already begun to occur, rapid research to identify these community-level thresholds is important in recognizing if or when an acceleration of out-migration is likely to occur.

The preceding examples are just two of at least six different types of critical thresholds in the progression from the time at which a community experiences a climate hazard to the point at which the short- and long-term impacts on migration and displacement have been realized (McLeman 2018). These include thresholds at which:

- Adaptation to specific climate hazards becomes necessary
- *In situ* adaptation approaches start to become less effective
- Substantive changes in land use and livelihoods become necessary
- Migration begins to occur
- Migration rates accelerate and/or become nonlinear
- Migration rates slow and stabilize

The identification and analysis of each of these in the context of specific communities and regions presents important opportunities for future research to help us better understand how migration and displacement is likely to occur in the future, and to devise strategies that ensure that migration that does occur takes place under the highest levels of agency possible under the circumstances.

7.6 Known Unknowns and Unknown Unknowns: Foreseeable and Unforeseeable Futures in Climate-Related Migration

The late Donald Rumsfeld once famously said that when describing military intelligence at a press briefing during his time as US Secretary of Defense:

… as we know, there are known knowns; there are things we know we know. We also know there are known unknowns; that is to say we know there are some things we do not know. But there are also unknown unknowns – the ones we don't know we don't know. And if one looks throughout the history of our country and other free countries, it is the latter category that tends to be the difficult ones.

Rumsfeld was at the time discussing whether or not the government of Iraq possessed weapons of mass destruction (it turns out they did not), but his categories might also be used to describe what we know and don't know about the future in terms of anthropogenic climate change and how it will affect migration and displacement. Our "known knowns" in terms of demonstrable facts can be summarized as follows:

- Greenhouse gasses such as carbon dioxide and methane trap heat in the atmosphere
- Human activity – specifically combustion of fossil fuels, industrial processes including concrete and steel production, and deforestation – is causing carbon dioxide, methane, and other heat-trapping gasses to accumulate rapidly in the atmosphere, and are currently at concentrations not seen in the past 800,000 years
- Average global temperatures have warmed by approximately 1.2°C since the late 1800s
- At current rates of emissions, average global temperatures will exceed pre-industrial levels by approximately 2.4°C or more by the end of this century
- Climate hazards including extreme storms, floods, and wildfire droughts currently displace on average over 20 million people per year, and the frequency and/or severity of those events increase positively with additional atmospheric warming

We also know a number of key unknowns that will eventually determine how many people will be on the move in the future due to climate change, which we can phrase as questions:

- Will we reduce our GHG emissions? If so, how soon will we do so? If not, will our emissions increase – and by how much and how soon – or will they stay roughly the same for decades to come?
- Will the human population continue to grow, and if so, how many more people will there be? Or, will population growth rates plateau and/or eventually fall? If so, when? Will population density be high in areas with high levels of exposure to climate hazards?
- What will be the economic development trajectories of countries and communities that are currently low- and middle-income and have high levels of vulnerability to climate hazards? Will poor countries and poor people still be poor in the year 2050 or 2080, or will the international community work collectively to

meet the Sustainable Development Goals, thereby reducing vulnerability and improving adaptive capacity?
- What future trajectory will international migration policy take? Will wealthy countries continue their current attempts to reduce and restrict migration? Or, will they facilitate freer movements of people, thereby allowing labor markets and social networks to play greater roles in shaping future international migration flows?

These "known unknowns" are in essence the result of uncertainty about how known relationships and connections will specifically play out in the future. We cannot predict exactly what will happen, but we can identify scenarios and outcome pathways that might arise under various combinations and permutations of causal drivers, and perhaps even assign probabilities to each. We can then even try to influence the outcomes.

Three decades ago, ecologist Norman Myers (1995) published a thoughtful essay on the topic of environmental unknowns, warning that environmental systems can respond in very unpredictable ways to human pressure. He was particularly concerned about threshold effects we discussed in the preceding section of this book, which Myers subdivided into *discontinuities* (sudden changes as opposed to gradual ones) and *synergistic outcomes*, which occur when two or more environmental processes interact in such a way that the outcome causes an exponential change. Among the many examples given in his article is one that is relevant for imagining the future of climate-related migration: What if ocean circulation patterns are disrupted by global warming?

As described in Chapter 1, the oceans and the atmosphere are linked systems: changes in one affect the other. The oceans are not bathtubs of still water, they have currents at the surface and below the surface that redistribute heat around the globe. One of the best-known ocean currents is the Gulf Stream, a surface current that flows from the warm waters of the Gulf of Mexico in a northeast direction toward the United Kingdom, Scandinavia, and the west coast of Europe. It is part of a larger ocean process known formally as Atlantic meridional overturning circulation (AMOC). The surface current of the Gulf Stream that warms Europe is matched by a current that flows deep beneath the surface in the opposite direction, returning cool water to the tropics and flowing on into the southern hemisphere. The whole system is like a conveyor belt, and the process that drives it is called *thermohaline circulation* (National Ocean Service 2024). It works like this: Wind currents in the higher latitudes of the northern hemisphere blow from west to east, passing across northern Canada and Greenland before they reach the open waters of the North Atlantic Ocean. Because they have passed over land, these winds are dry, containing little moisture. As they pass over the open ocean they collect moisture that evaporates from the sea surface. The water that evaporates leaves its salt behind, causing the sea surface layer to become saltier and denser. At the same time, cold air causes the surface temperature of the water to drop and sea ice forms close to the poles. This process also causes the ocean water to become saltier and denser, because the ice that forms on top of the sea is composed of freshwater, and the salt remains in the liquid water below. The dense cold surface layer water formed through the combination of evaporation and sea ice formation begins to sink down in the water column and flow southward; meanwhile, less dense warmer water at the surface is drawn from south to north to replace the sinking water. It is the downward sinking mechanism of cold, salty water that powers the currents that bring warm water to the coast of Europe. Similar processes operate to drive currents in other oceans.

Here is where an important known unknown lurks: Recall from Chapter 4 that ice and snow on Greenland is melting rapidly due to rising air temperatures. The resulting meltwaters that drain to the ocean are composed of freshwater, which is less dense than salt water and floats on the ocean surface. At some threshold, if enough freshwater pools on the surface of the North Atlantic, it could cause thermohaline circulation to abruptly stop, like putting a stick into the spokes of a moving bicycle. The reason is that, if freshwater is pooled on the ocean surface, it is this water that will evaporate and will form into sea ice. Because it carries no salt, there would no longer be any change in the density of water in the upper layers of the ocean, and so the downward sinking mechanism would cease. This is not a hypothetical event; there is evidence that abrupt shutdowns in thermohaline circulation have occurred in the long past in association with glacial/interglacial cycles (Pena and Goldstein 2014) and some ocean-atmosphere circulation models and early warning signals suggest AMOC has already begun slowing and is moving toward a tipping point, perhaps as soon as the middle of this century (van Westen et al. 2024, Ditlevsen and Ditlevsen 2023). The key uncertainty is when anthropogenic warming would reach the point where it triggers a serious AMOC slowdown or shutdown. Most scientists believe that such an event is unlikely to occur this century (IPCC 2019), but the possibility cannot be ruled out entirely given the complexity of atmosphere–ocean interactions involved and scientific uncertainty about them (Lohmann et al. 2024, Orbe et al. 2023).

Let's imagine what would happen if the AMOC abruptly halted. The heat transported northward by AMOC makes average temperatures in Western Europe disproportionately warmer than other locations at similar latitudes. For example, the Danish capital of Copenhagen is at approximately the same latitude as Goose Bay, the largest town on the east coast of Labrador, Canada, on the directly opposite side of the North Atlantic from Denmark. The metropolitan area of Copenhagen is home to approximately 1.4 million people and has a mild climate in which no month of the year has average daily temperatures below 0°C. Goose Bay has 8,000 people and average daily temperatures in January of -12°C (10°F). Without the warmth of the Gulf Stream, Copenhagen – and Oslo, Dublin, Glasgow, and other northern European capitals and large cities – would experience climates more like Goose Bay. Would the infrastructure in those cities be able to support such large populations in a suddenly harsher climate, and would people even want to live in them at that point? Further, the impacts would not just be felt in Europe. An abrupt AMOC shutdown, or even a significant slowing would trigger a wide range of cascading impacts on precipitation patterns, temperature regimes, and the geographic distribution of extreme climate events throughout the Atlantic basin (Bouwer et al. 2022). The impacts on global, regional, and local patterns of climate-related migration and displacement would be tremendous. An abrupt halt of the AMOC is unlikely in the near future, but its mere possibility makes it a worrisome known unknown. It is also a reminder that when planning for future migration and displacement in a changing climate, we should also consider other low probability/high consequence events, as was done, for example, by Gemenne (2011) in a study that considered the migration and displacement implications of a +4°C warming by 2100 (i.e. warming level roughly 1.5°C greater than is currently projected).

The big challenge for anyone concerned about migration in a changing climate is the likelihood of unknown unknowns, the inevitable unexpected surprises that might occur. More than two decades ago, Streets and Glantz (2000) wrote extensively about the idea of *climate surprise* – that is, the gap between what we expect is likely to occur because of climate change and what actually will end up happening. Myers (1995) referred to the gap between what we know about future environmental risks and what might potentially happen as being, "a black hole of knowledge and understanding." How do we shine light into this black hole? How do we identify unknown unknowns and describe them and thus convert them into known unknowns? Is this even possible?

In some cases, it is indeed possible. Some unknowns are unknown simply because we haven't looked carefully enough, or we haven't paid attention to facts that are already apparent. Myers (1995) notes that many environmental problems that were unknown unknowns in 1972 (significant because it was the year of the first big UN environmental conference) became known once concerted efforts were made to look for them. For example, consider the depletion of stratospheric ozone (also described as a "hole" in the ozone layer). Readers will recall from Chapter 1 that stratospheric ozone prevents ultraviolet radiation emitted by the sun – which causes damage to cells in living tissues – from reaching the Earth's surface. Scientists began collecting data on stratospheric ozone concentrations in the late 1950s, and noticed these were gradually declining. By 1974, scientists realized that human-produced chemicals known as chlorofluorocarbons (CFCs) caused stratospheric ozone to decompose (Molina and Rowland 1974). In 1980, the rate of ozone depletion accelerated sharply – a critical threshold or tipping point had been crossed – but it was not until 1985 that scientific studies were published that warned a rapid thinning of the ozone layer over Antarctica was underway (Farman et al. 1985). Once this was recognized, the international community acted quickly to implement a treaty to curb the production of CFCs, and we are the beneficiaries of that action today. Myers (1995) points to this example as showing how all the information we needed to know about the risks of ozone depletion was available to us by 1974, but it took another decade (and the crossing of a critical threshold) before we actually recognized the danger. The unknown unknown became known once we started looking for it.

However, there are some unknowns that are unknowable simply because the knowledge, skills, and observational tools at our disposal are simply inadequate. To illustrate, imagine we had a time machine and could travel to the year 1825 and speak to a scientist of the day. Imagine trying to explain to them mobile phones, social media, AI, cryptocurrency, and genetically modified organisms. Technologies that we use on a daily basis and which shape our world today would be unknown unknowns to a scientist of 1825. Many of the precursor technologies had yet to be invented, and even the basic relationships between power, voltage, current, and resistance necessary to develop electrical transmission systems were not known until 1826. Money then was scarce, and mostly coins; the idea of bitcoin would be very difficult to explain. Genes had not been discovered, and the very idea that humans might clone organisms and modify the physiology of living things in a godlike fashion would be intellectually troubling for many American and European scientists given that Creationism was a dominant scholarly paradigm (Darwin did not publish the *Origin of Species* until 1859). Now hop back into the time machine and move forward to 1925 and explain present-day technologies to a scientist of that year. They would probably marvel at what we describe to them, but they would not be nearly as surprised as their counterparts from a century before. By 1925, necessary precursor technologies such as electricity, telephones, and simple mechanical computers were already in use, Darwinian evolutionary theory was widely accepted, and research on molecular genetics was being done. Many technologies we use in 2025 would be knowable and predictable to any scientists of 1925, who paused to imagine them. Unknown unknowns in 1825 had become known unknowns by 1925. If we could set our time machine to the year 2125, we would no doubt find that the technologies people use are, for people living in 2025, a mix of unknown unknowns and known unknowns.

Since we don't have a time machine, what should our research agenda be to discover unknown unknowns in the context of future climate-related migration? Here are two ideas. First, we need to make sure we are not overlooking or discounting any relevant facts or possibilities. All of the information provided in this book until this very sentence is based on common assumptions and empirical evidence of causal connections between climate, adaptation, and migration. It may well be that we the authors, and other scholars working in this field are overlooking something obvious in much the same way that scientists in the early 1970s were slow to recognize that CFC emissions would rapidly degrade the ozone layer. Perhaps the risk of large-scale population displacements numbering in the hundreds of millions is more imminent than we have suggested in this book. Or perhaps it is the opposite – maybe we are on the verge of a global society that will be more peaceful, cooperative, and environmentally minded than ever before, and in no more than a few years the risks of climate-related displacement will dissolve (we certainly hope so). What would climate-related migration look like in a world in which everyone has enough to eat and everyone leads long, happy, healthy lives? That would make a wonderful topic for research, perhaps even the next book (the chapter on "how to make it happen" would be especially fun to write). Maybe everything we need to know to make those outcomes happen is right in front of us, we only need to look a little harder. Or perhaps we need to broaden our collaborations, so that we look at things from a more diverse set of perspectives.

Second, to discover the genuinely unknown unknowns, Myers (1995) suggests that we need to ask new questions. We can think of a few; readers can no doubt think of many more. We suggest asking, what technologies are currently in their infancy that might lead to fundamental changes in social, economic, and political systems 20, 50, and 100 years from now? For example, what would mobility look like in a world where we begin genetically modifying ourselves? If virtual reality technology improves and becomes widely accessible, will people lose interest in being physically mobile so long as their basic biological needs can be met where they are? Will individuals and households continue to be the dominant living arrangements in most societies in future decades, or might we be more likely to live communally or in some other type of social structure? These questions emerged from simply reflecting on some of the basic assumptions underpinning the climate-adaptation-migration diagram in Figure 1.13 and imagining how those assumptions could conceivably change in the coming decades. To conclude, we encourage readers to not simply accept at face value the assertions and assumptions made about future climate-related migration, but to imagine how things might be different.

Glossary

Adaptation – In the context of households, communities, and societies refers to the "process of adjustment to actual or expected climate change and its effects, to moderate or avoid harm or to exploit beneficial opportunities." From IPCC AR6 Glossary.

Adaptive capacity – The ability of systems, institutions, humans, and other organisms to adjust to potential damage, to take advantage of opportunities, or to respond to consequences of climatic variability and change. From IPCC AR6 Glossary.

Agency – In the context of migration refers to the extent to the degree of freedom in which a person's decision to migrate (or not) is made, as well as consequent decisions regarding the timing, destination, duration, and other relevant aspects of the act of migration. A person with low or no agency has little control over the migration decision (e.g. a refugee forced to flee an armed conflict), whereas a person who is able to choose at their leisure when and where to move, such as for lifestyle reasons or to pursue job opportunities, would be said to have high agency.

Agent-based model – A computer simulation method used to study the interactions between individuals (agents) assigned with certain attributes and decision-making methods. One important goal of agent-based modeling is to understand what system-level dynamics emerge from the individual behavior of many agents.

Anthropogenic – Resulting from or produced by human activities. From IPCC AR6 Glossary.

Climate – Average weather conditions experienced over an extended period of time that may range from months to millions of years, depending on context. Common measures used in describing climate are the mean and variability of temperature, precipitation, and wind as measured at the Earth's surface. The most common time scale used for describing average climatic conditions by the World Meteorological Organization (WMO) is a period of 30 years. Definition adapted from IPCC AR6 Glossary.

Climate change (general definition) – Refers to long-term shifts in average weather patterns and conditions. A change in the state of the climate can be identified (for example, using statistical tests) by changes in the mean and/or the variability of its properties, and that persists for an extended period, typically decades or longer. Climate change may be due to natural processes, such as fluctuations in solar radiation, volcanic eruptions, and variations in the Earth's orbit around the sun. Climate change may also be anthropogenic, caused by human modification of the composition of the Earth's atmosphere and by changes in deforestation and other changes in land use and land cover. Adapted from IPCC AR6 Glossary

Climate change (legal definition) – Article 1 of the United Nations Framework Convention on Climate Change, defines climate change as being, "a change of climate which is attributed directly or indirectly to human activity that alters the composition of the global atmosphere and which is in addition to natural climate variability observed over comparable time periods." In other words, under international law, "climate change" refers explicitly to anthropogenic climate change unless otherwise specified.

Climate migrant – In this book, unless otherwise clarified, this term is used in the generic sense to refer to any person moving for reasons directly or indirectly related to climatic variability or change, and may refer to people whose movements are voluntary or involuntary.

Climate migration or climate-related migration – Refers generally to migration for reasons directly or indirectly related to climatic variability or change. More specifically, according to the IOM Glossary, *climate migration* is defined as being, "The movement of a person or groups of persons who, predominantly for reasons of sudden or progressive change in the environment due to climate change, are obliged to leave their habitual place of residence, or choose to do so, either temporarily or permanently, within a State or across an international border."

Climate refugee – A term that is widely used in popular media to describe people who are displaced or move involuntarily for reasons related to extreme weather events or climate change. People displaced by environmental events are not eligible for international protection under the United Nations Convention Relating to the Status of Refugees and so this term is not used or recognized under international law. This term is not used in the present book.

Climate-related displacement – Displacement directly or indirectly resulting from the impacts of climatic variability and/or change. See definition of *displacement* further.

Climate system – The global climate system consists of five major components: the atmosphere, the hydrosphere, the cryosphere, the lithosphere, and the biosphere and the interactions between them. From IPCC AR6 Glossary.

Climate variability – Deviations of climate variables (temperature, precipitation, wind, etc.) from their mean in terms of frequency, severity, or spatial scale. This does not refer to individual weather events. Variability may be natural or intrinsic to the climate system (referred to as "internal variability" or may be the result of human activity ("forced variability"). Definition adapted from IPCC AR6 Glossary.

Disaster – A serious disruption to the functioning of a community or a society at any scale due to hazardous events interacting with conditions of exposure, vulnerability, and capacity, leading to one or more of the following: human, material, economic, and environmental losses and impacts. From IPCC AR6 Glossary.

Displacement – "The movement of persons who have been forced or obliged to flee or to leave their homes or places of habitual residence, in particular as a result of or in order to avoid the effects of armed conflict, situations of generalized violence, violations of human rights or natural or human-made disasters." Taken directly from IOM Glossary on Migration

Environmental migration/migrant – Similar to definitions of climate migrant/climate migration earlier, but referring to movements of people directly or indirectly influenced by a wider range of events or conditions not necessarily linked to climate change (e.g. people moving due to the impacts of earthquakes or volcanic eruptions).

Exposure – The presence of people; livelihoods; species or ecosystems; environmental functions, services, and resources; infrastructure; or economic, social, or cultural assets in places and settings that could be adversely affected by extreme weather events or by the impacts of climatic variability or change. Adapted from IPCC AR6 Glossary. See also *risk*.

Extreme weather event – Event that is rare at a particular place and time of year. Definitions of rare vary, but an extreme weather event would normally fall in the 10th or 90th percentile of a probability density function estimated from observations. The characteristics of extreme weather vary from place to place in an absolute sense. When a pattern of extreme weather persists for some time, such as a season, it may be classified as an extreme climate event, especially if it yields an average or total that is itself extreme (for example, drought or heavy rainfall over a season). From IPCC AR6 Glossary.

Flood – The overflowing of the normal confines of a stream or other water body, or the accumulation of water over areas that are not normally submerged. Floods can be caused by unusually heavy rain, for example during storms and cyclones. Floods include river (fluvial) floods, flash floods, urban floods, rain (pluvial) floods, sewer floods, coastal floods, and glacial lake outburst floods (GLOF). From IPCC AR6 Glossary.

Forced migration – A migratory movement which, although the drivers can be diverse, involves force, compulsion, or coercion. From IOM Glossary for Migration. Also described as involuntary migration. The reasons for forced migration may include, but are not limited to, threats to life and livelihood that may be of natural or human origin. The distinction between forced migration and voluntary migration is not always clear-cut, and any individual migration decision may occur along a continuum of compulsion and personal choice. See also *agency*.

Global warming – The increase in average global surface temperatures relative to a baseline reference period, measured over a period long enough to reduce the influence of short-term variability. The baseline period used in IPCC reporting is typically the average of temperatures between 1850 and 1900 (the earliest period of reliable observations with sufficient geographic coverage). At the time of publication of this book, global warming is approximately 1.1°C (i.e. current average global temperatures are approximately 1.1°C warmer than the baseline period). The term is distinct from *climate change* in that the latter encompasses a much wider range of variables and impacts than simply temperature change. Adapted from IPCC AR6 Glossary.

Gravity model – Models that draw on gravitational theory to estimate flows of people between two or more locations based on distance and population.

Greenhouse gas – Gaseous constituents of the atmosphere, both natural and anthropogenic, that absorb and emit radiation at specific wavelengths within the electromagnetic spectrum. By doing so, greenhouse gasses (GHGs) trap heat in the atmosphere, causing temperatures to be warmer than would occur in their absence (known as the "greenhouse effect"). The most common GHGs in order of abundance in the atmosphere are water vapor (H_2O), carbon dioxide (CO_2), nitrous oxide (N_2O), methane (CH_4) and ozone (O_3). In addition to these naturally occurring compounds are GHGs manufactured by humans, including sulfur hexafluoride (SF6), hydrofluorocarbons (HFCs), chlorofluorocarbons (CFCs) and perfluorocarbons (PFCs). Adapted from IPCC AR6 Glossary.

Hazard – Any physical event, process, or phenomenon that may cause loss of life, injury, or other health impacts; damage and/or loss to property, infrastructure, livelihoods, service provision, ecosystems, and environmental resources; and/or social and economic disruption. or environmental degradation. Hazards may occur due to natural processes (e.g. volcanic eruption), direct human activities (e.g. spillage of toxic waste into a waterway), or human modification of natural processes, including specific impacts of climate change. Definition adapted from IOM Glossary and IPCC AR6 Glossary. See also *risk*.

Immigration – International migration as viewed from the perspective of the country of arrival (*emigration* = international migration from the perspective of the country of origin). The act of moving into a country other than one's country of nationality or usual residence, so that the country of destination effectively becomes the new country of usual residence. From IOM Glossary.

Immobility – In the context of climate-related migration, an unwillingness or inability to move away from a location exposed to hazards, climate variability, or where a disaster has occurred. The distinction is clarified in this book through use of the terms *voluntary immobility* to refer to people who choose to remain in hazardous locations and *involuntary mobility* to refer to people that wish to move but lack the ability to do so for social, economic, or other reasons.

In situ **adaptation** – Refers to adaptation that does not require migration or relocation on the part of the person, community, or group in question.

Internal displacement/internally displaced persons – Describes people or groups forced or obliged to flee or to leave their homes or places of habitual residence, in particular as a result of or in order to avoid the effects of armed conflict, situations of generalized violence, violations of human rights or natural or human-made disasters, and who have not crossed an internationally recognized State border. Adapted from IOM Glossary

Internal migration – The movement of people within a State involving the establishment of a new temporary or permanent residence. From IOM Glossary.

International migration – The movement of people away from their place of usual residence and across an international border to a country of which they are not nationals. From IOM Glossary.

Involuntary migration – See *forced migration*.

Involuntary non-migration – See *immobility*.

Irregular migration – Movements of people that take place outside the laws, regulations, or international agreements governing the entry into or exit from the State of origin, transit, or destination. From IOM Glossary. Note that in this book we deliberately avoid the term *illegal migration* that is often used erroneously as a synonym for irregular migration; not all irregular migration is automatically illegal (e.g. under the UN Refugee Convention, a person fleeing persecution has a legal right to seek protection of signatory State, regardless of their migration status in the receiving State).

Machine learning – A broad subset of artificial intelligence (AI) used to identify patterns within data to make predictions.

Maladaptation – Actions (typically unintended) that may lead to increased risk of adverse climate-related outcomes. Adapted from IPCC AR6 Glossary

Managed retreat – The purposeful, planned, and coordinated movement of people and infrastructure away from areas of high risk.

Migration – The movement of persons away from their place of usual residence, either across an international border or within a State. From IOM Glossary for Migration.

Migration-as-adaptation – Adaptation that entails the voluntary movement of people on a temporary or permanent basis. See also *in situ adaptation.*

Mitigation – Any steps taken to reduce greenhouse gas emissions or to remove greenhouse gasses from the atmosphere (such as by increasing forest cover). Adapted from IPCC AR6 Glossary

Mobility – An umbrella term that refers to *all* aspects of the movement of people, regardless of the duration, timescale, destination, or reasons for moving.

Pastoralism – A livelihood strategy based on moving livestock to seasonal pastures primarily in order to convert grasses, tree leaves, or crop residues into human food. The search for feed is however not the only reason for mobility; people and livestock may move to avoid various natural and/or social hazards, to avoid competition with others, or to seek more favorable conditions. From IOM Glossary.

Planned relocation – In the context of disasters or environmental degradation, including when due to the effects of climate change, a planned process in which persons or groups of persons move or are assisted to move away from their homes or place of temporary residence, are settled in a new location and provided with the conditions for rebuilding their lives. From IOM Glossary.

Refugee – A person who qualifies for protection under the United Nations Convention Relating to the Status of Refugees, defined under Article 1 of that Convention as being any person that, "owing to well-founded fear of being persecuted for reasons of race, religion, nationality, membership of a particular social group or political opinion, is outside the country of [their] nationality and is unable or, owing to such fear, is unwilling to avail [them]self of the protection of that country; or who, not having a nationality and being outside the country of [their] former habitual residence as a result of such events, is unable or, owing to such fear, is unwilling to return to it." (pronouns changed from him/his to them/their by authors). By definition, a person that remains within their country of nationality is not legally considered to be a *refugee*, and is instead described as an *internally displaced person*. Also by definition, a person that moves across international borders for reasons related to climatic variability or change is not legally considered to be a *refugee*.

Relocation – In the context of humanitarian emergencies, relocations are to be considered as internal humanitarian evacuations and are understood as large-scale movements of civilians, who face an immediate threat to life in a conflict setting, to locations within the same country where they can be more effectively protected. From IOM Glossary. In the context of climate change, relocations may be organized in anticipation of hazards that may emerge (see *planned relocation*).

Remittances – Personal monetary transfers, cross border or within the same country, made by migrants to individuals or communities with whom the migrant has links. From IOM Glossary.

Resilience – The ability of a community, society, and/or environmental systems to cope with and recover from hazardous events, trends, or disturbances by responding or reorganizing in ways that maintain their essential function, identity, and structure while maintaining the capacity for adaptation, learning, and transformation. Definition combined from IOM and IPCC AR6 glossaries.

Return migration – In the context of international migration, the movement of persons returning to their country of origin after having moved away from their place of habitual residence and crossed an international border. In the context of internal migration, the movement of persons returning to their place of habitual residence after having moved away from it. From IOM Glossary.

Risk – The potential for adverse consequences for human or ecological systems. In the context of climate change, risks can arise from potential impacts of climate change as well as human responses to climate change. Relevant adverse consequences include those on lives, livelihoods, health and well-being, economic, social, and cultural assets and investments, infrastructure, services (including ecosystem services), ecosystems, and species. In IPCC framing, risk is a function of *hazard*, *vulnerability*, and *exposure*, with adaptation having the potential to moderate risk. From IPCC AR6 Glossary

Sea level change – Change to the height of sea level, which can be observed over time at a range of spatial scales, from global to local. Key causes are changes in the amount of water in the oceans, such as caused by the melting of glaciers and ice sheets on land; changes in the density of ocean water (water becomes less dense and rises in volume as its temperature increases); and, changes in the height of land relative to the ocean (such as by the subsidence of coastal or tectonic processes). Adapted from IPCC AR6 Glossary.

Slow-onset impacts of climate change – Impacts that emerge gradually, over extended periods of time, such as sea level rise and rising average temperatures. In this book, drought is treated as a slow-onset event that may unfold over weeks, months or years. Slow-onset impacts are predictable and foreseeable, creating the opportunity for mitigation of greenhouse gas emissions to prevent their occurrence and for adaptation measures to be put into place.

Temporary migration – Migration for a specific motivation and purpose with the intention to return to the country of origin or habitual residence after a limited period of time or to undertake an onward movement. From IOM Glossary.

Trapped populations – See *immobility*.

Voluntary migration – Migration that results from the deliberate choice to move.

Vulnerability – Propensity or predisposition to be adversely affected. Vulnerability encompasses a variety of concepts and elements including sensitivity or susceptibility to harm and lack of capacity to cope and adapt. From IPCC AR6 Glossary. See also *risk*.

References

Abdelgadir, A. and Fouka, V. (2020) 'Political secularism and Muslim integration in the West: assessing the effects of the French headscarf ban', *American Political Science Review*, 114(3), pp. 707–723. https://doi.org/10.1017/S0003055420000106.

Abel, G. J., Brottrager, M., Crespo Cuaresma, J. and Muttarak, R. (2019) 'Climate, conflict and forced migration', *Global Environmental Change*, 54, pp. 239–249. https://doi.org/10.1016/j.gloenvcha.2018.12.003.

Abrams, J. B., Gosnell, H., Gill, N. J. and Klepeis, P. J. (2012) 'Re-creating the rural, reconstructing nature: an international literature review of the environmental implications of amenity migration', *Conservation and Society*, 10(3), p. 270. https://doi.org/10.4103/0972-4923.101837.

Acharya, A. K. (2021) 'Caste-based migration and exposure to abuse and exploitation: Dadan labour migration in India', *Contemporary Social Science*, 16(3), pp. 371–383. https://doi.org/10.1080/21582041.2020.1855467.

Acosta, R. J., Kishore, N., Irizarry, R. A. and Buckee, C. O. (2020) 'Quantifying the dynamics of migration after Hurricane Maria in Puerto Rico', *Proceedings of the National Academy of Sciences*, 117(51), pp. 32772–32778. https://doi.org/10.1073/pnas.2001671117.

Adams, H. (2016) 'Why populations persist: mobility, place attachment and climate change', *Population and Environment*, 37(4), pp. 429–448. https://doi.org/10.1007/s11111-015-0246-3.

Adger, W. N., Crépin, A.-S., Folke, C., Ospina, D., Chapin, F. S., Segerson, K., Seto, K. C., Anderies, J. M., Barrett, S., Bennett, E. M., Daily, G., Elmqvist, T., Fischer, J., Kautsky, N., Levin, S. A., Shogren, J. F., van den Bergh, J., Walker, B. and Wilen, J. (2020) 'Urbanization, migration, and adaptation to climate change', *One Earth*, 3(4), pp. 396–399. https://doi.org/10.1016/j.oneear.2020.09.016.

Adger, W. N., Kelly, P. M., Winkels, A., Huy, L. Q. and Locke, C. (2002) 'Migration, remittances, livelihood trajectories, and social resilience', *Ambio: A Journal of the Human Environment*, 31(4), pp. 358–366. https://doi.org/10.1579/0044-7447-31.4.358.

Adger, W. N., Lorenzoni, I. and O'Brien, K. L. (eds.) (2009) 'Adaptation Now', in *Adapting to Climate Change: Thresholds, Values, Governance*. Cambridge; New York: Cambridge University Press, pp. 1–22. https://assets.cambridge.org/97805217/64858/excerpt/9780521764858_excerpt.pdf.

Adhikari, P., Hong, Y., Douglas, K. R., Kirschbaum, D. B., Gourley, J., Adler, R. and Robert Brakenridge, G. (2010) 'A digitized global flood inventory (1998–2008): compilation and preliminary results', *Natural Hazards*, 55(2), pp. 405–422. https://doi.org/10.1007/s11069-010-9537-2.

African Union (1974) *OAU Convention Governing the Specific Aspects of Refugee Problems in Africa.* https://au.int/en/treaties/oau-convention-governing-specific-aspects-refugee-problems-africa (Accessed: 27 May 2024).

African Union (2012) 'Kampala Convention: African Union Convention for the Protection and Assistance of Internally Displaced Persons in Africa [Treaty]'. African Union. https://au.int/sites/default/files/treaties/36846-treaty-kampala_convention.pdf.

African Union (2024) *About the African Union* (n.d.). https://au.int/en/overview (Accessed: 27 May 2024).

Ahmad, D. and Afzal, M. (2021) 'Flood hazards, human displacement and food insecurity in rural riverine areas of Punjab, Pakistan: policy implications', *Environmental Science and Pollution Research*, 28(8), pp. 10125–10139. https://doi.org/10.1007/s11356-020-11430-7.

Ahmed, B., Kelman, I., Kamruzzaman, Md., Mohiuddin, H., Rahman, Md. M., Das, A., Fordham, M. and Shamsudduha, M. (2019) 'Indigenous people's responses to drought in northwest Bangladesh', *Environmental Development*, 29, pp. 55–66. https://doi.org/10.1016/j.envdev.2018.11.004.

Ahmed, I. and Ledger, K. (2023) 'Lessons from the 2019/2020 "Black Summer Bushfires" in Australia', *International Journal of Disaster Risk Reduction*, 96, p. 103947. https://doi.org/10.1016/j.ijdrr.2023.103947.

Ahmed, K. J., Haq, S. M. A. and Bartiaux, F. (2019) 'The nexus between extreme weather events, sexual violence, and early marriage: a study of vulnerable populations in Bangladesh', *Population and Environment*, 40(3), pp. 303–324. https://doi.org/10.1007/s11111-019-0312-3.

Ahmed, S. and Eklund, E. (2021) 'Climate change impacts in coastal Bangladesh: migration, gender and environmental injustice', *Asian Affairs*, 52(1), pp. 155–174. https://doi.org/10.1080/03068374.2021.1880213.

Ahsan Ullah, A. K. M. and Chattoraj, D. (2023) 'International marriage migration: the predicament of culture and its negotiations', *International Migration*, 61(6), pp. 262–278. https://doi.org/10.1111/imig.13172.

Ajani, A. and van der Geest, K. (2021) 'Climate change in rural Pakistan: evidence and experiences from a people-centered perspective', *Sustainability Science*, 16(6), pp. 1999–2011. https://doi.org/10.1007/s11625-021-01036-4.

Ajibade, I., Sullivan, M., Lower, C., Yarina, L. and Reilly, A. (2022) 'Are managed retreat programs successful and just? A global mapping of success typologies, justice dimensions, and trade-offs', *Global Environmental Change*, 76, p. 102576. https://doi.org/10.1016/j.gloenvcha.2022.102576.

Akber, Md. A., Patwary, M. M., Islam, Md. A. and Rahman, M. R. (2018) 'Storm protection service of the Sundarbans mangrove forest, Bangladesh', *Natural Hazards*, 94(1), pp. 405–418. https://doi.org/10.1007/s11069-018-3395-8.

Akther, H. and Ahmad, M. M. (2021) 'Livelihood under stress: the case of urban poor during and post-flood in Dhaka, Bangladesh', *The Geographical Journal*, 187(3), pp. 186–199. https://doi.org/10.1111/geoj.12379.

Akyelken, N. (2020) 'Living with urban floods in Metro Manila: a gender approach to mobilities, work and climatic events', *Gender, Place & Culture*, 27(11), pp. 1580–1601. https://doi.org/10.1080/0966369X.2020.1726880.

Alam, Md. Z. and Mamun, A. A. (2022) 'Dynamics of internal migration in Bangladesh: trends, patterns, determinants, and causes', *PLoS ONE*, 17(2), p. e0263878. https://doi.org/10.1371/journal.pone.0263878.

Aldunce, P., Araya, D., Sapiain, R., Ramos, I., Lillo, G., Urquiza, A. and Garreaud, R. (2017) 'Local perception of drought impacts in a changing climate: the mega-drought in central Chile', *Sustainability*, 9(11), p. 2053. https://doi.org/10.3390/su9112053.

Aleksandrova, M. (2020) 'Principles and considerations for mainstreaming climate change risk into national social protection frameworks in developing countries', *Climate and Development*, 12(6), pp. 511–520. https://doi.org/10.1080/17565529.2019.1642180.

Alex, R. (2014) 'Don't call them "refugees": why climate-change victims need a different label', *The Guardian*, 18 September. www.theguardian.com/vital-signs/2014/sep/18/refugee-camps-climate-change-victims-migration-pacific-islands (Accessed: 27 May 2024).

Alexander, M., Polimis, K. and Zagheni, E. (2019) 'The impact of Hurricane Maria on out-migration from Puerto Rico: evidence from Facebook data', *Population and Development Review*, 45(3), pp. 617–630. https://doi.org/10.1111/padr.12289.

Al-Maruf, A., Pervez, A. K. M. K., Sarker, P. K., Rahman, M. S. and Ruiz-Menjivar, J. (2022) 'Exploring the factors of farmers' rural–urban migration decisions in Bangladesh', *Agriculture*, 12(5). https://doi.org/10.3390/agriculture12050722.

Alston, M. (2014) 'Gender mainstreaming and climate change', *Women's Studies International Forum*, 47, pp. 287–294. https://doi.org/10.1016/j.wsif.2013.01.016.

Alston, M. and Whittenbury, K. (eds.) (2012) *Research, Action and Policy: Addressing the Gendered Impacts of Climate Change*. Netherlands: Springer Science and Business Media (EBL ebooks online). https://doi.org/10.1007/978-94-007-5518-5.

Amri, I. and Giyarsih, S. R. (2022) 'Monitoring urban physical growth in tsunami-affected areas: a case study of Banda Aceh City, Indonesia', *GeoJournal*, 87(3), pp. 1929–1944. https://doi.org/10.1007/s10708-020-10362-6.

Andrews, O., Quéré, C. L., Kjellstrom, T., Lemke, B. and Haines, A. (2018) 'Implications for workability and survivability in populations exposed to extreme heat under climate change: a modelling study', *The Lancet Planetary Health*, 2(12), pp. e540–e547. https://doi.org/10.1016/S2542-5196(18)30240-7.

Anisi, A. (2020) 'Addressing challenges in climate change adaptation: learning from the Narikoso community relocation in Fiji', *Toda Peace Institute*, (Policy Brief No. 84), p. 21.

Antunes, L., Silva, S. P., Marques, J., Nunes, B. and Antunes, S. (2017) 'The effect of extreme cold temperatures on the risk of death in the two major Portuguese cities', *International Journal of Biometeorology*, 61(1), pp. 127–135. https://doi.org/10.1007/s00484-016-1196-x.

Antwi-Agyei, P. and Nyantakyi-Frimpong, H. (2021) 'Evidence of climate change coping and adaptation practices by smallholder farmers in northern Ghana', *Sustainability*, 13(3), p. 1308. https://doi.org/10.3390/su13031308.

Arboleda, E. (1995) 'The Cartagena Declaration of 1984 and its similarities to the 1969 OAU Convention – a comparative perspective', *International Journal of Refugee Law*, 7(Special_Issue), pp. 87–101. https://doi.org/10.1093/reflaw/7.Special_Issue.87.

Arctic Climate Impact Assessment (ACIA) (2005) *Arctic Climate Impact Assessment*. New York, NY: Cambridge University Press. www.acia.uaf.edu (Accessed: 29 April 2024).

Arias, P., Bellouin, N., Coppola, E., Jones, R., Krinner, G., Marotzke, J., Naik, V., Palmer, M., Plattner, G.-K., Rogelj, J., Rojas, M., Sillmann, J., Storelvmo, T., Thorne, P., Trewin, B., Achutarao, K., Adhikary, B., Allan, R., Armour, K., Bala, G., Barimalala, R., Berger, S., Canadell, J. G., Cassou, C., Cherchi, A., Collins, W. D., Collins, W. J., Connors, S., Corti, S., Cruz, F., Dentener, F. J., Dereczynski, C., Di Luca, A., Diongue Niang, A., Doblas-Reyes, P., Dosio, A., Douville, H., Engelbrecht, F., Eyring, V., Fischer, E. M., Forster, P., Fox-Kemper, B., Fuglestvedt, J., Fyfe, J., Gillett, N., Goldfarb, L., Gorodetskaya, I., Gutierrez, J. M., Hamdi, R., Hawkins, E., Hewitt, H., Hope, P., Islam, A. S., Jones, C., Kaufmann, D., Kopp, R., Kosaka, Y., Kossin, J., Krakovska, S., Li, J., Lee, J.-Y., Masson-Delmotte, V., Mauritsen,

T., Maycock, T., Meinshausen, M., Min, S., Ngo Duc, T., Otto, F., Pinto, I., Pirani, A., Raghavan, K., Ranasighe, R., Ruane, A., Ruiz, L., Sallée, J.-B., Samset, B. H., Sathyendranath, S., Monteiro, P. S., Seneviratne, S. I., Sörensson, A. A., Szopa, S., Takayabu, I., Treguier, A.-M., van den Hurk, B., Vautard, R., Von Schuckmann, K., Zaehle, S., Zhang, X. and Zickfeld, K. (2021) 'Climate Change 2021: The Physical Science Basis. Contribution of Working Group I to the Sixth Assessment Report of the Intergovernmental Panel on Climate Change; Technical Summary', in V. Masson-Delmotte, P. Zhai, A. Pirani, S. L. Conners, C. Péan, S. Berger, N. Caud, Y. Chen, L. Goldfarb, M. I. Gomis, M. Huang, K. Leitzell, E. Lonnoy, J. B. R. Matthews, T. K. Maycock, T. Waterfield, O. Yelekçi, R. Yu, and B. Zhou (eds.). *The Intergovernmental Panel on Climate Change AR6*, Remote, p. 33–144. www.ipcc .ch/report/ar6/wg1/ (Accessed: 11 April 2024).

Arnold, C. (2019) 'Death, statistics and a disaster zone: the struggle to count the dead after Hurricane Maria', *Nature*, 566(7742), pp. 22–25. https://doi.org/10.1038/ d41586-019-00442-0.

Asadi Zarch, M. A., Sivakumar, B., Malekinezhad, H. and Sharma, A. (2017) 'Future aridity under conditions of global climate change', *Journal of Hydrology*, 554, pp. 451–469. https://doi.org/10.1016/j.jhydrol.2017.08.043.

Asefawu, G. S. (2022) 'Seasonal migration and household food security status in the drought-prone areas of Northeast Ethiopia', *Environmental Challenges*, 8, p. 100566. https://doi.org/10.1016/j.envc.2022.100566.

Ashraf, M., Routray, J. K. and Saeed, M. (2014) 'Determinants of farmers' choice of coping and adaptation measures to the drought hazard in northwest Balochistan, Pakistan', *Natural Hazards*, 73(3), pp. 1451–1473. https://doi.org/10.1007/s11069-014-1149-9.

Atif, S., Umar, M. and Ullah, F. (2021) 'Investigating the flood damages in Lower Indus Basin since 2000: spatiotemporal analyses of the major flood events', *Natural Hazards*, 108(2), pp. 2357–2383. https://doi.org/10.1007/s11069-021-04783-w.

Aune, K. T., Gesch, D. and Smith, G. S. (2020) 'A spatial analysis of climate gentrification in Orleans Parish, Louisiana post-Hurricane Katrina', *Environmental Research*, 185, p. 109384. https://doi.org/10.1016/j.envres.2020.109384.

Ayeb-Karlsson, S., Smith, C. D. and Kniveton, D. (2018) 'A discursive review of the textual use of "trapped" in environmental migration studies: the conceptual birth and troubled teenage years of trapped populations', *Ambio*, 47(5), pp. 557–573. https:// doi.org/10.1007/s13280-017-1007-6.

Azumah, S. B. and Ahmed, A. (2023) 'Climate-induced migration among maize farmers in Ghana: a reality or an illusion?', *Environmental Development*, 45, p. 100808. https:// doi.org/10.1016/j.envdev.2023.100808.

Baada, J. N., Baruah, B., Sano, Y. and Luginaah, I. (2021) 'Mothers in a "strange land": migrant women farmers' reproductive health in the Brong-Ahafo region of Ghana', *Journal of Health Care for the Poor and Underserved*, 32(2), pp. 910–930. https:// doi.org/10.1353/hpu.2021.0071.

Bach, A. J. E., Cunningham, S. J. K., Morris, N. R., Xu, Z., Rutherford, S., Binnewies, S. and Meade, R. D. (2024) 'Experimental research in environmentally induced hyperthermic older persons: a systematic quantitative literature review mapping the available evidence', *Temperature*, 11(1), pp. 4–26. https://doi.org/10.1080/23328940.20 23.2242062.

Baez, J., Caruso, G., Mueller, V. and Niu, C. (2017a) 'Droughts augment youth migration in Northern Latin America and the Caribbean', *Climatic Change*, 140(3), pp. 423–435. https://doi.org/10.1007/s10584-016-1863-2.

Baez, J., Caruso, G., Mueller, V. and Niu, C. (2017b) 'Heat Exposure and Youth Migration in Central America and the Caribbean', *American Economic Review*, 107(5), pp. 446–450. https://doi.org/10.1257/aer.p20171053.

Balachandran, B., Olshansky, R. B. and Johnson, L. A. (2022) 'Planning for disaster-induced relocation of communities', *Journal of the American Planning Association*, 88(3), pp. 288–304. https://doi.org/10.1080/01944363.2021.1978855.

Baldauf, M., Garlappi, L. and Yannelis, C. (2020) 'Does climate change affect real estate prices? Only if you believe in it', *The Review of Financial Studies*, 33(3), pp. 1256–1295. https://doi.org/10.1093/rfs/hhz073.

Banerjee, S., Yuves-Gerlitz, J. and Kniveton, D. (2011) 'Remittances as a Key to Adaptation: Perspectives on Remittances from Communities Exposed to Water Hazards in the Hindu Kush-Himalayan Region', in *Labour Migration: Opportunities and challenges for mountain livelihoods*. Patan, Nepal: The International Centre for Integrated Mountain Development (ICIMOD) (59), pp. 24–27.

Barbosa, L. M. and Coates, R. (2021) 'Resisting disaster chronopolitics: Favelas and forced displacement in Rio de Janeiro, Brazil', *International Journal of Disaster Risk Reduction*, 63, p. 102447. https://doi.org/10.1016/j.ijdrr.2021.102447.

Barclay, J., Few, R., Armijos, M. T., Phillips, J. C., Pyle, D. M., Hicks, A., Brown, S. K. and Robertson, R. E. A. (2019) 'Livelihoods, wellbeing and the risk to life during volcanic eruptions', *Frontiers in Earth Science*, 7. https://doi.org/10.3389/feart.2019.00205.

Bardsley, D. K. and Hugo, G. J. (2010) 'Migration and climate change: examining thresholds of change to guide effective adaptation decision-making', *Population and Environment*, 32(2), pp. 238–262. https://doi.org/10.1007/s11111-010-0126-9.

Barkat, A. and Manzuma, A. (2014) Gender and Migration from Bangladesh: Mainstreaming Migration into the National Development Plans from a Gender Perspective. Edited by International Labour Organisation. Dhaka: ILO Country Office for Bangladesh. https://webapps.ilo.org/wcmsp5/groups/public/---asia/---ro-bangkok/---ilo-dhaka/documents/publication/wcms_310411.pdf.

Barnwell, G., Makaulule, M., Stroud, L., Watson, M. and Dima, M. (2021). 'The Lived Experiences of Place Severing and Decolonial Resurgence in Vhembe District, South Africa. *Awry: Journal of Critical Psychology*,2(1), Article 1.

Basile, R., Centofanti, F., Giallonardo, L. and Licari, F. (2024) 'Migration responses to earthquakes: evidence from Italy', *Italian Economic Journal*, 10(1), pp. 269–291. https://doi.org/10.1007/s40797-023-00226-6.

Baylie, M. M. and Fogarassy, C. (2022) 'Decision analysis of the adaptation of households to extreme floods using an extended protection motivation framework – a case study from Ethiopia', *Land*, 11(10). https://doi.org/10.3390/land11101755.

Bayrak, M. M., Van Hieu, T., Tran, T. A., Hsu, Y.-Y., Nien, T. and Quynh, D. T. T. (2023) 'Climate change adaptation responses and human mobility in the Mekong Delta: local perspectives from rural households in An Giang Province, Vietnam', *Humanities and Social Sciences Communications*, 10(1), pp. 1–14. https://doi.org/10.1057/s41599-023-01817-5.

BBC News (2015) 'New Zealand deports climate change asylum seeker to Kiribati', 24 September. www.bbc.com/news/world-asia-34344513 (Accessed: 27 May 2024).

Beduschi, A. (2021). 'International migration management in the age of artificial intelligence'. *Migration Studies*, 9(3), pp. 576–596. https://doi.org/10.1093/migration/mnaa003.

Begum, A., Lempert, R., Ali, E., Benjaminsen, T. A., Bernauer, T., Cramer, W., Cui, X., Mach, K., Nagy, G., Stenseth, N. C., Sukumar, R. and Wester, P. (2022a) 'Point of Departure and Key Concepts', in H.-O. Pörtner, D. C. Roberts, M. Tignor, E. S. Poloczanska, K. Mintenbeck, A. Algeria, M. Craig, S. Langsdorf, S. Löschke, V. Möller, A. Okem, and B. Rama (eds.) *Climate Change 2022 – Impacts, Adaptation and Vulnerability: Working Group II Contribution to the Sixth Assessment Report of the Intergovernmental Panel on Climate Change*. 1st ed. Cambridge, UK and

New York, NY, USA: Cambridge University Press, pp. 121–196. https://doi.org/10.1017/9781009325844.

Begum, A., Lempert, R., Ali, E., Benjaminsen, T. A., Bernauer, T., Cramer, W., Cui, X., Mach, K., Nagy, G., Stenseth, N. C., Sukumar, R. and Wester, P. (2022b) 'Point of Departure and Key Concepts, in H.-O. Pörtner, D. C. Roberts, M. Tignor, E. S. Poloczanska, K. Mintenbeck, A. Algeria, M. Craig, S. Langsdorf, S. Löschke, V. Möller, A. Okem, and B. Rama (eds.) *Climate Change 2022 – Impacts, Adaptation and Vulnerability*. Contribution of Working Group II to the Sixth Assessment Report of the Intergovernmental Panel on Climate Change. Cambridge, UK and New York, NY, USA: Cambridge University Press, pp. 121–196. doi:10.1017/9781009325844.003.

Bell, A., Ward, P., Tamal, Md. E. H. and Killilea, M. (2019) 'Assessing recall bias and measurement error in high-frequency social data collection for human-environment research', *Population and Environment*, 40(3), pp. 325–345. https://doi.org/10.1007/s11111-019-0314-1.

Bell, A. R. (2017) 'Informing decisions in agent-based models – A mobile update', *Environmental Modelling & Software*, 93, pp. 310–321. https://doi.org/10.1016/j.envsoft.2017.03.028.

Bell, A. R., Calvo-Hernandez, C. and Oppenheimer, M. (2019) 'Migration, intensification, and diversification as adaptive strategies', *Socio-Environmental Systems Modelling*, 1, pp. 16102–16102. https://doi.org/10.18174/sesmo.2019a16102.

Bell, A. R., Ward, P. S., Killilea, M. E. and Tamal, M. E. H. (2016) 'Real-time social data collection in rural Bangladesh via a "microtasks for micropayments" platform on android smartphones', *PLOS ONE*, 11(11), p. e0165924. https://doi.org/10.1371/journal.pone.0165924.

Bell, A. R., Wrathall, D. J., Mueller, V., Chen, J., Oppenheimer, M., Hauer, M., Adams, H., Kulp, S., Clark, P. U., Fussell, E., Magliocca, N., Xiao, T., Gilmore, E. A., Abel, K., Call, M. and Slangen, A. B. A. (2021) 'Migration towards Bangladesh coastlines projected to increase with sea-level rise through 2100', *Environmental Research Letters*, 16(2), p. 024045. https://doi.org/10.1088/1748-9326/abdc5b.

Bell, C. and Keys, P. W. (2018) 'Conditional relationships between drought and civil conflict in sub-Saharan Africa', *Foreign Policy Analysis*, 14(1), pp. 1–23. https://doi.org/10.1093/fpa/orw002.

Beniston, M. and Stephenson, D. B. (2004) 'Extreme climatic events and their evolution under changing climatic conditions', *Global and Planetary Change*, 44(1), pp. 1–9. https://doi.org/10.1016/j.gloplacha.2004.06.001.

Berdugo, M., Delgado-Baquerizo, M., Soliveres, S., Hernández-Clemente, R., Zhao, Y., Gaitán, J. J., Gross, N., Saiz, H., Maire, V., Lehmann, A., Rillig, M. C., Solé, R. V. and Maestre, F. T. (2020) 'Global ecosystem thresholds driven by aridity', *Science*, 367(6479), pp. 787–790. https://doi.org/10.1126/science.aay5958.

Berlemann, M. and Tran, T. X. (2020) 'Climate-related hazards and internal migration empirical evidence for rural Vietnam', *Economics of Disasters and Climate Change*, 4(2), pp. 385–409. https://doi.org/10.1007/s41885-020-00062-3.

Berlemann, M. and Tran, T. X. (2021) 'Tropical storms and temporary migration in Vietnam', *Population and Development Review*, 47(4), pp. 1107–1142. https://doi.org/10.1111/padr.12455.

Berlin Rubin, N. and Wong-Parodi, G. (2022) 'As California burns: the psychology of wildfire- and wildfire smoke-related migration intentions', *Population and Environment*, 44(1), pp. 15–45. https://doi.org/10.1007/s11111-022-00409-w.

Bernzen, A., Jenkins, J. C. and Braun, B. (2019) 'Climate change-induced migration in coastal Bangladesh? A critical assessment of migration drivers in rural households under economic and environmental stress', *Geosciences*, 9(1). https://doi.org/10.3390/geosciences9010051.

Bertana, A. (2020) 'The role of power in community participation: relocation as climate change adaptation in Fiji', *Environment and Planning C: Politics and Space*, 38, p. 239965442090939. https://doi.org/10.1177/2399654420909394.

Best, K., Gilligan, J., Baroud, H., Carrico, A., Donato, K. and Mallick, B. (2022) 'Applying machine learning to social datasets: a study of migration in southwestern Bangladesh using random forests', *Regional Environmental Change*, 22(2), p. 52. https://doi.org/10.1007/s10113-022-01915-1.

Best, K., Gilligan, J. and Qu, A. (2021) 'Modeling multi-level patterns of environmental migration in Bangladesh: an agent-based approach', in *2021 Winter Simulation Conference (WSC). 2021 Winter Simulation Conference (WSC)*, Phoenix, AZ, USA: IEEE, pp. 1–12. https://doi.org/10.1109/WSC52266.2021.9715380.

Best, K. and Jouzi, Z. (2022) 'Climate gentrification: methods, gaps, and framework for future research', *Frontiers in Climate*, 4. https://doi.org/10.3389/fclim.2022.828067.

Best, K. B., Gilligan, J. M., Baroud, H., Carrico, A. R., Donato, K. M., Ackerly, B. A. and Mallick, B. (2021) 'Random forest analysis of two household surveys can identify important predictors of migration in Bangladesh', *Journal of Computational Social Science*, 4(1), pp. 77–100. https://doi.org/10.1007/s42001-020-00066-9.

Bezgrebelna, M., McKenzie, K., Wells, S., Ravindran, A., Kral, M., Christensen, J., Stergiopoulos, V., Gaetz, S. and Kidd, S. A. (2021) 'Climate change, weather, housing precarity, and homelessness: a systematic review of reviews', *International Journal of Environmental Research and Public Health*, 18(11), p. 5812. https://doi.org/10.3390/ijerph18115812.

Bhusal, P., Kimengsi, J. N. and Raj Awasthi, K. (2021) 'What drives environmental (non-)migration around the Himalayan region? Evidence from rural Nepal', *World Development Perspectives*, 23, p. 100350. https://doi.org/10.1016/j.wdp.2021.100350.

Bhutta, Z. A., Bhutta, S. Z., Raza, S. and Sheikh, A. T. (2022) 'Addressing the human costs and consequences of the Pakistan flood disaster', *The Lancet*, 400(10360), pp. 1287–1289. https://doi.org/10.1016/S0140-6736(22)01874-8.

Billson, J. M. (1990) 'Opportunity or tragedy: the impact of Canadian resettlement policy on Inuit families', *American Review of Canadian Studies*, 20(2), pp. 187–218. https://doi.org/10.1080/02722019009481650.

Black, R., Bennett, S. R. G., Thomas, S. M. and Beddington, J. R. (2011) 'Migration as adaptation', *Nature*, 478(7370), pp. 447–449. https://doi.org/10.1038/478477a.

Black, R. and Collyer, M. (2014) '"Trapped" Populations: Limits on Mobility at Times of Crisis', in S. Martin, *Humanitarian Crises and Migration: Causes, Consequences and Responses*. 1st ed. London: Routledge, pp. 287–305. https://doi.org/10.4324/9780203797860-14.

Blaikie, P., Cannon, T., Davies, I. and Wisner, B. (1994) *At Risk: Natural Hazards, People's Vulnerability, and Disasters*. 1st ed. London: Routledge.

Bloomberg, M. (2023) 'Cyclone Gabrielle triggered more destructive forestry "slash" – NZ must change how it grows trees on fragile land', *The Conversation*, 16 February. http://theconversation.com/cyclone-gabrielle-triggered-more-destructive-forestry-slash-nz-must-change-how-it-grows-trees-on-fragile-land-200059 (Accessed: 11 April 2024).

Boas, I. (2020) 'Social networking in a digital and mobile world: the case of environmentally-related migration in Bangladesh', *Journal of Ethnic and Migration Studies*, 46(7), pp. 1330–1347. https://doi.org/10.1080/1369183X.2019.1605891.

Boege, V. and Rakova, U. (2019) 'Climate change-induced relocation: problems and achievements – the Carterets case', *Toda Peace Institute*, (Policy Brief No.33), p. 18.

Bohra-Mishra, P., Oppenheimer, M. and Hsiang, S. M. (2014) 'Nonlinear permanent migration response to climatic variations but minimal response to disasters', *Proceedings of the National Academy of Sciences*, 111(27), pp. 9780–9785. https://doi.org/10.1073/pnas.1317166111.

Bonabeau, E. (2002) 'Agent-based modeling: methods and techniques for simulating human systems', *Proceedings of the National Academy of Sciences*, 99(suppl_3), pp. 7280–7287. https://doi.org/10.1073/pnas.082080899.

Bonfils, C. J. W., Santer, B. D., Fyfe, J. C., Marvel, K., Phillips, T. J. and Zimmerman, S. R. H. (2020) 'Human influence on joint changes in temperature, rainfall and continental aridity', *Nature Climate Change*, 10(8), pp. 726–731. https://doi.org/10.1038/s41558-020-0821-1.

Borderon, M., Best, K. B., Bailey, K., Hopping, D. L., Dove, M. and Cervantes de Blois, C. L. (2021) 'The risks of invisibilization of populations and places in environment-migration research', *Humanities and Social Sciences Communications*, 8(1), pp. 1–11. https://doi.org/10.1057/s41599-021-00999-0.

Borger, J. and editor, diplomatic (2007) 'Darfur conflict heralds era of wars triggered by climate change, UN report warns', *The Guardian*, 23 June. www.theguardian.com/environment/2007/jun/23/sudan.climatechange (Accessed: 23 April 2024).

Borjas, G. J. (2001) 'The economics of immigration', in *The New Immigrant in the American Economy*. 1st ed. London: Routledge, p. 52.

Boston, J., Panda, A. and Surminski, S. (2021) 'Designing a funding framework for the impacts of slow-onset climate change – insights from recent experiences with planned relocation', *Current Opinion in Environmental Sustainability*, 50, pp. 159–168. https://doi.org/10.1016/j.cosust.2021.04.001.

Bouwer, L. M., Cheong, S.-M., Jacot Des Combes, H., Frölicher, T. L., McInnes, K. L., Ratter, B. M. W. and Rivera-Arriaga, E. (2022) 'Risk management and adaptation for extremes and abrupt changes in climate and oceans: current knowledge gaps', *Frontiers in Climate*, 3. https://doi.org/10.3389/fclim.2021.785641.

Bowser, G. C. and Cutter, S. L. (2015) 'Stay or go? Examining decision making and behavior in hurricane evacuations', *Environment: Science and Policy for Sustainable Development*, 57(6), pp. 28–41. https://doi.org/10.1080/00139157.2015.1089145.

Box, J. E., Hubbard, A., Bahr, D. B., Colgan, W. T., Fettweis, X., Mankoff, K. D., Wehrlé, A., Noël, B., van den Broeke, M. R., Wouters, B., Bjørk, A. A. and Fausto, R. S. (2022) 'Greenland ice sheet climate disequilibrium and committed sea-level rise', *Nature Climate Change*, 12(9), pp. 808–813. https://doi.org/10.1038/s41558-022-01441-2.

BRAC (2023) 'Climate-resilient and migrant-friendly towns through locally-led adaptation [News and Press Release]', *BRAC*, 10 August. https://reliefweb.int/report/bangladesh/climate-resilient-and-migrant-friendly-towns-through-locally-led-adaptation#:~:text=In%20pursuit%20of%20this%20Goal,communities%20living%20in%20informal%20settlements.

Bradshaw, C. J. A., Sodhi, N. S., Peh, K. S.-H. and Brook, B. W. (2007) 'Global evidence that deforestation amplifies flood risk and severity in the developing world', *Global Change Biology*, 13(11), pp. 2379–2395. https://doi.org/10.1111/j.1365-2486.2007.01446.x.

Bragg, C., Gibson, G., King, H., Lefler, A. A. and Ntoubandi, F. (2018) 'Remittances as aid following major sudden-onset natural disasters', *Disasters*, 42(1), pp. 3–18. https://doi.org/10.1111/disa.12229.

Braunschweiger, D. and Pütz, M. (2021) 'Climate adaptation in practice: how mainstreaming strategies matter for policy integration', *Environmental Policy and Governance*, 31(4), pp. 361–373. https://doi.org/10.1002/eet.1936.

Brennan, M., Srini, T., Steil, J., Mazereeuw, M. and Ovalles, L. (2022) 'A perfect storm? Disasters and evictions', *Housing Policy Debate*, 32(1), pp. 52–83. https://doi.org/1 0.1080/10511482.2021.1942131.

Bronen, R. and Chapin, F. S. (2013) 'Adaptive governance and institutional strategies for climate-induced community relocations in Alaska', *Proceedings of the National Academy of Sciences*, 110(23), pp. 9320–9325. https://doi.org/10.1073/pnas.1210508110.

Bronfman, N. C., Cisternas, P. C., Repetto, P. B., Castañeda, J. V. and Guic, E. (2020) 'Understanding the relationship between direct experience and risk perception of natural hazards', *Risk Analysis*, 40(10), pp. 2057–2070. https://doi.org/10.1111/ risa.13526.

Bronfman, N. C., Repetto, P. B., Guerrero, N., Castañeda, J. V. and Cisternas, P. C. (2021) 'Temporal evolution in social vulnerability to natural hazards in Chile', *Natural Hazards*, 107(2), pp. 1757–1784. https://doi.org/10.1007/s11069-021-04657-1.

Broughton, C. (2008) 'Migration as engendered practice: Mexican men, masculinity, and northward migration', *Gender & Society*, 22(5), pp. 568–589. https://doi .org/10.1177/0891243208321275.

Brown, G., Raymond, C. M. and Corcoran, J. (2015) 'Mapping and measuring place attachment', *Applied Geography*, 57, pp. 42–53. https://doi.org/10.1016/j .apgeog.2014.12.011.

Brown, I. A. (2010) 'Assessing eco-scarcity as a cause of the outbreak of conflict in Darfur: a remote sensing approach', *International Journal of Remote Sensing*, 31(10), pp. 2513–2520. https://doi.org/10.1080/01431161003674592.

Brown, O., Hammill, A. and McLeman, R. (2007) 'Climate change as the "new" security threat: implications for Africa', *International Affairs*, 83(6), pp. 1141–1154. https:// doi.org/10.1111/j.1468-2346.2007.00678.x.

Brown, O. and McLeman, R. (2009) 'A recurring anarchy? The emergence of climate change as a threat to international peace and security: analysis', *Conflict, Security & Development*, 9(3), pp. 289–305. https://doi.org/10.1080/14678800903142680.

Brown, P., Vélez Vega, C. M., Murphy, C. B., Welton, M., Torres, H., Rosario, Z., Alshawabkeh, A., Cordero, J. F., Padilla, I. Y. and Meeker, J. D. (2018) 'Hurricanes and the environmental justice island: Irma and Maria in Puerto Rico', *Environmental Justice (Print)*, 11(4), pp. 148–153. https://doi.org/10.1089/env.2018.0003.

Brown, S., Jenkins, K., Goodwin, P., Lincke, D., Vafeidis, A. T., Tol, R. S. J., Jenkins, R., Warren, R., Nicholls, R. J., Jevrejeva, S., Arcilla, A. S. and Haigh, I. D. (2021) 'Global costs of protecting against sea-level rise at 1.5 to 4.0 °C', *Climatic Change*, 167(1), p. 4. https://doi.org/10.1007/s10584-021-03130-z.

Brown, S. and Nicholls, R. J. (2015) 'Subsidence and human influences in mega deltas: The case of the Ganges–Brahmaputra–Meghna', *Science of The Total Environment*, 527–528, pp. 362–374. https://doi.org/10.1016/j.scitotenv.2015.04.124.

Brown, S., Nicholls, R. J., Bloodworth, A., Bragg, O., Clauss, A., Field, S., Gibbons, L., Pladaitė, M., Szuplewski, M., Watling, J., Shareef, A. and Khaleel, Z. (2023) 'Pathways to sustain atolls under rising sea levels through land claim and island raising', *Environmental Research: Climate*, 2(1), p. 015005. https://doi .org/10.1088/2752-5295/acb4b3.

Brown, S., Nicholls, R. J., Lowe, J. A. and Hinkel, J. (2016) 'Spatial variations of sea-level rise and impacts: An application of DIVA', *Climatic Change*, 134(3), pp. 403–416. https://doi.org/10.1007/s10584-013-0925-y.

Brown, S., Wadey, M. P., Nicholls, R. J., Shareef, A., Khaleel, Z., Hinkel, J., Lincke, D. and McCabe, M. V. (2020) 'Land raising as a solution to sea-level rise: an analysis of coastal flooding on an artificial island in the Maldives', *Journal of Flood Risk Management*, 13(S1), p. e12567. https://doi.org/10.1111/jfr3.12567.

de Bruijn, J. A., de Moel, H., Jongman, B., de Ruiter, M. C., Wagemaker, J. and Aerts, J. C. J. H. (2019) 'A global database of historic and real-time flood events based on social media', *Scientific Data*, 6(1), p. 311. https://doi.org/10.1038/s41597-019-0326-9.

Burke, M., Driscoll, A., Heft-Neal, S., Xue, J., Burney, J. and Wara, M. (2021) 'The changing risk and burden of wildfire in the United States', *Proceedings of the National Academy of Sciences*, 118(2), p. e2011048118. https://doi.org/10.1073/pnas.2011048118.

Burton, I., Kates, R. W. and White, G. F. (1978) *The Environment as Hazard*. Oxford University Press.

Bylander, M. (2016) 'Cambodian migration to Thailand: The role of environmental shocks and stress'. (The KNOMAD *Working Paper Series*, 7).

C40 Cities Leadership Group and Mayors Migration Council (2021) *Global Mayors Action Agenda on Climate and Migration*, p. 52. www.mayorsmigrationcouncil.org.

Call, M. and Gray, C. (2020) 'Climate anomalies, land degradation, and rural out-migration in Uganda', *Population and Environment*, 41(4), pp. 507–528. https://doi.org/10.1007/s11111-020-00349-3.

Call, M. A., Gray, C., Yunus, M. and Emch, M. (2017) 'Disruption, not displacement: environmental variability and temporary migration in Bangladesh', *Global Environmental Change*, 46, pp. 157–165. https://doi.org/10.1016/j.gloenvcha.2017.08.008.

Callahan, C. W. and Mankin, J. S. (2022) 'Globally unequal effect of extreme heat on economic growth', *Science Advances*, 8(43), p. eadd3726. https://doi.org/10.1126/sciadv.add3726.

Calliari, E., Serdeczny, O. and Vanhala, L. (2020) 'Making sense of the politics in the climate change loss and damage debate', *Global Environmental Change*, 64, p. 102133. https://doi.org/10.1016/j.gloenvcha.2020.102133.

Calverley, C. M. and Walther, S. C. (2022) 'Drought, water management, and social equity: analyzing Cape Town, South Africa's water crisis', *Frontiers in Water*, 4. www.frontiersin.org/journals/water/articles/10.3389/frwa.2022.910149.

Cano, G. and Délano, A. (2007) 'The Mexican government and organised Mexican immigrants in the United States: a historical analysis of political transnationalism (1848–2005)', *Journal of Ethnic and Migration Studies*, 33(5), pp. 695–725. https://doi.org/10.1080/13691830701359157.

Cantor, D. (2021) 'Environment, mobility, and international law: a new approach in the Americas', *Chicago Journal of International Law*, 21(2, Article 3). https://chicagounbound.uchicago.edu/cjil/vol21/iss2/3.

Cantor, D., Swartz, J., Roberts, B., Abbara, A., Ager, A., Bhutta, Z. A., Blanchet, K., Madoro Bunte, D., Chukwuorji, J. C., Daoud, N., Ekezie, W., Jimenez-Damary, C., Jobanputra, K., Makhashvili, N., Rayes, D., Restrepo-Espinosa, M. H., Rodriguez-Morales, A. J., Salami, B. and Smith, J. (2021) 'Understanding the health needs of internally displaced persons: a scoping review', *Journal of Migration and Health*, 4, p. 100071. https://doi.org/10.1016/j.jmh.2021.100071.

Cárdenas-Vélez, M., Barrott, J., Betancur Jaramillo, J. C., Hernández-Orozco, E., Maestre-Másmela, D. and Lobos-Alva, I. (2024) 'A combined cognitive and spatial model to map and understand climate-induced migration', *Environment, Development and Sustainability*, 26(3), pp. 6781–6807. https://doi.org/10.1007/s10668-023-02987-7.

Carlot, J., Vousdoukas, M., Rovere, A., Karambas, T., Lenihan, H. S., Kayal, M., Adjeroud, M., Pérez-Rosales, G., Hedouin, L. and Parravicini, V. (2023) 'Coral reef structural complexity loss exposes coastlines to waves', *Scientific Reports*, 13(1), p. 1683. https://doi.org/10.1038/s41598-023-28945-x.

Carlyle, W. J. (1994) 'Rural population change on the Canadian prairies', *Great Plains Research*, 4(1), pp. 65–87.

Carrico, A. R. and Donato, K. (2019) 'Extreme weather and migration: evidence from Bangladesh', *Population and Environment*, 41(1), pp. 1–31. https://doi.org/10.1007/s11111-019-00322-9.

Carrico, A. R., Donato, K. M., Best, K. B. and Gilligan, J. (2020) 'Extreme weather and marriage among girls and women in Bangladesh', *Global Environmental Change*, 65, p. 102160. https://doi.org/10.1016/j.gloenvcha.2020.102160.

Case, D., Fowler, D., Morgan, H., Schwellenbach, S. and Culbertson, K. (2008) 'Moving to the mountains: Amenity migration in the Sierra and Southern Appalachian mountains', in J. L. Wescoat and D. M. Johnston (eds.) *Political Economies of Landscape Change: Places of Integrative Power*. Dordrecht: Springer Netherlands, pp. 77–88. https://doi.org/10.1007/978-1-4020-5849-3_4.

Castillo, C. M. (2015) 'The Cartagena process: 30 years of innovation and solidarity', *Refugee Studies Centre, Oxford Department of International Development*, (49), pp. 89–91.

Cattaneo, C. and Bosetti, V. (2017) 'Climate-induced international migration and conflicts', *CESifo Economic Studies*, 63(4), pp. 500–528. https://doi.org/10.1093/cesifo/ifx010.

Cattaneo, C. and Peri, G. (2016) 'The migration response to increasing temperatures', *Journal of Development Economics*, 122, pp. 127–146. https://doi.org/10.1016/j.jdeveco.2016.05.004.

Chakraborty, J., McAfee, A. A., Collins, T. W. and Grineski, S. E. (2021) 'Exposure to Hurricane Harvey flooding for subsidized housing residents of Harris County, Texas', *Natural Hazards*, 106(3), pp. 2185–2205. https://doi.org/10.1007/s11069-021-04536-9.

Chamlee-Wright, E. and Storr, V. H. (2009) '"There's no place like New Orleans": sense of place and community recovery in the ninth ward after Hurricane Katrina', *Journal of Urban Affairs*, 31(5), pp. 615–634. https://doi.org/10.1111/j.1467-9906.2009.00479.x.

Chang, D. (2015) 'From global factory to continent of labour: labour and development in Asia', *Asian Labour Review*, 1(1), pp. 5–48.

Chase, J. and Hansen, P. (2021) 'Displacement after the camp fire: where are the most vulnerable?', *Society & Natural Resources*, 34(12), pp. 1566–1583. https://doi.org/10.1080/08941920.2021.1977879.

Chaulagain, S., Li, J., Baniya, R. and Pizam, A. (2023) 'Should I stay or should I go back? Factors influencing snowbirds' permanent relocation intention', *Journal of Hospitality and Tourism Insights*, ahead-of-print(ahead-of-print). https://doi.org/10.1108/JHTI-07-2023-0485.

Chen, D., Rojas, M., Samset, B. H., Cobb, K., Diongue Niang, A., Edwards, P., Emori, S., Faria, S. H., Hawkins, E., Hope, P., Huybrechts, P., Meinshausen, M., Mustafa, S. K., Plattner, G.-K. and Tréguier, A.-M. (2021) 'Framing, Context, and Methods', in V. Masson-Delmotte, P. Zhai, A. Pirani, S. L. Connors, C. Péan, S. Berger, N. Caud, Y. Chen, L. Goldfarb, M. I. Gomis, M. Huang, K. Leitzell, E. Lonnoy, J. B. R. Matthews, T. K. Maycock, T. Waterfield, O. Yelekçi, R. Yu, and B. Zhou (eds.) *Climate Change 2021 – The Physical Science Basis: Working Group I Contribution to the Sixth Assessment Report of the Intergovernmental Panel on Climate Change*. Cambridge, United Kingdom and New York, NY, USA: Cambridge University Press, pp. 147–286. https://doi.org/10.1017/9781009157896.003.

Chen, J. (2020) 'Displacement, emplacement and the lifestyles of Chinese "snowbirds" and local residents in tropical Sanya', *Social & Cultural Geography*, 21(5), pp. 629–650. https://doi.org/10.1080/14649365.2018.1509113.

Chen, J. and Mueller, V. (2018) 'Coastal climate change, soil salinity and human migration in Bangladesh', *Nature Climate Change*, 8(11), pp. 981–985. https://doi.org/10.1038/s41558-018-0313-8.

Cheng, M. and Duan, C. (2021) 'The changing trends of internal migration and urbanization in China: new evidence from the seventh National Population Census', *China Population and Development Studies*, 5(3), pp. 275–295. https://doi.org/10.1007/s42379-021-00093-7.

Chevtaeva, E. and Denizci-Guillet, B. (2021) 'Digital nomads' lifestyles and coworkation', *Journal of Destination Marketing & Management*, 21, p. 100633. https://doi.org/10.1016/j.jdmm.2021.100633.

Chim, M. M., Aubry, T. J., Abraham, N. L., Marshall, L., Mulcahy, J., Walton, J. and Schmidt, A. (2023) 'Climate projections very likely underestimate future volcanic forcing and its climatic effects', *Geophysical Research Letters*, 50(12), p. e2023GL103743. https://doi.org/10.1029/2023GL103743.

Choi, C., Hwang, M. C. and Parreñas, R. S. (2018) 'Women on the Move: Stalled Gender Revolution in Global Migration', in B. J. Risman, C. M. Froyum, and W. J. Scarborough (eds.) *Handbook of the Sociology of Gender*. Cham: Springer International Publishing (Handbooks of Sociology and Social Research), pp. 493–506. https://doi.org/10.1007/978-3-319-76333-0_36.

Christ, S., Schwarz, N. and Sliuzas, R. (2023) 'Understanding residential choice under risk: A case study of settlement fire and wildfire risk', *Habitat International*, 136, p. 102815. https://doi.org/10.1016/j.habitatint.2023.102815.

Church, J. A. and White, N. J. (2011) 'Sea-level rise from the late 19th to the early 21st century', *Surveys in Geophysics*, 32(4), pp. 585–602. https://doi.org/10.1007/s10712-011-9119-1.

Cissé, G., McLeman, R., Adams, H., Aldunce, P., Bowen, K., Campbell-Lendrum, D., Clayton, S., Ebi, K. L., Hess, J., Huang, C., Liu, Q., McGregor, G., Semenza, J. and Tirado, M. C. (2022) 'Health, Wellbeing, and the Changing Structure of Communities', in H.-O. Pörtner, D. C. Roberts, M. Tignor, E. S. Poloczanska, K. Mintenbeck, A. Alegría, M. Craig, S. Langsdorf, S. Löschke, V. Möller, A. Okem, and B. Rama (eds.) *Climate Change 2022: Impacts, Adaptation and Vulnerability. Contribution of Working Group II to the Sixth Assessment Report of the Intergovernmental Panel on Climate Change*. 1st ed. Cambridge, UK and New York, NY, USA: Cambridge University Press, pp. 1041–1170. https://doi.org/10.1017/9781009325844.009.

Clark, G. E., Moser, S. C., Ratick, S. J., Dow, K., Meyer, W. B., Emani, S., Jin, W., Kasperson, J. X., Kasperson, R. E. and Schwarz, H. E. (1998) 'Assessing the vulnerability of coastal communities to extreme storms: the case of Revere, MA., USA', *Mitigation and Adaptation Strategies for Global Change*, 3(1), pp. 59–82. https://doi.org/10.1023/A:1009609710795.

Clarke, M., Rendell, H., Tastet, J.-P., Clave, B. and Masse, L. (2002) 'Late-Holocene sand invasion and North Atlantic storminess along the Aquitaine Coast, southwest France', *The Holocene*, 12(2), pp. 231–238. https://doi.org/10.1191/0959683602hl539rr.

Cleaveland, C. and Kirsch, V. (2020) '"They took all my clothes and made me walk naked for two days so I couldn't escape": Latina immigrant experiences of human smuggling in Mexico', *Qualitative Social Work*, 19(2), pp. 213–228. https://doi.org/10.1177/1473325018816362.

Clement, V., Rigaud, K. K., de Sherbinin, A., Jones, B., Adamo, S., Schewe, J., Sadiq, N. and Shabahat, E. (2021) *Groundswell Part 2: Acting on Internal Climate Migration. Publications & Research*. Washington, DC: World Bank. https://hdl.handle.net/10986/36248.

Climate change impacts and cross-border displacement: International refugee law and UNHCR's mandate (2023). UN High Commissioner for Refugees (UNHCR), p. 6.

www.unhcr.org/sites/default/files/2023-12/UNHCR%20note%20on%20climate%20 change%20international%20protection%20UNHCRs%20mandate%20Dec%20 2023.pdf.

Cloos, P., Belloiseau, M., Mc Pherson, N., Harris-Glenville, F., Joseph, D. D. and Zinszer, K. (2023) 'Discussing linkages between climate change, human mobility and health in the Caribbean: the case of Dominica, a qualitative study', *The Journal of Climate Change and Health*, 11, p. 100237. https://doi.org/10.1016/j.joclim.2023.100237.

Cockel, I., Zani, B. and Parhusip, J. S. (2023) '"There will be no law, or people to protect us": irregular Southeast Asian seasonal workers in Taiwan before and during the pandemic', *Journal of Agrarian Change*, 23(3), pp. 634–644. https://doi.org/10.1111/ joac.12550.

Cohen, J. P., Barr, J. and Kim, E. (2021) 'Storm surges, informational shocks, and the price of urban real estate: An application to the case of Hurricane Sandy', *Regional Science and Urban Economics*, 90, p. 103694. https://doi.org/10.1016/j .regsciurbeco.2021.103694.

Collins, J., Polen, A., McSweeney, K., Colón-Burgos, D. and Jernigan, I. (2021) 'Hurricane risk perceptions and evacuation decision-making in the age of Covid-19', *Bulletin of the American Meteorological Society*, 102(4), pp. E836–E848. https://doi .org/10.1175/BAMS-D-20-0229.1.

Collins, T. W. (2010) 'Marginalization, facilitation, and the production of unequal risk: the 2006 Paso del Norte floods', *Antipode*, 42(2), pp. 258–288. https://doi .org/10.1111/j.1467-8330.2009.00755.x.

Collins, T. W., Grineski, S. E., Chakraborty, J. and Flores, A. B. (2019) 'Environmental injustice and Hurricane Harvey: a household-level study of socially disparate flood exposures in Greater Houston, Texas, USA', *Environmental Research*, 179, p. 108772. https://doi.org/10.1016/j.envres.2019.108772.

Connell, J. (2018) 'Nothing there atoll? "Farewell to the Carteret islands": Living climate change in Oceania', in *Pacific Climate Cultures*, pp. 73–87. https://doi .org/10.2478/9783110591415-007.

Connell, J. and Conway, D. (2000) 'Migration and remittances in island microstates: a comparative perspective on the South Pacific and the Caribbean', *International Journal of Urban and Regional Research*, 24(1), pp. 52–78. https://doi .org/10.1111/1468-2427.00235.

Conrad, C. P. and Husson, L. (2009) 'Influence of dynamic topography on sea level and its rate of change', *Lithosphere*, 1(2), pp. 110–120. https://doi.org/10.1130/L32.1.

Cook, B. I., Seager, R. and Smerdon, J. E. (2014) 'The worst North American drought year of the last millennium: 1934', *Geophysical Research Letters*, 41(20), pp. 7298–7305. https://doi.org/10.1002/2014GL061661.

Corfidi, S. F., Evans, J. S. and Johns, R. H. (2024) *Facts About Derechos – Very Damaging Windstorms, About Derechos*. www.spc.noaa.gov/misc/AbtDerechos/derechofacts .htm (Accessed: 21 April 2024).

Cortés Arbués, I., Chatzivasileiadis, T., Ivanova, O., Storm, S., Bosello, F. and Filatova, T. (2024) 'Distribution of economic damages due to climate-driven sea-level rise across European regions and sectors', *Scientific Reports*, 14(1), p. 126. https://doi .org/10.1038/s41598-023-48136-y.

Corwin, D. L. (2021) 'Climate change impacts on soil salinity in agricultural areas', *European Journal of Soil Science*, 72(2), pp. 842–862. https://doi.org/10.1111/ ejss.13010.

Costanza, R., Anderson, S. J., Sutton, P., Mulder, K., Mulder, O., Kubiszewski, I., Wang, X., Liu, X., Pérez-Maqueo, O., Luisa Martinez, M., Jarvis, D. and Dee, G. (2021)

'The global value of coastal wetlands for storm protection', *Global Environmental Change*, 70, p. 102328. https://doi.org/10.1016/j.gloenvcha.2021.102328.

Cottier, F. and Salehyan, I. (2021) 'Climate variability and irregular migration to the European Union', *Global Environmental Change*, 69, p. 102275. https://doi.org/10.1016/j.gloenvcha.2021.102275.

Coupland, K. (2020) Disaster displacement: Examining the post-Dorian experience on Eleuthera. IMRC Policy Points, Issue XVIII. International Migration Research Centre, Waterloo. https://scholars.wlu.ca/imrc/43/

Coutin, S. B. (2010) 'Confined within: national territories as zones of confinement', *Political Geography*, 29(4), pp. 200–208. https://doi.org/10.1016/j.polgeo.2010.03.005.

Cowan, T., Hegerl, G. C., Colfescu, I., Bollasina, M., Purich, A. and Boschat, G. (2017) 'Factors contributing to record-breaking heat waves over the Great Plains during the 1930s Dust Bowl', *Journal of climate*, 30(7), pp. 2437–2461. https://doi.org/10.1175/jcli-d-16-0436.1.

Crawley, H. (2016) 'Managing the unmanageable? Understanding Europe's response to the migration "crisis"', *Human Geography*, 9(2), pp. 13–23. https://doi.org/10.1177/194277861600900202.

Cremins, D., Far, N., Finegold, C., Francis, A., Neusner, J. and Salazar, D. (2023) *Climate of Coercion: Environmental and Other Drivers of Cross-Border Displacement in Central America and Mexico.* Central America North America (central)/ Mexico: U.S. Committee for Refugees and Immigrants, International Refugee Assistance Project (IRAP), and Human Security Initiative (HUMSI), p. 14. https://refugees.org/wp-content/uploads/2023/03/Climate-of-Coercion-Report.pdf.

Crenshaw, K. W. (2017) *On Intersectionality: Essential Writings.* New York, NY: The New Press (Faculty Books). https://scholarship.law.columbia.edu/books/255.

Crépeau, F. and Vezmar, M. (2021) 'Words matter: "illegal", "irregular", "unauthorized", "undocumented". Policy Note 15. KNOMAD. www.knomad.org/sites/default/files/2021-07/Policy%20Brief%2015%20%20-%20Words%20matter%20illegal%20irregular%20unauthorized%20undocumented%20-%20June%202021.pdf.

Cunsolo, A. and Ellis, N. R. (2018) 'Ecological grief as a mental health response to climate change-related loss', *Nature Climate Change*, 8(4), pp. 275–281. https://doi.org/10.1038/s41558-018-0092-2.

Curtis, K. J., Fussell, E. and DeWaard, J. (2015) 'Recovery migration after hurricanes Katrina and Rita: spatial concentration and intensification in the migration system', *Demography*, 52(4), pp. 1269–1293. https://doi.org/10.1007/s13524-015-0400-7.

Cutter, S. L. and Finch, C. (2008) 'Temporal and spatial changes in social vulnerability to natural hazards', *Proceedings of the National Academy of Sciences*, 105(7), pp. 2301–2306. https://doi.org/10.1073/pnas.0710375105.

Cutting CO_2 (2024) *World Meteorological Organization.* https://wmo.int/site/frontline-of-climate-action/climate-change-mitigation/cutting-co2 (Accessed: 27 May 2024).

van Daalen, K. R., Dada, S., Issa, R., Chowdhury, M., Jung, L., Singh, L., Stokes, D., Orcutt, M. and Singh, N. S. (2021) 'A scoping review to assess sexual and reproductive health outcomes, challenges and recommendations in the context of climate migration', *Frontiers in Global Women's Health*, 2. https://doi.org/10.3389/fgwh.2021.757153.

Dahri, G. N., Mangan, T., Nangraj, G. M., Talpur, B. A., Jarwar, I.A., Sial, M., Nangraj, A. N. and Aamir, M. (2021) 'Socio-economic impact and migration due to water shortage in district Badin Sindh province of Pakistan', *Journal of Economic Impact*, 3(1), pp. 01–11. https://doi.org/10.52223/jei3012101.

Dallmann, I. and Millock, K. (2017) 'Climate variability and inter-state migration in India', *CESifo Economic Studies*, 63(4), pp. 560–594. https://doi.org/10.1093/cesifo/ifx014.

Dannecker, P. (2005) 'Transnational migration and the transformation of gender relations: the case of Bangladeshi labour migrants', *Current Sociology*, 53(4), pp. 655–674. https://doi.org/10.1177/0011392105052720.

Dasgupta, S., Laplante, B., Meisner, C., Wheeler, D. and Yan, J. (2009) 'The impact of sea level rise on developing countries: a comparative analysis', *Climatic Change*, 93(3), pp. 379–388. https://doi.org/10.1007/s10584-008-9499-5.

Davis, J. and Lopez-Carr, D. (2014) 'Migration, remittances and smallholder decision-making: implications for land use and livelihood change in Central America', *Land Use Policy*, 36, pp. 319–329. https://doi.org/10.1016/j.landusepol.2013.09.001.

Davis, K. F., Bhattachan, A., D'Odorico, P. and Suweis, S. (2018) 'A universal model for predicting human migration under climate change: examining future sea level rise in Bangladesh', *Environmental Research Letters*, 13(6), p. 064030. https://doi.org/10.1088/1748-9326/aac4d4.

Davis, L., Gertler, P., Jarvis, S. and Wolfram, C. (2021) 'Air conditioning and global inequality', *Global Environmental Change*, 69, p. 102299. https://doi.org/10.1016/j.gloenvcha.2021.102299.

De Lellis, P., Ruiz Marín, M. and Porfiri, M. (2021) 'Modeling human migration under environmental change: a case study of the effect of sea level rise in Bangladesh', *Earth's Future*, 9(4), p. e2020EF001931. https://doi.org/10.1029/2020EF001931.

De Silva, D. G., Kruse, J. B. and Wang, Y. (2008) 'Spatial dependencies in wind-related housing damage', *Natural Hazards*, 47(3), pp. 317–330. https://doi.org/10.1007/s11069-008-9221-y.

DeAngelis, D. L. and Diaz, S. G. (2019) 'Decision-making in agent-based modeling: a current review and future prospectus', *Frontiers in Ecology and Evolution*, 6. https://doi.org/10.3389/fevo.2018.00237.

Debnath, M. and Nayak, D. K. (2022) 'Assessing drought-induced temporary migration as an adaptation strategy: evidence from rural India', *Migration and Development*, 11(3), pp. 521–542. https://doi.org/10.1080/21632324.2020.1797458.

Dedekorkut-Howes, A., Torabi, E. and Howes, M. (2020) 'When the tide gets high: a review of adaptive responses to sea level rise and coastal flooding', *Journal of Environmental Planning and Management*, 63(12), pp. 2102–2143. https://doi.org/10.1080/09640568.2019.1708709.

Defrance, D., Delesalle, E. and Gubert, F. (2023) 'Migration response to drought in Mali. An analysis using panel data on Malian localities over the 1987–2009 period', *Environment and Development Economics*, 28(2), pp. 171–190. https://doi.org/10.1017/S1355770X22000183.

Delazeri, L. M. M., Da Cunha, D. A. and Oliveira, L. R. (2022) 'Climate change and rural–urban migration in the Brazilian Northeast region', *GeoJournal*, 87(3), pp. 2159–2179. https://doi.org/10.1007/s10708-020-10349-3.

Denys, P. H., Beavan, R. J., Hannah, J., Pearson, C. F., Palmer, N., Denham, M. and Hreinsdottir, S. (2020) 'Sea level rise in New Zealand: the effect of vertical land motion on century-long tide gauge records in a tectonically active region', *Journal of Geophysical Research: Solid Earth*, 125(1), p. e2019JB018055. https://doi.org/10.1029/2019JB018055.

Depsky, N. and Pons, D. (2020) 'Meteorological droughts are projected to worsen in Central America's dry corridor throughout the 21st century', *Environmental Research Letters*, 16(1), p. 014001. https://doi.org/10.1088/1748-9326/abc5e2.

Derickson, K. D. (2022) 'Disrupting displacements: making knowledges for futures otherwise in Gullah/Geechee nation', *Annals of the American Association of Geographers*, 112(3), pp. 838–846. https://doi.org/10.1080/24694452.2021.1996219.

Designing a future Tuktoyaktuk (n.d.) *Future Tuktoyaktuk*. [Website]. https://futuretuktoyaktuk.org (Accessed: 29 April 2024).

Dessalegn, M., Debevec, L., Nicol, A. and Ludi, E. (2023) 'A critical examination of rural out-migration studies in Ethiopia: considering impacts on agriculture in the sending communities', *Land*, 12(1), p. 176. https://doi.org/10.3390/land12010176.

Devine-Wright, P. (2013) 'Think global, act local? The relevance of place attachments and place identities in a climate changed world', *Global Environmental Change*, 23(1), pp. 61–69. https://doi.org/10.1016/j.gloenvcha.2012.08.003.

DeWaard, J., Johnson, J. and Whitaker, S. (2019) 'Internal migration in the United States: a comprehensive comparative assessment of the consumer credit panel', *Demographic Research*, 41, pp. 953–1006. https://doi.org/10.4054/demres.2019.41.33.

DeWaard, J., Johnson, J. E. and Whitaker, S. D. (2020) 'Out-migration from and return migration to Puerto Rico after Hurricane Maria: evidence from the Consumer Credit Panel', *Population and Environment*, 42(1), pp. 28–42. https://doi.org/10.1007/s11111-020-00339-5.

Di Virgilio, G., Evans, J. P., Blake, S. A. P., Armstrong, M., Dowdy, A. J., Sharples, J. and McRae, R. (2019) 'Climate change increases the potential for extreme wildfires', *Geophysical Research Letters*, 46(14), pp. 8517–8526. https://doi.org/10.1029/2019GL083699.

Dillon, R. H. (1967) 'Stephen long's great American desert', *Proceedings of the American Philosophical Society*, 111(2), pp. 93–108.

Ditlevsen, P. and Ditlevsen, S. (2023) 'Warning of a forthcoming collapse of the Atlantic meridional overturning circulation', *Nature Communications*, 14(1), p. 4254. https://doi.org/10.1038/s41467-023-39810-w.

Doberstein, B., Tadgell, A. and Rutledge, A. (2020) 'Managed retreat for climate change adaptation in coastal megacities: a comparison of policy and practice in Manila and Vancouver', *Journal of Environmental Management*, 253, p. 109753. https://doi.org/10.1016/j.jenvman.2019.109753.

Dobler-Morales, C. and Bocco, G. (2021) 'Social and environmental dimensions of drought in Mexico: an integrative review', *International Journal of Disaster Risk Reduction*, 55, p. 102067. https://doi.org/10.1016/j.ijdrr.2021.102067.

Dominguez, F., Dall'erba, S., Huang, S., Avelino, A., Mehran, A., Hu, H., Schmidt, A., Schick, L. and Lettenmaier, D. (2018) 'Tracking an atmospheric river in a warmer climate: from water vapor to economic impacts', *Earth System Dynamics*, 9(1), pp. 249–266. https://doi.org/10.5194/esd-9-249-2018.

Donato, K. M., Enriquez, L. E. and Llewellyn, C. (2017) 'Frozen and stalled? Gender and migration scholarship in the 21st century', *American Behavioral Scientist*, 61(10), pp. 1079–1085. https://doi.org/10.1177/0002764217734259.

Dong, S., Wen, L., Liu, S., Zhang, X., Lassoie, J. P., Yi, S., Li, X., Li, J. and Li, Y. (2011) 'Vulnerability of worldwide pastoralism to global changes and interdisciplinary strategies for sustainable pastoralism', *Ecology and Society*, 16(2). www.jstor.org/stable/26268902 (Accessed: 23 April 2024).

Douben, K.-J. (2006) 'Characteristics of river floods and flooding: a global overview, 1985–2003', *Irrigation and Drainage*, 55(S1), pp. S9–S21. https://doi.org/10.1002/ird.239.

Drought Summary – Met Éireann – The Irish Meteorological Service (2021). www.met.ie/drought-summary (Accessed: 23 April 2024).

Du, S., Scussolini, P., Ward, P. J., Zhang, M., Wen, J., Wang, L., Koks, E., Diaz-Loaiza, A., Gao, J., Ke, Q. and Aerts, J. C. J. H. (2020) 'Hard or soft flood adaptation?

Advantages of a hybrid strategy for Shanghai', *Global Environmental Change*, 61, p. 102037. https://doi.org/10.1016/j.gloenvcha.2020.102037.

Duan, Y. and Wilmsen, B. (2012) 'Addressing the resettlement challenges at the Three Gorges Project', *International Journal of Environmental Studies*, 69(3), pp. 461–474. https://doi.org/10.1080/00207233.2012.676374.

Dugmore, A. J., McGovern, T. H., Vésteinsson, O., Arneborg, J., Streeter, R. and Keller, C. (2012) 'Cultural adaptation, compounding vulnerabilities and conjunctures in Norse Greenland', *Proceedings of the National Academy of Sciences*, 109(10), pp. 3658–3663. https://doi.org/10.1073/pnas.1115292109.

Duke, N. C., Meynecke, J.-O., Dittmann, S., Ellison, A. M., Anger, K., Berger, U., Cannicci, S., Diele, K., Ewel, K. C., Field, C. D., Koedam, N., Lee, S. Y., Marchand, C., Nordhaus, I. and Dahdouh-Guebas, F. (2007) 'A world without mangroves?', *Science*, 317(5834), pp. 41–42. https://doi.org/10.1126/science.317.5834.41b.

Dundon, L. A. and Abkowitz, M. (2021) 'Climate-induced managed retreat in the U.S.: a review of current research', *Climate Risk Management*, 33, p. 100337. https://doi.org/10.1016/j.crm.2021.100337.

Dupuy, J., Fargeon, H., Martin-StPaul, N., Pimont, F., Ruffault, J., Guijarro, M., Hernando, C., Madrigal, J. and Fernandes, P. (2020) 'Climate change impact on future wildfire danger and activity in southern Europe: a review', *Annals of Forest Science*, 77(2), p. 35. https://doi.org/10.1007/s13595-020-00933-5.

Durand, J. and Massey, D. S. (eds.) (2004) *Crossing the Border: Research from the Mexican Migration Project*. Russell Sage Foundation.

Durand, J. and Massey, D. S. (2019) 'Evolution of the Mexico-U.S. Migration System: insights from the Mexican Migration Project', *The ANNALS of the American Academy of Political and Social Science*, 684(1), pp. 21–42. https://doi.org/10.1177/0002716219857667.

Ebi, K. L. and McLeman, R. (2022) 'Climate related migration and displacement', *BMJ*, 379, p. o2389. https://doi.org/10.1136/bmj.o2389.

Echendu, A. J. (2022) 'Flooding in Nigeria and Ghana: opportunities for partnerships in disaster-risk reduction', *Sustainability: Science, Practice and Policy*, 18(1), pp. 1–15. https://doi.org/10.1080/15487733.2021.2004742.

Edwards, J. B. (2013) 'The logistics of climate-induced resettlement: lessons from the Carteret Islands, Papua New Guinea', *Refugee Survey Quarterly*, 32(3), pp. 52–78. https://doi.org/10.1093/rsq/hdt011.

Electromagnetic Spectrum Diagram | MyNASAData (2024). https://mynasadata.larc.nasa.gov/basic-page/electromagnetic-spectrum-diagram (Accessed: 27 May 2024).

El-Hinnawi, E. (1985) *Environmental Refugees*. Nairobi, Kenya: UNEP – United Nations Environment Programme. https://digitallibrary.un.org/record/121267 (Accessed: 23 April 2024).

Elliott, J. R. and Pais, J. (2006) 'Race, class, and Hurricane Katrina: social differences in human responses to disaster', *Social Science Research*, 35(2), pp. 295–321. https://doi.org/10.1016/j.ssresearch.2006.02.003.

Enarson, E., Fothergill, A. and Peek, L. (2018) 'Gender and Disaster: Foundations and New Directions for Research and Practice', in H. Rodríguez, W. Donner, and J.E. Trainor (eds.) *Handbook of Disaster Research*. Cham: Springer International Publishing, pp. 205–223. https://doi.org/10.1007/978-3-319-63254-4_11.

Engler, P., MacDonald, M., Piazza, R. and Sher, G. (2020) 'Migration to advanced economies can raise growth', *IMF Blog*, 19 June. https://meetings.imf.org/en/IMF/Home/Blogs/Articles/2020/06/19/blog-weo-chapter4-migration-to-advanced-economies-can-raise-growth.

Epstein, A., Treleaven, E., Ghimire, D.J. and Diamond-Smith, N. (2022) 'Drought and migration: an analysis of the effects of drought on temporary labor and return migration from a migrant-sending area in Nepal', *Population and Environment*, 44(3), pp. 145–167. https://doi.org/10.1007/s11111-022-00406-z.

Erdal, M. B. (2022) 'Migrant Transnationalism, Remittances and Development', in *Handbook on Transnationalism*. Edward Elgar Publishing, pp. 356–370. www.elgaronline.com/edcollchap/book/9781789904017/book-part-9781789904017-32.xml (Accessed: 27 May 2024).

Eswar, D., Karuppusamy, R. and Chellamuthu, S. (2021) 'Drivers of soil salinity and their correlation with climate change', *Current Opinion in Environmental Sustainability*, 50, pp. 310–318. https://doi.org/10.1016/j.cosust.2020.10.015.

Etana, D., Snelder, D. J. R. M., Wesenbeeck, C. F. A. van and Buning, T. D. C. (2022) 'Climate change, in-situ adaptation, and migration decisions of smallholder farmers in central Ethiopia', *Migration and Development*, 11(3), pp. 737–761. https://doi.org/10.1080/21632324.2020.1827538.

European Commission Joint Research Centre (2019) *World Atlas of Desertification: Global Urbanisation and Aridity*. https://wad.jrc.ec.europa.eu/globalurbanisation (Accessed: 23 April 2024).

European Environment Agency (EEA) (2024) *Global and European sea level rise*. www.eea.europa.eu/en/analysis/indicators/global-and-european-sea-level-rise (Accessed: 28 April 2024).

Eyer, J., Dinterman, R., Miller, N. and Rose, A. (2018) 'The effect of disasters on migration destinations: evidence from Hurricane Katrina', *Economics of Disasters and Climate Change*, 2(1), pp. 91–106. https://doi.org/10.1007/s41885-017-0020-3.

Ezra, M. and Kiros, G.-E. (2001) 'Rural out-migration in the drought prone areas of Ethiopia: a multilevel analysis1', *International Migration Review*, 35(3), pp. 749–771. https://doi.org/10.1111/j.1747-7379.2001.tb00039.x.

Facts + Statistics: Hurricanes | III (2022) *Insurance Information Institute*. www.iii.org/fact-statistic/facts-statistics-hurricanes (Accessed: 11 April 2024).

Fagan, B. (2004) *The Long Summer: How Climate Changed Civilization*. New York, NY: Basic Books.

Fagherazzi, S., Mariotti, G., Leonardi, N., Canestrelli, A., Nardin, W. and Kearney, W. S. (2020) 'Salt marsh dynamics in a period of accelerated sea level rise', *Journal of Geophysical Research: Earth Surface*, 125(8), p. e2019JF005200. https://doi.org/10.1029/2019JF005200.

Fahad, S., Hossain, M. S., Huong, N. T. L., Nassani, A. A., Haffar, M. and Naeem, M. R. (2023) 'An assessment of rural household vulnerability and resilience in natural hazards: evidence from flood prone areas', *Environment, Development and Sustainability*, 25(6), pp. 5561–5577. https://doi.org/10.1007/s10668-022-02280-z.

Fainu, K. (2021) '"My father will go down like the captain of the Titanic": life on the Pacific's disappearing islands', *The Guardian*, 15 October. www.theguardian.com/world/2021/oct/16/my-father-will-go-down-like-the-captain-of-the-titanic-life-on-the-pacifics-disappearing-islands (Accessed: 30 April 2024).

Faist, T. (2010) 'Towards transnational studies: world theories, transnationalisation and changing institutions', *Journal of Ethnic and Migration Studies*, 36(10), pp. 1665–1687. https://doi.org/10.1080/1369183X.2010.489365.

Falco, C., Donzelli, F. and Olper, A. (2018) 'Climate change, agriculture and migration: a survey', *Sustainability*, 10(5), p. 1405. https://doi.org/10.3390/su10051405.

Farman, J. C., Gardiner, B. G. and Shanklin, J. D. (1985) 'Large losses of total ozone in Antarctica reveal seasonal ClOx/NOx interaction', *Nature*, 315(6016), pp. 207–210. https://doi.org/10.1038/315207a0.

Farrell, J., Burow, P. B., McConnell, K., Bayham, J., Whyte, K. and Koss, G. (2021) 'Effects of land dispossession and forced migration on Indigenous peoples in North America', *Science*, 374(6567), p. eabe4943. https://doi.org/10.1126/science.abe4943.

Faust, E.C. (1951). 'The History of Malaria in the United States', *American Scientist*, 39(1), pp. 121–130. www.jstor.org/stable/27826354

Favell, A. (2019) 'Integration: twelve propositions after Schinkel', *Comparative Migration Studies*, 7(1), p. 21. https://doi.org/10.1186/s40878-019-0125-7.

FEMA (2018) Hurricanes Irma and Maria in Puerto Rico: Building Performance, Observations, Recommendations, and Technical Guidance (FEMA P-2020 / October *2018*) – Puerto Rico (The United States of America) | ReliefWeb (2018). https://reliefweb.int/report/puerto-rico-united-states-america/hurricanes-irma-and-maria-puerto-rico-building-performance-observations-recommendations-and-technical-guidance-fema-p-2020-october-2018 (Accessed: 21 April 2024).

Feng, S., Krueger, A. B. and Oppenheimer, M. (2010) 'Linkages among climate change, crop yields and Mexico–US cross-border migration', *Proceedings of the National Academy of Sciences*, 107(32), pp. 14257–14262. https://doi.org/10.1073/pnas.1002632107.

Fernandez Milan, B. and Creutzig, F. (2015) 'Reducing urban heat wave risk in the 21st century', *Current Opinion in Environmental Sustainability*, 14, pp. 221–231. https://doi.org/10.1016/j.cosust.2015.08.002.

Fernandez-Gimenez, M. E. and Le Febre, S. (2006) 'Mobility in pastoral systems: dynamic flux or downward trend?', *International Journal of Sustainable Development & World Ecology*, 13(5), pp. 341–362. https://doi.org/10.1080/13504500609469685.

Fernández-Sánchez, H., Salma, J., Márquez-Vargas, P. M. and Salami, B. (2020) 'Left-behind women in the context of international migration: a scoping review', *Journal of Transcultural Nursing*, 31(6), pp. 606–616. https://doi.org/10.1177/1043659620935962.

Fiedler, J. W. and Conrad, C. P. (2010) 'Spatial variability of sea level rise due to water impoundment behind dams', *Geophysical Research Letters*, 37(12). https://doi.org/10.1029/2010GL043462.

Filippova, O., Nguyen, C., Noy, I. and Rehm, M. (2020) 'Who cares? Future sea level rise and house prices', *Land Economics*, 96(2), pp. 207–224. https://doi.org/10.3368/le.96.2.207.

Findlay, A. M. (2011) 'Migrant destinations in an era of environmental change', *Global Environmental Change*, 21, pp. S50–S58. https://doi.org/10.1016/j.gloenvcha.2011.09.004.

Finch, C., Emrich, C. T., and Cutter, S. L. (2010). 'Disaster disparities and differential recovery in New Orleans'. *Population and Environment*, 31(4), pp. 179–202. https://doi.org/10.1007/s11111-009-0099-8.

Fishman, R. and Li, S. (2022) 'Agriculture, irrigation and drought induced international migration: evidence from Mexico', *Global Environmental Change*, 75, p. 102548. https://doi.org/10.1016/j.gloenvcha.2022.102548.

Food and Agriculture Organization of the United Nations (FAO) (2024) What are drylands? www.fao.org/dryland-forestry/background/what-are-drylands/en/ (Accessed: 7 November 2024).

Flörke, M., Schneider, C. and Mcdonald, R. (2018) 'Water competition between cities and agriculture driven by climate change and urban growth', *Nature Sustainability*, 1. https://doi.org/10.1038/s41893-017-0006-8.

Ford, J. D., King, N., Galappaththi, E. K., Pearce, T., McDowell, G. and Harper, S. L. (2020) 'The resilience of Indigenous peoples to environmental change', *One Earth*, 2(6), pp. 532–543. https://doi.org/10.1016/j.oneear.2020.05.014.

Foresight (2011) *Migration and Global Environmental Change: Future Challenges and Opportunities. Final Project Report.* London: The Government Office for Science, p. 234. https://assets.publishing.service.gov.uk/media/5a74b18840f0b61df4777b6c/11-11 16-migration-and-global-environmental-change.pdf.

Fowler, T., Southgate, R. J., Waite, T., Harrell, R., Kovats, S., Bone, A., Doyle, Y. and Murray, V. (2015) 'Excess Winter Deaths in Europe: a multi-country descriptive analysis', *European Journal of Public Health*, 25(2), pp. 339–345. https://doi.org/10.1093/eurpub/cku073.

Fox-Kemper, B., Hewitt, H. T., Xiao, C., Aðalgeirsdóttir, G., Drijfhout, S. S., Edwards, T. L., Golledge, N. R., Hemer, M., Kopp, R. E., Krinner, G., Mix, A., Notz, D., Nowicki, S., Nurhati, I. S., Ruiz, L., Sallée, J.-B., Slangen, A. B. A. and Yu, Y. (2021) 'Ocean, cryosphere and sea level change', in *IPCC, Climate Change 2021: The Physical Science Basis. Working Group I Contribution to the Sixth Assessment Report of the Intergovernmental Panel on Climate Change.* 1st ed. Edited by V. Masson-Delmotte, P. Zhai, A. Pirani, S. L. Connors, C. Péan, S. Berger, N. Caud, Y. Chen, L. Goldfarb, M. I. Gomis, M. Huang, K. Leitzell, E. Lonnoy, J. B. R. Matthews, T. K. Maycock, T. Waterfield, O. Yelekçi, R. Yu, and B. Zhou. Cambridge, United Kingdom and New York, NY, USA: Cambridge University Press, pp. 1211–1362. https://doi.org/10.1017/9781009157896.011.

Francis, A. (2019) '*Free movement agreements & climate-induced migration: a Caribbean case study [Case Study]*'. Sabin Center for Climate Change Law, Columbia Law School. http://columbiaclimatelaw.com/files/2019/09/FMAs-Climate-Induced-Migration-AFrancis.pdf.

Frank, T. and EE News (2024) 'Demolishing Homes That Sustain Hurricane Damage Can Improve Local Economy: Buying out and razing homes harmed by Hurricane Sandy Boosted Business Development, Jobs and Property Values in Nearby Neighborhoods', *Scientific American*, 5 January. www.scientificamerican.com/article/demolishing-homes-that-sustain-hurricane-damage-can-improve-local-economy/.

Frank-Vitale, A. (2023) 'Coyotes, caravans, and the Central American migrant smuggling continuum', *Trends in Organized Crime*, 26(1), pp. 64–79. https://doi.org/10.1007/s12117-022-09480-z.

Fransen, S., Werntges, A., Hunns, A., Sirenko, M. and Comes, T. (2024) 'Refugee settlements are highly exposed to extreme weather conditions', *Proceedings of the National Academy of Sciences*, 121(3), p. e2206189120. https://doi.org/10.1073/pnas.2206189120.

Frederikse, T., Landerer, F., Caron, L., Adhikari, S., Parkes, D., Humphrey, V. W., Dangendorf, S., Hogarth, P., Zanna, L., Cheng, L. and Wu, Y.-H. (2020) 'The causes of sea-level rise since 1900', *Nature*, 584(7821), pp. 393–397. https://doi.org/10.1038/s41586-020-2591-3.

Freier, L. F., Berganza, I. and Blouin, C. (2022) 'The Cartagena Refugee definition and Venezuelan displacement in Latin America', *International Migration*, 60(1), pp. 18–36. https://doi.org/10.1111/imig.12791.

Friend, R., Jarvie, J., Reed, S. O., Sutarto, R., Thinphanga, P. and Toan, V. C. (2014) 'Mainstreaming urban climate resilience into policy and planning; reflections from Asia', *Urban Climate*, 7, pp. 6–19. https://doi.org/10.1016/j.uclim.2013.08.001.

Friesen, T. M., Finkelstein, S. A. and Medeiros, A. S. (2020) 'Climate variability of the Common Era (AD 1–2000) in the Eastern North American Arctic: impacts on human migrations', *Quaternary International*, 549, pp. 142–154. https://doi.org/10.1016/j.quaint.2019.06.002.

Frowd, P. M., Apard, É. and Dele-Adedeji, I. (2023) 'Quasilegality and migrant smuggling in Northern Niger', *Trends in Organized Crime*, 26(4), pp. 379–396. https://doi.org/10.1007/s12117-023-09492-3.

Funkhouser, E. (2009) 'The choice of migration destination: a longitudinal approach using pre-migration outcomes', *Review of Development Economics*, 13(4), pp. 626–640. https://doi.org/10.1111/j.1467-9361.2009.00523.x.

Fussell, E. (2018) 'Population Displacements and Migration Patterns in Response to Hurricane Katrina', in R. McLeman and F. Gemenne (eds.) *Routledge Handbook of Environmental Displacement and Migration*. 1st ed. London: Routledge, pp. 277–288. www.taylorfrancis.com/chapters/edit/10.4324/9781315638843-22/population-displacements-migration-patterns-response-hurricane-katrina-elizabeth-fussell.

Fussell, E., Hunter, L. M. and Gray, C. L. (2014) 'Measuring the environmental dimensions of human migration: the demographer's toolkit', *Global environmental change : human and policy dimensions*, 28, pp. 182–191. https://doi.org/10.1016/j.gloenvcha.2014.07.001.

Fussell, E., Sastry, N. and VanLandingham, M. (2010) 'Race, socioeconomic status, and return migration to New Orleans after Hurricane Katrina', *Population and Environment*, 31(1), pp. 20–42. https://doi.org/10.1007/s11111-009-0092-2.

Gabriel, M. (2006) 'Youth migration and social advancement: how young people manage emerging differences between themselves and their hometown', *Journal of Youth Studies*, 9(1), pp. 33–46. https://doi.org/10.1080/13676260500523622.

García-Hernández, C. (2022) 'Disaster, demographics, and vulnerability: interrogating the long-term effects of an extreme weather event', *Geographical Research*, 60(4), pp. 549–562. https://doi.org/10.1111/1745-5871.12550.

García-López, G. A. (2018) 'The multiple layers of environmental injustice in contexts of (un)natural disasters: the case of Puerto Rico post-Hurricane Maria', *Environmental Justice*, 11(3), pp. 101–108. https://doi.org/10.1089/env.2017.0045.

Garland, A., Bukvic, A. and Maton-Mosurska, A. (2022) 'Capturing complexity: environmental change and relocation in the North Slope Borough, Alaska', *Climate Risk Management*, 38, p. 100460. https://doi.org/10.1016/j.crm.2022.100460.

Gaur, A., Bénichou, N., Armstrong, M. and Hill, F. (2021) 'Potential future changes in wildfire weather and behavior around 11 Canadian cities', *Urban Climate*, 35, p. 100735. https://doi.org/10.1016/j.uclim.2020.100735.

Gaurav, K., Sinha, R. and Panda, P.K. (2011) 'The Indus flood of 2010 in Pakistan: a perspective analysis using remote sensing data'. *Natural Hazards*, 59, pp. 1815–1826. https://doi.org/10.1007/s11069-011-9869-6

Gautam, Y. (2017) 'Seasonal migration and livelihood resilience in the face of climate change in Nepal', *Mountain Research and Development*, 37(4), pp. 436–445. https://doi.org/10.1659/MRD-JOURNAL-D-17-00035.1.

Gautier, D., Denis, D. and Locatelli, B. (2016) 'Impacts of drought and responses of rural populations in West Africa: a systematic review', *WIREs Climate Change*, 7(5), pp. 666–681. https://doi.org/10.1002/wcc.411.

Geddes, A., Adger, W. N., Arnell, N. W., Black, R. and Thomas, D. S. G. (2012) 'Migration, environmental change, and the "challenges of governance"', *Environment and Planning C: Government and Policy*, 30(6), pp. 951–967. https://doi.org/10.1068/c3006ed.

Geisen, T. (2012) 'Understanding Cultural Differences as Social Limits to Learning: Migration Theory, Culture and Young Migrants', in Z. Bekerman and T. Geisen (eds.) *International Handbook of Migration, Minorities and Education: Understanding Cultural and Social Differences in Processes of Learning*. Dordrecht: Springer Netherlands, pp. 19–33. https://doi.org/10.1007/978-94-007-1466-3_3.

Gemenne, F. (2011) 'Climate-induced population displacements in a 4°C+ world', *Philosophical Transactions of the Royal Society A: Mathematical, Physical and Engineering Sciences*, 369(1934), pp. 182–195. https://doi.org/10.1098/rsta.2010.0287.

Gemenne, F. (2018) 'Qualitative Research Techniques: It's a Case-Studies World', in *Routledge Handbook of Environmental Displacement and Migration*. 1st ed. Routledge, pp. 117–124.

Gender, Migration, Environment and Climate Change | Environmental Migration Portal (2024) *IOM: UN Migration*. https://environmentalmigration.iom.int/gender-migration-environment-and-climate-change (Accessed: 2 May 2024).

Gender quotas increase the equality and effectiveness of climate policy interventions – Nature Climate Change (2019) Nature. www.nature.com/articles/s41558-019-0438-4 (Accessed: 2 May 2024).

Gensini, V. (2021) 'Severe Convective Storms in a Changing *Climate', in*, pp. 39–56. https://doi.org/10.1016/B978-0-12-822700-8.00007-X.

Gensini, V. A. and Brooks, H. E. (2018) 'Spatial trends in United States tornado frequency', *npj Climate and Atmospheric Science*, 1(1), p. 38. https://doi.org/10.1038/s41612-018-0048-2.

Georges, E. (1992) 'Gender, class, and migration in the Dominican Republic: women's experiences in a transnational community', *Annals of the New York Academy of Sciences*, 645, pp. 81–99. https://doi.org/10.1111/j.1749-6632.1992.tb33487.x.

Giannelli, G. C. and Canessa, E. (2022) 'After the flood: Migration and remittances as coping strategies of rural Bangladeshi household', *Economic Development and Cultural Change*, 70(3), pp. 1159–1195. https://doi.org/10.1086/713939.

Giczy, H. (2009) 'The Bum Blockade: Los Angeles and the great depression', *Voces Novae*, 1(1). https://digitalcommons.chapman.edu/vocesnovae/vol1/iss1/6.

Gidey, E., Mhangara, P., Gebregergs, T., Zeweld, W., Gebretsadik, H., Dikinya, O., Mussa, S., Zenebe, A., Girma, A., Fisseha, G., Addisu, A., Nasir, J., Zeleke, T. and Birhane, E. (2023) 'Analysis of drought coping strategies in northern Ethiopian highlands'. *SN Applied Sciences*, 5, 195. https://doi.org/10.1007/s42452-023-05409-5.

Gilbert, G. and McLeman, R. (2010) 'Household access to capital and its effects on drought adaptation and migration: a case study of rural Alberta in the 1930s', *Population and Environment*, 32(1), pp. 3–26. https://doi.org/10.1007/s11111-010-0112-2.

Gilmore, E. A., Wrathall, D., Adams, H., Buhaug, H., Castellanos, E., Hilmi, N., McLeman, R., Singh, C. and Adelekan, I. (2024) 'Defining severe risks related to mobility from climate change', *Climate Risk Management*, 44, p. 100601. https://doi.org/10.1016/j.crm.2024.100601.

Ginsburg, C., Bocquier, P., Béguy, D., Afolabi, S., Derra, K., Augusto, O., Otiende, M., Odhiambo, F., Zabré, P., Soura, A., White, M. J. and Collinson, M. A. (2016) 'Human capital on the move: education as a determinant of internal migration in selected INDEPTH surveillance populations in Africa', *Demographic research*, 34, pp. 845–884. https://doi.org/10.4054/DemRes.2016.34.30.

Gleick, P. H. (2014) 'Water, drought, climate change, and conflict in Syria', *Weather, Climate, and Society*, 6(3), pp. 331–340. https://doi.org/10.1175/WCAS-D-13-00059.1.

Glick, J. L., Lopez, A., Pollock, M. and Theall, K. P. (2019) '"Housing insecurity seems to almost go hand in hand with being trans": housing stress among transgender and gender non-conforming individuals in New Orleans', *Journal of Urban Health*, 96(5), pp. 751–759. https://doi.org/10.1007/s11524-019-00384-y.

Goelzer, H., Nowicki, S., Payne, A., Larour, E., Seroussi, H., Lipscomb, W. H., Gregory, J., Abe-Ouchi, A., Shepherd, A., Simon, E., Agosta, C., Alexander, P., Aschwanden, A., Barthel, A., Calov, R., Chambers, C., Choi, Y., Cuzzone, J., Dumas, C., Edwards, T., Felikson, D., Fettweis, X., Golledge, N.R., Greve, R., Humbert, A., Huybrechts, P., Le clec'h, S., Lee, V., Leguy, G., Little, C., Lowry, D.P., Morlighem, M., Nias,

I., Quiquet, A., Rückamp, M., Schlegel, N.-J., Slater, D. A., Smith, R. S., Straneo, F., Tarasov, L., van de Wal, R. and van den Broeke, M. (2020) 'The future sea-level contribution of the Greenland ice sheet: a multi-model ensemble study of ISMIP6', *The Cryosphere*, 14(9), pp. 3071–3096. https://doi.org/10.5194/tc-14-3071-2020.

Goldberg, L., Lagomasino, D., Thomas, N. and Fatoyinbo, T. (2020) 'Global declines in human-driven mangrove loss', *Global Change Biology*, 26(10), pp. 5844–5855. https://doi.org/10.1111/gcb.15275.

Goldman, A., Eggen, B., Golding, B. and Murray, V. (2014) 'The health impacts of windstorms: a systematic literature review', *Public Health*, 128(1), pp. 3–28. https://doi.org/10.1016/j.puhe.2013.09.022.

Goldsmith, L., Raditz, V. and Méndez, M. (2022) 'Queer and present danger: understanding the disparate impacts of disasters on LGBTQ+ communities', *Disasters*, 46(4), pp. 946–973. https://doi.org/10.1111/disa.12509.

Gonzalez, N. L. S. de (1961) 'Family organization in five types of migratory wage labor', *American Anthropologist*, 63(6), pp. 1264–1280.

Goss, M., Swain, D. L., Abatzoglou, J. T., Sarhadi, A., Kolden, C. A., Williams, A. P. and Diffenbaugh, N. S. (2020) 'Climate change is increasing the likelihood of extreme autumn wildfire conditions across California', *Environmental Research Letters*, 15(9), p. 094016. https://doi.org/10.1088/1748-9326/ab83a7.

Government of Canada (2023) 'The Government of Canada invests to protect Tuktoyaktuk from coastal erosion [News release]', *Canada.ca*, 27 July. www.canada.ca/en/office-infrastructure/news/2023/07/the-government-of-canada-invests-to-protect-tuktoyaktuk-from-coastal-erosion.html.

Government of Canada (2024) *Claim refugee status from inside Canada: Work and study permits.* www.canada.ca/en/immigration-refugees-citizenship/services/refugees/asylum/claim-protection-inside-canada/work-study.html (Accessed: 26 May 2024).

Gramlich, J. (2024) 'Migrant encounters at the U.S.-Mexico border hit a record high at the end of 2023', *Pew Research Center*, 15 February. www.pewresearch.org/short-reads/2024/02/15/migrant-encounters-at-the-us-mexico-border-hit-a-record-high-at-the-end-of-2023/ (Accessed: 28 May 2024).

Grant, M. J. (2002) *Down and Out on the Family Farm: Rural Rehabilitation in the Great Plains, 1929–1945.* Lincoln: University of Nebraska Press (Our Sustainable Future).

Graumann, A., Houston, T. G., Lawrimore, J. H. (Jay H.), Levinson, D. H. (David H.), Lott, N., McCown, S., Stephens, S. and Wuertz, D. B. (2006) 'Hurricane Katrina: a climatological perspective (preliminary report)'. Edited by National Climatic Data Center (U.S.). https://repository.library.noaa.gov/view/noaa/13833.

Graveline, M.-H. and Germain, D. (2022) 'Disaster risk resilience: conceptual evolution, key issues, and opportunities', *International Journal of Disaster Risk Science*, 13(3), pp. 330–341. https://doi.org/10.1007/s13753-022-00419-0.

Gray, C. and Call, M. (2023) 'Heat and drought reduce subnational population growth in the global tropics', *Population and Environment*, 45(2), p. 6. https://doi.org/10.1007/s11111-023-00420-9.

Gray, C. and Mueller, V. (2012) 'Drought and population mobility in rural Ethiopia', *World Development*, 40(1), pp. 134–145. https://doi.org/10.1016/j.worlddev.2011.05.023.

Gray, C. and Wise, E. (2016) 'Country-specific effects of climate variability on human migration', *Climatic Change*, 135(3), pp. 555–568. https://doi.org/10.1007/s10584-015-1592-y.

Gray, C. L. and Mueller, V. (2012) 'Natural disasters and population mobility in Bangladesh', *Proceedings of the National Academy of Sciences*, 109(16), pp. 6000–6005. https://doi.org/10.1073/pnas.1115944109.

Gray, I. (2021) 'Hazardous simulations: pricing climate risk in US coastal insurance markets', *Economy and Society*, 50(2), pp. 196–223. https://doi.org/10.1080/03085147.2020.1853358.

Gray, J. H. (2004) *The Winter Years: The Depression on the Prairies*. Calgary, AB: Fifth House Ltd.

Green, R., Bates, L. K. and Smyth, A. (2007) 'Impediments to recovery in New Orleans' Upper and Lower Ninth Ward: one year after Hurricane Katrina', *Disasters*, 31(4), pp. 311–335. https://doi.org/10.1111/j.1467-7717.2007.01011.x.

Greenfield, N. M. (2023) 'International students add US$40 billion to economy', *University World News*, 13 November. www.universityworldnews.com/post.php?story=20231113124228557.

Greer, A., Binder, S. B., Thiel, A., Jamali, M. and Nejat, A. (2020) 'Place attachment in disaster studies: measurement and the case of the 2013 Moore tornado', *Population and Environment*, 41(3), pp. 306–329. https://doi.org/10.1007/s11111-019-00332-7.

Gregory, J. N. (1989). *American Exodus: The Dust Bowl Migration and Okie Culture in California*. New York: Oxford University Press.

GRID (2015) *The Ganges river basin*. www.grida.no/resources/6685.

Grimm, V., Revilla, E., Berger, U., Jeltsch, F., Mooij, W. M., Railsback, S. F., Thulke, H.-H., Weiner, J., Wiegand, T. and DeAngelis, D. L. (2005) 'Pattern-oriented modeling of agent-based complex systems: lessons from ecology', *Science*, 310(5750), pp. 987–991. https://doi.org/10.1126/science.1116681.

Groen, J. A. and Polivka, A. E. (2010) 'Going home after Hurricane Katrina: determinants of return migration and changes in affected areas', *Demography*, 47(4), pp. 821–844. https://doi.org/10.1007/BF03214587.

Groth, J., Ide, T., Sakdapolrak, P., Kassa, E. and Hermans, K. (2020) 'Deciphering interwoven drivers of environment-related migration – A multisite case study from the Ethiopian highlands', *Global Environmental Change*, 63, p. 102094. https://doi.org/10.1016/j.gloenvcha.2020.102094.

Grove, J. M. (1988) *The Little Ice Age*. 1st ed. London: Routledge. https://doi.org/10.4324/9780203402863.

Grove, J. M. (2002) 'Climatic Change in Northern Europe Over the Last Two Thousand Years and its Possible Influence on Human Activity', in G. Wefer, W. H. Berger, K.-E. Behre, and E. Jansen (eds.) *Climate Development and History of the North Atlantic Realm*. Berlin, Heidelberg: Springer, pp. 313–326. https://doi.org/10.1007/978-3-662-04965-5_19.

Grzegorzewski, A. S., Cialone, M. A. and Wamsley, T. V. (2011) 'Interaction of barrier islands and storms: implications for flood risk reduction in Louisiana and Mississippi', *Journal of Coastal Research*, 2011(10059), pp. 156–164. https://doi.org/10.2112/SI59-016.1.

Grzymala-Kazlowska, A. and Phillimore, J. (2018) 'Introduction: rethinking integration. New perspectives on adaptation and settlement in the era of super-diversity', *Journal of Ethnic and Migration Studies*, 44(2), pp. 179–196. https://doi.org/10.1080/1369183X.2017.1341706.

Guérou, A., Meyssignac, B., Prandi, P., Ablain, M., Ribes, A. and Bignalet-Cazalet, F. (2023) 'Current observed global mean sea level rise and acceleration estimated from satellite altimetry and the associated measurement uncertainty', *Ocean Science*, 19(2), pp. 431–451. https://doi.org/10.5194/os-19-431-2023.

Guerry, A. D., Silver, J., Beagle, J., Wyatt, K., Arkema, K., Lowe, J., Hamel, P., Griffin, R., Wolny, S., Plane, E., Griswold, M., Papendick, H. and Sharma, J. (2022) 'Protection and restoration of coastal habitats yield multiple benefits for urban residents as sea levels rise', *npj Urban Sustainability*, 2(1), pp. 1–12. https://doi.org/10.1038/s42949-022-00056-y.

Gunarathna, U., Bandara, C. S., Dissanayake, R. and Munasinghe, H. (2023) 'Tsunami-resilient building guidelines for Sri Lankan coastal belt: a critical review and consolidation based on significant institutional perceptions', *International Journal of Disaster Resilience in the Built Environment*, 14(4), pp. 453–470. https://doi.org/10.1108/IJDRBE-06-2022-0058.

Gutmann, M. P. and Field, V. (2010) 'Katrina in historical context: environment and migration in the U.S.', *Population and Environment*, 31(1), pp. 3–19. https://doi.org/10.1007/s11111-009-0088-y.

Haasnoot, M., Lawrence, J. and Magnan, A. K. (2021) 'Pathways to coastal retreat', *Science*, 372(6548), pp. 1287–1290. https://doi.org/10.1126/science.abi6594.

Haberlie, A. M., Ashley, W. S., Battisto, C. M. and Gensini, V. A. (2022) 'Thunderstorm activity under intermediate and extreme climate change scenarios', *Geophysical Research Letters*, 49(14), p. e2022GL098779. https://doi.org/10.1029/2022GL098779.

Hagger, V., Worthington, T. A., Lovelock, C. E., Adame, M. F., Amano, T., Brown, B. M., Friess, D. A., Landis, E., Mumby, P. J., Morrison, T. H., O'Brien, K. R., Wilson, K. A., Zganjar, C. and Saunders, M. I. (2022) 'Drivers of global mangrove loss and gain in social-ecological systems', *Nature Communications*, 13(1), p. 6373. https://doi.org/10.1038/s41467-022-33962-x.

Hague, B. S., McGregor, S., Jones, D. A., Reef, R., Jakob, D. and Murphy, B. F. (2023) 'The global drivers of chronic coastal flood hazards under sea-level rise', *Earth's Future*, 11(8), p. e2023EF003784. https://doi.org/10.1029/2023EF003784.

Haile, M. G., Wossen, T., Tesfaye, K. and von Braun, J. (2017) 'Impact of climate change, weather extremes, and price risk on global food supply', *Economics of Disasters and Climate Change*, 1(1), pp. 55–75. https://doi.org/10.1007/s41885-017-0005-2.

Hall, N. (2015) 'Money or mandate? Why international organizations engage with the climate change regime', *Global Environmental Politics*, 15(2), pp. 79–97. https://doi.org/10.1162/GLEP_a_00299.

Hall, S. (2023) Youth, Migration and Development: A New Lens for Critical Times: Case Studies from Colombia, Jordan, Kenya, Mexico, Pakistan, Senegal, Thailand, and Tunisia. KNOMAD Paper 53. Global Knowledge Partnership on Migration and Development (KNOMAD), p. 66. www.knomad.org/sites/default/files/publication-doc/knomad_paper_53_youth_migration_and_development_samuel_hall_nov_2023_0.pdf.

Hamideh, S. and Sen, P. (2022) 'Experiences of vulnerable households in low-attention disasters: Marshalltown, Iowa (United States) after the EF3 Tornado', *Global Environmental Change*, 77, p. 102595. https://doi.org/10.1016/j.gloenvcha.2022.102595.

Harding, B. and Pohle, C. (2022) 'The next five years are crucial for Bougainville's independence bid: Bougainville wants full sovereignty, while Papua New Guinea is unlikely to let it secede. Is compromise possible?', *United States Institute of Peace*, 12 August. www.usip.org/publications/2022/08/next-five-years-are-crucial-bougainvilles-independence-bid.

Hartmann, H. and Andresky, L. (2013) 'Flooding in the Indus River basin – A spatio-temporal analysis of precipitation records', *Global and Planetary Change*, 107, pp. 25–35. https://doi.org/10.1016/j.gloplacha.2013.04.002.

Hassani-Mahmooei, B. and Parris, B. W. (2012) 'Climate change and internal migration patterns in Bangladesh: an agent-based model', *Environment and Development Economics*, 17(6), pp. 763–780. https://doi.org/10.1017/S1355770X12000290.

Hauer, M. E., Fussell, E., Mueller, V., Burkett, M., Call, M., Abel, K., McLeman, R. and Wrathall, D. (2020) 'Sea-level rise and human migration', *Nature Reviews Earth & Environment*, 1(1), pp. 28–39. https://doi.org/10.1038/s43017-019-0002-9.

Hauer, M. E., Hardy, D., Kulp, S. A., Mueller, V., Wrathall, D. J. and Clark, P. U. (2021) 'Assessing population exposure to coastal flooding due to sea level rise', *Nature Communications*, 12(1), p. 6900. https://doi.org/10.1038/s41467-021-27260-1.

Haug, G. H., Günther, D., Peterson, L. C., Sigman, D.M., Hughen, K. A. and Aeschlimann, B. (2003) 'Climate and the collapse of Maya civilization', *Science*, 299(5613), pp. 1731–1735. https://doi.org/10.1126/science.1080444.

Heaney, A. K. and Winter, S. J. (2016) 'Climate-driven migration: an exploratory case study of Maasai health perceptions and help-seeking behaviors', *International Journal of Public Health*, 61(6), pp. 641–649. https://doi.org/10.1007/s00038-015-0759-7.

Heidari, H., Arabi, M. and Warziniack, T. (2021) 'Effects of climate change on natural-caused fire activity in Western U.S. national forests', *Atmosphere*, 12(8), p. 981. https://doi.org/10.3390/atmos12080981.

Hemmati, M., Ellingwood, B. R. and Mahmoud, H. N. (2020) 'The role of urban growth in resilience of communities under flood risk', *Earth's Future*, 8(3), p. e2019EF001382. https://doi.org/10.1029/2019EF001382.

Hermans, K. and Garbe, L. (2019) 'Droughts, livelihoods, and human migration in northern Ethiopia', *Regional Environmental Change*, 19(4), pp. 1101–1111. https://doi.org/10.1007/s10113-019-01473-z.

Hermans, K. and McLeman, R. (2021) 'Climate change, drought, land degradation and migration: exploring the linkages', *Current Opinion in Environmental Sustainability*, 50, pp. 236–244. https://doi.org/10.1016/j.cosust.2021.04.013.

Hermans, T. H. J., Gregory, J. M., Palmer, M. D., Ringer, M. A., Katsman, C. A. and Slangen, A. B. A. (2021) 'Projecting global mean sea-level change using CMIP6 models', *Geophysical Research Letters*, 48(5), p. e2020GL092064. https://doi.org/10.1029/2020GL092064.

Hermans-Neumann, K., Priess, J. and Herold, M. (2017) 'Human migration, climate variability, and land degradation: hotspots of socio-ecological pressure in Ethiopia', *Regional Environmental Change*, 17(5), pp. 1479–1492. https://doi.org/10.1007/s10113-017-1108-6.

Hernández, B., Hidalgo, M. C. and Ruiz, C. (2020) 'Theoretical and Methodological Aspects of Research on Place Attachment', in *Place Attachment*. 2nd ed. Routledge, pp. 94–110.

Herrero, M., Addison, J., Bedelian, C., Carabine, E., Havlík, P., Henderson, B., Van De Steeg, J. and Thornton, P.K. (2016) 'Climate change and pastoralism: impacts, consequences and adaptation', *Revue Scientifique Et Technique (International Office of Epizootics)*, 35(2), pp. 417–433. https://doi.org/10.20506/rst.35.2.2533.

Hewitt, K. (2013) 'Environmental disasters in social context: toward a preventive and precautionary approach', *Natural Hazards*, 66(1), pp. 3–14. https://doi.org/10.1007/s11069-012-0205-6.

Hewitt, K. (ed.) (2019) 'The Idea of Calamity in a Technocratic Age', in *Interpretations of Calamity: From the Viewpoint of Human Ecology*. London: Routledge, pp. 3–32. https://doi.org/10.4324/9780429329579.

Hill, R., Walsh, F. J., Davies, J., Sparrow, A., Mooney, M., Wise, R. M. and Tengö, M. (2020) 'Knowledge co-production for Indigenous adaptation pathways: transform post-colonial articulation complexes to empower local decision-making', *Global Environmental Change*, 65, p. 102161. https://doi.org/10.1016/j.gloenvcha.2020.102161.

Hillel, D. (1991) *Out of the Earth: Civilization and the Life of the Soil*. Los Angeles: UCLA Press.

Hinkel, J., Lincke, D., Vafeidis, A. T., Perrette, M., Nicholls, R. J., Tol, R. S. J., Marzeion, B., Fettweis, X., Ionescu, C. and Levermann, A. (2014) 'Coastal flood damage and adaptation costs under 21st century sea-level rise', *Proceedings of the National Academy of Sciences*, 111(9), pp. 3292–3297. https://doi.org/10.1073/pnas.1222469111.

Hino, M., Field, C. B. and Mach, K. J. (2017) 'Managed retreat as a response to natural hazard risk', *Nature Climate Change*, 7(5), pp. 364–370. https://doi.org/10.1038/nclimate3252.

Hobbie, S. E. and Grimm, N. B. (2020) 'Nature-based approaches to managing climate change impacts in cities', *Philosophical Transactions of the Royal Society B: Biological Sciences*, 375(1794), p. 20190124. https://doi.org/10.1098/rstb.2019.0124.

Hobbins, M., Jansma, T., Sarmiento, D. P., McNally, A., Magadzire, T., Jayanthi, H., Turner, W., Hoell, A., Husak, G., Senay, G., Boiko, O., Budde, M., Mogane, P. and Dewes, C. F. (2023) 'A global long-term daily reanalysis of reference evapotranspiration for drought and food-security monitoring', *Scientific Data*, 10(1), p. 746. https://doi.org/10.1038/s41597-023-02648-4.

Hoegh-Guldberg, O., Jacob, D., Taylor, M., Guillén Bolaños, T., Bindi, M., Brown, S., Camilloni, I. A., Diedhiou, A., Djalante, R., Ebi, K., Engelbrecht, F., Guiot, J., Hijioka, Y., Mehrotra, S., Hope, C. W., Payne, A.J., Pörtner, H.-O., Seneviratne, S. I., Thomas, A., Warren, R. and Zhou, G. (2019) 'The human imperative of stabilizing global climate change at 1.5°C', *Science*, 365(6459), p. eaaw6974. https://doi.org/10.1126/science.aaw6974.

Hoffmann, R., Dimitrova, A., Muttarak, R., Crespo Cuaresma, J. and Peisker, J. (2020) 'A meta-analysis of country-level studies on environmental change and migration', *Nature Climate Change*, 10(10), pp. 904–912. https://doi.org/10.1038/s41558-020-0898-6.

Hoffmann, R., Šedová, B. and Vinke, K. (2021). 'Improving the evidence base: a methodological review of the quantitative climate migration literature'. *Global Environmental Change*, 71, p. 102367. https://doi.org/10.1016/j.gloenvcha.2021.102367.

Hoffmann, R., Wiederkehr, C., Dimitrova, A. and Hermans, K. (2022) 'Agricultural livelihoods, adaptation, and environmental migration in sub-Saharan drylands: a meta-analytical review', *Environmental Research Letters*, 17(8), p. 083003. https://doi.org/10.1088/1748-9326/ac7d65.

Hoffmann, R., Abel, G., Malpede, M., Muttarak, R. and Percoco, M. (2023) 'Climate change, aridity, and internal migration: evidence from census microdata for 72 countries', *IIASA* [Preprint]. https://iiasa.dev.local/ (Accessed: 23 April 2024).

Holdaway, A., Ford, M. and Owen, S. (2021) 'Global-scale changes in the area of atoll islands during the 21st century', *Anthropocene*, 33, p. 100282. https://doi.org/10.1016/j.ancene.2021.100282.

Hondagneu-Sotelo, P. and Cranford, C. (2006) 'Gender and Migration', in *Handbook of the Sociology of Gender*. Springer US (Handbooks of *Sociology and Social Research*), pp. 105–126. https://doi.org/10.1007/0-387-36218-5_6.

Horton, B. P., Khan, N. S., Cahill, N., Lee, J. S. H., Shaw, T. A., Garner, A. J., Kemp, A. C., Engelhart, S. E. and Rahmstorf, S. (2020) 'Estimating global mean sea-level rise and its uncertainties by 2100 and 2300 from an expert survey', *npj Climate and Atmospheric Science*, 3(1), pp. 1–8. https://doi.org/10.1038/s41612-020-0121-5.

Horton, R. M., de Sherbinin, A., Wrathall, D. and Oppenheimer, M. (2021) 'Assessing human habitability and migration', *Science*, 372(6548), pp. 1279–1283. https://doi.org/10.1126/science.abi8603.

Hsiang, S. M., Burke, M. and Miguel, E. (2013) 'Quantifying the influence of climate on human conflict', *Science*, 341(6151), p. 1235367. https://doi.org/10.1126/science.1235367.

Huber, J., Madurga-Lopez, I., Murray, U., McKeown, P. C., Pacillo, G., Laderach, P. and Spillane, C. (2023) 'Climate-related migration and the climate-security-migration nexus in the Central American Dry Corridor', *Climatic Change*, 176(6), p. 79. https://doi.org/10.1007/s10584-023-03549-6.

Hunt, J. D. and Byers, E. (2019) 'Reducing sea level rise with submerged barriers and dams in Greenland', *Mitigation and Adaptation Strategies for Global Change*, 24(5), pp. 779–794. https://doi.org/10.1007/s11027-018-9831-y.

Hunter, L. and David, E. (2009) 'Climate change and migration: considering the gender dimensions'.

Hunter, L. M., Murray, S. and Riosmena, F. (2013) 'Rainfall patterns and U.S. Migration from rural Mexico', *International Migration Review*, 47(4), pp. 874–909. https://doi.org/10.1111/imre.12051.

Hunter, L. M. and Simon, D. H. (2017) 'Might climate change the "healthy migrant" effect?', *Global Environmental Change*, 47, pp. 133–142. https://doi.org/10.1016/j.gloenvcha.2017.10.003.

Huntley, I. D. (1957) 'The Thames in winter', *Weather*, 12(12), pp. 373–376. https://doi.org/10.1002/j.1477-8696.1957.tb00415.x.

Hurricanes Irma and Maria in Puerto Rico: Building Performance, Observations, Recommendations, and Technical Guidance (FEMA P-2020 / October 2018) – Puerto Rico (The United States of America) | ReliefWeb (2018). https://reliefweb.int/report/puerto-rico-united-states-america/hurricanes-irma-and-maria-puerto-rico-building-performance-observations-recommendations-and-technical-guidance-fema-p-2020-october-2018 (Accessed: 21 April 2024).

ICCCAD and Khan, M. R. (2024) 'Bangladesh's leadership of the Climate Vulnerable Forum: How has the world benefitted?', *International Centre for Climate Change and Development (ICCCAD), The Daily Star*, 18 February. www.icccad.net/daily-star-articles/bangladeshs-leadership-of-the-climate-vulnerable-forumhow-has-the-world-benefitted/.

İçduygu, A. (2021) 'Decentring migrant smuggling: reflections on the Eastern Mediterranean route to Europe', *Journal of Ethnic and Migration Studies*, 47(14), pp. 3293–3309. https://doi.org/10.1080/1369183X.2020.1804194.

Ide, T. (2018) 'Climate war in the Middle East? Drought, the Syrian civil war and the state of climate-conflict research', *Current Climate Change Reports*, 4(4), pp. 347–354. https://doi.org/10.1007/s40641-018-0115-0.

IDMC (2017) *Positioned for Action: Displacement in the Sendai Framework for Disaster Risk Reduction. Briefing Paper*. Geneva, Switzerland: Internal Displacement Monitoring Centre (IDMC), p. 10. https://api.internal-displacement.org/sites/default/files/publications/documents/20170216-idmc-briefing-paper-drr.pdf.

IDMC (2022) *Country Profile: Bangladesh, IDMC – Internal Displacement Monitoring Centre*. www.internal-displacement.orgundefined (Accessed: 2 May 2024).

IDMC (2024) *Global Report on Internal Displacement 2024*. Geneva: Internal Displacement Monitoring Centre (IDMC), p. 69. www.internal-displacement.org/global-report/grid2024/.

INED (2023) *L'Institut national d'études démographiques*, www.ined.fr/en/everything_about_population/data/all-countries/?lst_continent=900&lst_pays= (Accessed: 11 July 2024).

Insurance Information Institute (2022) *Facts + Statistics: Hurricanes | III. Insurance Information Institute*. www.iii.org/fact-statistic/facts-statistics-hurricanes (Accessed: 11 April 2024).

Intergovernmental Authority on Development (IGAD) (2020) *Protocol on Free Movement of Persons in the Igad Region*. Khartoum, Sudan: Office of the Executive Secretary,

IGAD Secretariat, Republic of Djibouti, p. 19. https://environmentalmigration.iom .int/sites/g/files/tmzbdl1411/files/event/file/Final%20IGAD%20PROTOCOL%20 ENDORSED%20BY%20IGAD%20Ambassadors%20and%20Ministers%20of%20 Interior%20and%20Labour%20Khartoum%2026%20Feb%202020.pdf.

International Labour Organization (ILO) (1996) 'Lobalization changes the face of textile, clothing and footwear industries', *ILO News*. 33rd ed., 28 October. www.ilo.org/ resource/news/globalization-changes-face-textile-clothing-and-footwear-industries.

International Labour Organization (ILO) (2024) *Labour Migration | International Labour Organization*. www.ilo.org/regions-and-countries/arab-states/ilo-arab-states/areas-work/labour-migration (Accessed: 27 May 2024).

International Labour Organization (IMO) (2022) International Organization for Migration (IOM), and Walk Free (September 2022) *Global Estimates of Modern Slavery: Forced Labour and Forced Marriage – Executive summary*, p. 14. www.ilo.org/ sites/default/files/wcmsp5/groups/public/@ed_norm/@ipec/documents/publication/ wcms_854795.pdf.

International Organization for Migration (IOM) (2024) 'Migration and Human Mobility: Key Global Figures'. www.migrationdataportal.org/sites/g/files/tmzbdl251/ files/2024-03/EN%20March%20update%20%281%29.pdf.

IOM (2023) *Snapshot: Remittance Inflows to Bangladesh (2019–2022)*. UN Migration, p. 6. https://bangladesh.iom.int/sites/g/files/tmzbdl1006/files/documents/2024-02/ remittance-snapshot-2022.pdf.

IOM (2024) *IOM Bangladesh: Rohingya Humanitarian Crisis Response – Monthly Situation Report*. Bangladesh: UN Migration, p. 2. https://bangladesh.iom.int/sites/g/files/ tmzbdl1006/files/documents/2024-03/iom-rohingya-crisis-response_external-sitrep_ january-2024.pdf.

IOM (n.d.) *Who We Are, International Organization for Migration*. www.iom.int/who-we-are (Accessed: 23 May 2024).

IPCC (2018) *Global Warming of 1.5°c. An IPCC Special Report on the Impacts of Global Warming of 1.5°c Above Pre-Industrial Levels and Related Global Greenhouse Gas Emission Pathways, in the Context of Strengthening the Global Response to the Threat of Climate Change, Sustainable Development, and Efforts to Eradicate Poverty*. 1st ed. Edited by V. Masson-Delmotte, P. Zhai, H.-O. Pörtner, D. Roberts, J. Skea, P. R. Shukla, A. Pirani, W. Moufouma-Okia, C. Péan, R. Pidcock, S. Connors, J. B. R. Matthews, Y. Chen, X. Zhou, M. I. Gomis, E. Lonnoy, T. Maycock, M. Tignor, and T. Waterfield. Cambridge, UK and New York, NY, USA: Cambridge University Press. https://doi.org/10.1017/9781009157940.

IPCC (2019) IPCC Special Report on the Ocean and Cryosphere in a Changing Climate [Pörtner, H.-O., Roberts, D.C., Masson-Delmotte, V., Zhai, P., Tignor, M., Poloczanska, E., Mintenbeck, K., Alegría, A., Nicolai, M., Okem, A., Petzold, J., Rama, B., Weyer, N.M. (eds.)]. Cambridge University Press, Cambridge, UK and New York, NY, USA, 755 pp. https://doi.org/10.1017/9781009157964.

IPCC (2021) *Climate Change 2021: The Physical Science Basis. Contribution of Working Group I to the Sixth Assessment Report of the Intergovernmental Panel on Climate Change*. [Masson-Delmotte, V., Zhai, P., Pirani, A., Connors, S.L., Péan, C., Berger, S., Caud, N., Chen, Y., Goldfarb, L., Gomis, M.I., Huang, M., Leitzell, K., Lonnoy, E., Matthews, J.B.R., Maycock, T.K., Waterfield, T., Yelekçi, O., Yu, R. and Zhou, B. (eds.)] Cambridge University Press, Cambridge, UK and New York, NY, In press. doi:10.1017/9781009157896.

IPCC (2021) Technical Summary, in *'Climate Change 2021: The Physical Science Basis. Contribution of Working Group I to the Sixth Assessment Report of the Intergovernmental Panel on Climate Change'*. (Chen, D., Rojas, M., Samset, B.H.,

Cobb, K., Diongue Niang, A., Edwards, P., Emori, S., Faria, S.H., Hawkins, E., Hope, P., Huybrechts, P., Meinshausen, M., Mustafa, S.K., Plattner, G.-K., and Tréguier, A.-M. (2021) Framing, Context, and Methods. InClimate Change 2021: The Physical Science Basis. Contribution of Working Group I to the Sixth Assessment Report of the Intergovernmental Panel on Climate Change [MassonDelmotte, V., P. Zhai, A. Pirani, S.L. Connors, C. Péan, S. Berger, N. Caud, Y. Chen, L. Goldfarb, M.I. Gomis, M. Huang, K. Leitzell, E. Lonnoy, J.B.R. Matthews, T.K. Maycock, T. Waterfield, O. Yelekçi, R. Yu, and B. Zhou (eds.)]. Cambridge University Press, Cambridge, United Kingdom and New York, NY, pp. 147–286. doi:10.1017/9781009157896.003.)

IPCC (2024) 'About – IPCC'. www.ipcc.ch/about/ (Accessed: 27 May 2024).

Irrgang, A. M., Bendixen, M., Farquharson, L. M., Baranskaya, A. V., Erikson, L. H., Gibbs, A. E., Ogorodov, S. A., Overduin, P. P., Lantuit, H., Grigoriev, M. N. and Jones, B. M. (2022) 'Drivers, dynamics and impacts of changing Arctic coasts', *Nature Reviews Earth & Environment*, 3(1), pp. 39–54. https://doi.org/10.1038/s43017-021-00232-1.

Islam, Z. and Singh, S. K. (2021) 'Geospatial analysis of the impact of flood and drought hazards on crop land and its relationship with human migration at the district level in Uttar Pradesh, India', *Geomatics and Environmental Engineering*, 15(4), pp. 117–127. https://doi.org/10.7494/geom.2021.15.4.117.

Iyer, G., Ou, Y., Edmonds, J., Fawcett, A. A., Hultman, N., McFarland, J., Fuhrman, J., Waldhoff, S. and McJeon, H. (2022) 'Ratcheting of climate pledges needed to limit peak global warming', *Nature Climate Change*, 12(12), pp. 1129–1135. https://doi.org/10.1038/s41558-022-01508-0.

James N., G. (1989) *American Exodus: The Dust Bowl Migration and Okie Culture in California*. New York: Oxford University Press. https://history.washington.edu/research/books/american-exodus-dust-bowl-migration-and-okie-culture-california (Accessed: 23 April 2024).

Jamshed, A., Birkmann, J., Feldmeyer, D. and Rana, I. A. (2020) 'A conceptual framework to understand the dynamics of rural–urban linkages for rural flood vulnerability', *Sustainability*, 12(7). https://doi.org/10.3390/su12072894.

Jarawura, F. X. (2021) 'Dynamics of drought-related migration among five villages in the Savannah of Ghana', *Ghana Journal of Geography*, 13(1), pp. 103–125. https://doi.org/10.4314/gjg.v13i1.6.

Jasechko, S., Perrone, D., Seybold, H., Fan, Y. and Kirchner, J. W. (2020) 'Groundwater level observations in 250,000 coastal US wells reveal scope of potential seawater intrusion', *Nature Communications*, 11(1), p. 3229. https://doi.org/10.1038/s41467-020-17038-2.

Jayasinghe, A. (2023) 'Maldives to battle rising seas by building fortress islands', *Phys.Org*, 21 November. https://phys.org/news/2023-11-maldives-seas-fortress-islands.html.

Jevrejeva, S., Williams, J., Vousdoukas, M. I. and Jackson, L. P. (2023) 'Future sea level rise dominates changes in worst case extreme sea levels along the global coastline by 2100', *Environmental Research Letters*, 18(2), p. 024037. https://doi.org/10.1088/1748-9326/acb504.

Jia, H., Chen, F., Pan, D., Du, E., Wang, L., Wang, N. and Yang, A. (2022) 'Flood risk management in the Yangtze River basin – Comparison of 1998 and 2020 events', *International Journal of Disaster Risk Reduction*, 68, p. 102724. https://doi.org/10.1016/j.ijdrr.2021.102724.

Joarder, M. A. M. and Miller, P. W. (2013) 'Factors affecting whether environmental migration is temporary or permanent: evidence from Bangladesh', *Global Environmental Change*, 23(6), pp. 1511–1524. https://doi.org/10.1016/j.gloenvcha.2013.07.026.

Johnson, K. M. (2011) 'Rural-urban population change', *Encyclopedia of the Great Plains*. Edited by D.J. Wishart. University of Nebraska–Lincoln.

Jones, D. C. (2002) *Empire of Dust: Settling and Abandoning the Prairie Dry Belt*. University of Calgary Press. https://books.google.ca/books?id=NbzWpHBAaHYC.

Jones, G. W. (2020) 'New patterns of female migration in South Asia', *Asian Population Studies*, 16(1), pp. 1–4. https://doi.org/10.1080/17441730.2019.1701802.

Jonkman, S. N., Godfroy, M., Sebastian, A. and Kolen, B. (2018) 'Brief communication: loss of life due to Hurricane Harvey', *Natural Hazards and Earth System Sciences*, 18(4), pp. 1073–1078. https://doi.org/10.5194/nhess-18-1073-2018.

Jonkman, S. N., Hillen, M. M., Nicholls, R. J., Kanning, W. and Ledden, M. van (2013) 'Costs of adapting coastal defences to sea-level rise – new estimates and their implications', *Journal of Coastal Research*, 29(5), pp. 1212–1226. https://doi.org/10.2112/JCOASTRES-D-12-00230.1.

Jonkman, S. N., Maaskant, B., Boyd, E. and Levitan, M. L. (2009) 'Loss of life caused by the flooding of New Orleans after Hurricane Katrina: analysis of the relationship between flood characteristics and mortality', *Risk Analysis*, 29(5), pp. 676–698. https://doi.org/10.1111/j.1539-6924.2008.01190.x.

Jubilut, L., Sombra Muiños de Andrade, C. and de Lima Madureira, A. (2016) 'Humanitarian visas: building on Brazil's experience', *Forced Migration Review (FMR)*, (53), p. 3.

Junod, A., Rivera, F. and Rogin, A. (2023) *Climate Migration and Receiving Community Institutional Capacity in the US Gulf Coast*. Houston-The Woodlands-Sugar Land, TX, New Orleans-Metairie, LA, Orlando-Kissimmee-Sanford, FL: Metropolitan Housing and Communities Policy Center, Urban Institute, p. 65. www.urban.org/research/publication/climate-migration-and-receiving-community-institutional-capacity-us-gulf-coast.

Kabir, M. E., Davey, P., Serrao-Neumann, S. and Hossain, M. (2018) 'Seasonal Drought Thresholds and Internal Migration for Adaptation: Lessons from Northern Bangladesh', in M. Hossain, R. Hales, and T. Sarker (eds.) *Pathways to a Sustainable Economy: Bridging the Gap between Paris Climate Change Commitments and Net Zero Emissions*. Cham: Springer International Publishing, pp. 167–189. https://doi.org/10.1007/978-3-319-67702-6_10.

Kabir, M. E., Serrao-Neumann, S., Davey, P., Hossain, M. and Alam, Md. T. (2018) 'Drivers and temporality of internal migration in the context of slow-onset natural hazards: Insights from north-west rural Bangladesh', *International Journal of Disaster Risk Reduction*, 31, pp. 617–626. https://doi.org/10.1016/j.ijdrr.2018.06.010.

Kaijser, A. and Kronsell, A. (2014) 'Climate change through the lens of intersectionality', *Environmental Politics*, 23(3), pp. 417–433. https://doi.org/10.1080/09644016.2013.835203.

Kakinuma, K., Puma, M. J., Hirabayashi, Y., Tanoue, M., Baptista, E. A. and Kanae, S. (2020) 'Flood-induced population displacements in the world', *Environmental Research Letters*, 15(12), p. 124029. https://doi.org/10.1088/1748-9326/abc586.

Kam, P. M., Aznar-Siguan, G., Schewe, J., Milano, L., Ginnetti, J., Willner, S., McCaughey, J. W. and Bresch, D. N. (2021) 'Global warming and population change both heighten future risk of human displacement due to river floods', *Environmental Research Letters*, 16(4), p. 044026. https://doi.org/10.1088/1748-9326/abd26c.

Kane, H. H. and Fletcher, C. H. (2020) 'Rethinking reef island stability in relation to Anthropogenic sea level rise', *Earth's Future*, 8(10), p. e2020EF001525. https://doi.org/10.1029/2020EF001525.

Kaplan, R. D. (1994) 'The coming anarchy', *The Atlantic*, 1 February. www.theatlantic.com/magazine/archive/1994/02/the-coming-anarchy/304670/ (Accessed: 23 April 2024).

Karemera, D., Oguledo, V. I. and Davis, B. (2000) 'A gravity model analysis of international migration to North America', *Applied Economics*, 32(13), pp. 1745–1755. https://doi.org/10.1080/000368400421093.

Karnieli, A., Shtein, A., Panov, N., Weisbrod, N. and Tal, A. (2019) 'Was drought really the trigger behind the Syrian civil war in 2011?', *Water*, 11(8), p. 1564. https://doi.org/10.3390/w11081564.

Karras, G. and Chiswick, C. U. (1999) 'Macroeconomic determinants of migration: the case of Germany 1964–1988', *International Migration*, 37(4), pp. 657–677. https://doi.org/10.1111/1468-2435.00089.

Kartiki, K. (2011) 'Climate change and migration: a case study from rural Bangladesh', *Gender & Development*, 19(1), pp. 23–38. https://doi.org/10.1080/13552074.2011.554017.

Kashian, R.D., Buchman, T. and Drago, R. (2021) 'Tornadoes, poverty and race in the USA: A five-decade analysis', *Journal of Economic Studies*, 49(7), pp. 1304–1319. https://doi.org/10.1108/JES-06-2021-0287.

Kaufmann, B. A., Hülsebusch, C. G. and Krätli, S. (2019) 'Pastoral Livestock Systems', in P. Ferranti, E. M. Berry, and J. R. Anderson (eds.) *Encyclopedia of Food Security and Sustainability*. Oxford: Elsevier, pp. 354–360. https://doi.org/10.1016/B978-0-08-100596-5.22179-3.

Keellings, D. and Hernández Ayala, J. J. (2019) 'Extreme rainfall associated with Hurricane Maria over Puerto Rico and its connections to climate variability and change', *Geophysical Research Letters*, 46(5), pp. 2964–2973. https://doi.org/10.1029/2019GL082077.

Keenan, J. M., Hill, T. and Gumber, A. (2018) 'Climate gentrification: from theory to empiricism in Miami-Dade County, Florida', *Environmental Research Letters*, 13(5), p. 054001. https://doi.org/10.1088/1748-9326/aabb32.

Kefi, M., Mishra, B. K., Masago, Y. and Fukushi, K. (2020) 'Analysis of flood damage and influencing factors in urban catchments: case studies in Manila, Philippines, and Jakarta, Indonesia', *Natural Hazards*, 104(3), pp. 2461–2487. https://doi.org/10.1007/s11069-020-04281-5.

Keim, B. D., Muller, R. A. and Stone, G. W. (2007) 'Spatiotemporal patterns and return periods of tropical storm and hurricane strikes from Texas to Maine', *Journal of Climate*, 20(14), pp. 3498–3509. https://doi.org/10.1175/JCLI4187.1.

Kelley, C. P., Mohtadi, S., Cane, M. A., Seager, R. and Kushnir, Y. (2015) 'Climate change in the Fertile Crescent and implications of the recent Syrian drought', *Proceedings of the National Academy of Sciences*, 112(11), pp. 3241–3246. https://doi.org/10.1073/pnas.1421533112.

Kelly, M. (2023) 'Snowbirds and snowflakes: Mobility and aging across the Canada-United States border', *Canadian Geographies / Géographies canadiennes*, 67(2), pp. 217–225. https://doi.org/10.1111/cag.12798.

Kenner, B., Russell, D., Valdes, C., Sowell, A., Pham, X., Terán, A. and Kaufman, J. (2023) *Puerto Rico's agricultural economy in the aftermath of hurricanes Irma and Maria: A brief overview*. AP-114. Economic Research Service, U.S. Department of Agriculture. www.ers.usda.gov/webdocs/publications/106261/ap-114.pdf.

Kevane, M. and Gray, L. (2008) 'Darfur: rainfall and conflict', *Environmental Research Letters*, 3(3), p. 034006. https://doi.org/10.1088/1748-9326/3/3/034006.

Khajehei, S. and Hamideh, S. (2024) 'Post-disaster recovery challenges of public housing residents: Lumberton, North Carolina after Hurricane Matthew', *Urban Affairs Review*, 60(1), pp. 232–271. https://doi.org/10.1177/10780874231167570.

Kim, D. and Tajima, Y. (2022) 'Smuggling and border enforcement', *International Organization*, 76(4), pp. 830–867. https://doi.org/10.1017/S002081832200011X.

King, R., Cela, E. and Fokkema, T. (2021) 'New frontiers in international retirement migration', *Ageing & Society*, 41(6), pp. 1205–1220. https://doi.org/10.1017/S0144686X21000179.

Kirezci, E., Young, I. R., Ranasinghe, R., Muis, S., Nicholls, R. J., Lincke, D. and Hinkel, J. (2020) 'Projections of global-scale extreme sea levels and resulting episodic coastal flooding over the 21st Century', *Scientific Reports*, 10(1), p. 11629. https://doi.org/10.1038/s41598-020-67736-6.

Kirsch, T. D., Wadhwani, C., Sauer, L., Doocy, S. and Catlett, C. (2012) 'Impact of the 2010 Pakistan floods on rural and urban populations at six months', *PLoS Currents*, 4, p. e4fdfb212d2432. https://doi.org/10.1371/4fdfb212d2432.

Klabunde, A. and Willekens, F. (2016) 'Decision-making in agent-based models of migration: state of the art and challenges', *European Journal of Population = Revue Europeenne De Demographie*, 32(1), pp. 73–97. https://doi.org/10.1007/s10680-015-9362-0.

Kniveton, D., Smith, C. and Wood, S. (2011) 'Agent-based model simulations of future changes in migration flows for Burkina Faso', *Global Environmental Change*, 21, pp. S34–S40. https://doi.org/10.1016/j.gloenvcha.2011.09.006.

Knutson, T. R., Chung, M. V., Vecchi, G., Sun, J., Hsieh, T.-L. and Smith, A. J. P. (2021) *Climate change is probably increasing the intensity of tropical cyclones. Zenodo.* https://doi.org/10.5281/zenodo.4570334.

de Koning, K., Filatova, T., Need, A. and Bin, O. (2019) 'Avoiding or mitigating flooding: bottom-up drivers of urban resilience to climate change in the USA', *Global Environmental Change*, 59, p. 101981. https://doi.org/10.1016/j.gloenvcha.2019.101981.

Kopp, R. E., Simons, F. J., Mitrovica, J. X., Maloof, A. C. and Oppenheimer, M. (2009) 'Probabilistic assessment of sea level during the last interglacial stage', *Nature*, 462(7275), pp. 863–867. https://doi.org/10.1038/nature08686.

Koslov, L., Merdjanoff, A., Sulakshana, E. and Klinenberg, E. (2021) 'When rebuilding no longer means recovery: the stress of staying put after Hurricane Sandy', *Climatic Change*, 165(3), p. 59. https://doi.org/10.1007/s10584-021-03069-1.

Kou, L., Xu, H. and Kwan, M.-P. (2018) 'Seasonal mobility and well-being of older people: the case of "Snowbirds" to Sanya, China', *Health & Place*, 54, pp. 155–163. https://doi.org/10.1016/j.healthplace.2018.08.008.

Koubi, V., Nguyen, Q., Spilker, G. and Böhmelt, T. (2021) 'Environmental migrants and social-movement participation', *Journal of Peace Research*, 58(1), pp. 18–32. https://doi.org/10.1177/0022343320972153.

Koubi, V., Spilker, G., Schaffer, L. and Böhmelt, T. (2016) 'The role of environmental perceptions in migration decision-making: evidence from both migrants and non-migrants in five developing countries', *Population and Environment*, 38(2), pp. 134–163. https://doi.org/10.1007/s11111-016-0258-7.

Kremmidas, T. (2024) 'Flocking to the south this winter? Canadian snowbirds and the dollar', *Canadian Business Journal (Burlington, ON)* [Preprint]. www.cbj.ca/flocking-to-the-south-this-winter-canadian-snowbirds-and-the-dollar/.

Kubwarugira, G., Mayoussi, M. and El Khalki, Y. (2019) 'Assessing flood exposure in informal districts: a case study of Bujumbura, Burundi', *Journal of Applied Water Engineering and Research*, 7(3), pp. 207–215. https://doi.org/10.1080/23249676.2019.1611494.

Kugler, A. and Yuksel, M. (2008) 'Effects of low-skilled immigration on U.S. natives: evidence from Hurricane Mitch'. National Bureau of Economic Research (*Working Paper Series*). https://doi.org/10.3386/w14293.

Kulp, S. and Strauss, B. H. (2017) 'Rapid escalation of coastal flood exposure in US municipalities from sea level rise', *Climatic Change*, 142(3), pp. 477–489. https://doi.org/10.1007/s10584-017-1963-7.

Kulp, S. A. and Strauss, B. H. (2019) 'New elevation data triple estimates of global vulnerability to sea-level rise and coastal flooding', *Nature Communications*, 10(1), p. 4844. https://doi.org/10.1038/s41467-019-12808-z.

Kura, Y., Joffre, O., Laplante, B. and Sengvilaykham, B. (2017) 'Coping with resettlement: a livelihood adaptation analysis in the Mekong River basin', *Land Use Policy*, 60, pp. 139–149. https://doi.org/10.1016/j.landusepol.2016.10.017.

Kwarteng, J., Perfumi, S. C., Farrell, T., Third, A. and Fernandez, M. (2022) 'Misogynoir: challenges in detecting intersectional hate', *Social Network Analysis and Mining*, 12(1), p. 166. https://doi.org/10.1007/s13278-022-00993-7.

Läderach, P., Kommerell, V., Schapendonk, F., Van Loon, J., Martinez-Baron, D., Castellanos, A., González, C. E., Vega Lira, D., Ramirez Villegas, J., Achicanoy, H., Madurga-Lopez, I., Dutta Gupta, T., Carneiro, B., Resce, G., Ruscica, G. and Pacillo, G. (2021) 'Climate security in the dry corridor of Latin America', *CIMMYT*, Position Paper No. 2021/2, p. 54.

Lamb, H.H. (1995) *Climate, History and the Modern World*. 2nd ed. London: Routledge. https://doi.org/10.4324/9780203433652.

Lawanson, O. I., Proverbs, D. and Ibrahim, R. L. (2023) 'The impact of flooding on poor communities in Lagos State, Nigeria: the case of the Makoko urban settlement', *Journal of Flood Risk Management*, 16(1), p. e12838. https://doi.org/10.1111/jfr3.12838.

Le De, L., Gaillard, J. C. and Friesen, W. (2013) 'Remittances and disaster: a review', *International Journal of Disaster Risk Reduction*, 4, pp. 34–43. https://doi.org/10.1016/j.ijdrr.2013.03.007.

Le De, L., Gaillard, J. C., Friesen, W. and Smith, F. M. (2015) 'Remittances in the face of disasters: a case study of rural Samoa', *Environment, Development and Sustainability*, 17(3), pp. 653–672. https://doi.org/10.1007/s10668-014-9559-0.

Lee, C.-C., Chou, C. and Mostafavi, A. (2022) 'Specifying evacuation return and home-switch stability during short-term disaster recovery using location-based data', *Scientific Reports*, 12(1), p. 15987. https://doi.org/10.1038/s41598-022-20384-4.

Lee, E. S. (1966) 'A theory of migration', *Demography*, 3(1), pp. 47–57.

Lekson, S.H. and Cameron, C.M. (1995) 'The abandonment of Chaco Canyon, the Mesa Verde migrations, and the reorganization of the Pueblo world', *Journal of Anthropological Archaeology*, 14(2), pp. 184–202. https://doi.org/10.1006/jaar.1995.1010.

Leong, K. J., Airriess, C. A., Li, W., Chen, A. C.-C. and Keith, V. M. (2007) 'Resilient history and the rebuilding of a community: the Vietnamese American community in New Orleans East', *Journal of American History*, 94(3), pp. 770–779. https://doi.org/10.2307/25095138.

Levy, B. S. and Patz, J. A. (2015) 'Climate change, human rights, and social justice', *Annals of Global Health*, 81(3), pp. 310–322. https://doi.org/10.1016/j.aogh.2015.08.008.

Li, D., Yuan, J. and Kopp, R. E. (2020) 'Escalating global exposure to compound heat-humidity extremes with warming', *Environmental Research Letters*, 15(6), p. 064003. https://doi.org/10.1088/1748-9326/ab7d04.

Li, X., Bellerby, R., Craft, C. and Widney, S. E. (2018) 'Coastal wetland loss, consequences, and challenges for restoration', *Anthropocene Coasts*, 1(1), pp. 1–15. https://doi.org/10.1139/anc-2017-0001.

Libecap, G. D. and Hansen, Z. K. (2002) '"Rain follows the plow" and dryfarming doctrine: the climate information problem and homestead failure in the upper great plains, 1890–1925', *The Journal of Economic History*. 2002/05/22 ed., 62(1), pp. 86–120. https://doi.org/10.1017/S0022050702044042.

Lincke, D. and Hinkel, J. (2021) 'Coastal migration due to 21st century sea-level rise', *Earth's Future*, 9(5), p. e2020EF001965. https://doi.org/10.1029/2020EF001965.

Linke, A., Leutert, S., Busby, J., Duque, M., Shawcroft, M. and Brewer, S. (2023) 'Dry growing seasons predicted Central American migration to the US from 2012 to 2018', *Scientific Reports*, 13(1), p. 18400. https://doi.org/10.1038/s41598-023-43668-9.

Linke, A. M. and Ruether, B. (2021) 'Weather, wheat, and war: security implications of climate variability for conflict in Syria', *Journal of Peace Research*, 58(1), pp. 114–131. https://doi.org/10.1177/0022343320973070.

Liu, J., Feng, S., Gu, X., Zhang, Y., Beck, H. E., Zhang, J. and Yan, S. (2022) 'Global changes in floods and their drivers', *Journal of Hydrology*, 614, p. 128553. https://doi.org/10.1016/j.jhydrol.2022.128553.

Liu, P. R. and Raftery, A. E. (2021) 'Country-based rate of emissions reductions should increase by 80% beyond nationally determined contributions to meet the 2°C target', *Communications Earth & Environment*, 2(1), pp. 1–29. https://doi.org/10.1038/s43247-021-00097-8.

Lohmann, J., Dijkstra, H. A., Jochum, M., Lucarini, V. and Ditlevsen, P. D. (2024) 'Multistability and intermediate tipping of the Atlantic Ocean circulation', *Science Advances*, 10(12), p. eadi4253. https://doi.org/10.1126/sciadv.adi4253.

Lopez, G. (2023) 'Finding climate havens', *The New York Times*, 23 August. www.nytimes.com/2023/08/23/briefing/finding-climate-havens.html (Accessed: 16 May 2024).

López, P., Rodríguez, A. C. and Escapa, S. (2022) 'Psychosocial effects of gentrification on elderly people in Barcelona from the perspective of bereavement', *Emotion, Space and Society*, 43, p. 100880. https://doi.org/10.1016/j.emospa.2022.100880.

Lu, X., Wrathall, D. J., Sundsøy, P. R., Nadıruzzaman, Md., Wetter, E., Iqbal, A., Qureshi, T., Tatem, A. J., Canright, G. S., Engø-Monsen, K. and Bengtsson, L. (2016) 'Detecting climate adaptation with mobile network data in Bangladesh: anomalies in communication, mobility and consumption patterns during cyclone Mahasen', *Climatic Change*, 138(3), pp. 505–519. https://doi.org/10.1007/s10584-016-1753-7.

Lund, D. (2021) 'Navigating slow-onset risks through foresight and flexibility in Fiji: emerging recommendations for the planned relocation of climate-vulnerable communities', *Current Opinion in Environmental Sustainability*, 50, pp. 12–20. https://doi.org/10.1016/j.cosust.2020.12.004.

Lustgarten, A. (2020) 'The great climate migration has begun', *The New York Times*, 23 July. www.nytimes.com/interactive/2020/07/23/magazine/climate-migration.html (Accessed: 23 April 2024).

Lyons, K. (2022) 'How to move a country: Fiji's radical plan to escape rising sea levels', *The Guardian*, 8 November. www.theguardian.com/environment/2022/nov/08/how-to-move-a-country-fiji-radical-plan-escape-rising-seas-climate-crisis (Accessed: 30 April 2024).

Maamoun, N. (2019) 'The Kyoto protocol: empirical evidence of a hidden success', *Journal of Environmental Economics and Management*, 95, pp. 227–256. https://doi.org/10.1016/j.jeem.2019.04.001.

van Maarseveen, R. (2021) 'Urbanization and educational attainment: evidence from *Africa'*. Rochester, NY. https://doi.org/10.2139/ssrn.3836097.

Mabrouk, F. and Mekni, M. M. (2018) 'Remittances and Food Security in African Countries', *African Development Review*, 30(3), pp. 252–263. https://doi.org/10.1111/1467-8268.12334.

Mach, K. J. and Siders, A. R. (2021) 'Reframing strategic, managed retreat for transformative climate adaptation', *Science*, 372(6548), pp. 1294–1299. https://doi.org/10.1126/science.abh1894.

Maestre, F. T., Benito, B. M., Berdugo, M., Concostrina-Zubiri, L., Delgado-Baquerizo, M., Eldridge, D. J., Guirado, E., Gross, N., Kéfi, S., Le Bagousse-Pinguet, Y., Ochoa-Hueso, R. and Soliveres, S. (2021) 'Biogeography of global drylands', *New Phytologist*, 231(2), pp. 540–558. https://doi.org/10.1111/nph.17395.

Maharjan, A., de Campos, R. S., Singh, C., Das, S., Srinivas, A., Bhuiyan, M. R. A., Ishaq, S., Umar, M. A., Dilshad, T., Shrestha, K., Bhadwal, S., Ghosh, T., Suckall, N. and Vincent, K. (2020) 'Migration and household adaptation in climate-sensitive hotspots in South Asia', *Current Climate Change Reports*, 6(1), pp. 1–16. https://doi.org/10.1007/s40641-020-00153-z.

Maharjan, A., Tuladhar, S., Hussain, A., Mishra, A., Bhadwal, S., Ishaq, S., Saeed, B. A., Sachdeva, I., Ahmad, B., Ferdous, J. and Hassan, S. M. T. (2021) 'Can labour migration help households adapt to climate change? Evidence from four river basins in South Asia', *Climate and Development*, 13(10), pp. 879–894. https://doi.org/10.1080/17565529.2020.1867044.

Makaske, B., Vries, E. de, Tainter, J. A. and McIntosh, R. J. (2007) 'Aeolian and fluvio-lacustrine landforms and prehistoric humanoccupation on a technically influenced floodplain margin, the Méma, central Mali', *Netherlands Journal of Geosciences*, 86(3), pp. 241–256. https://doi.org/10.1017/S0016774600077830.

Malak, Md. A., Hossain, N. J., Quader, M. A., Akter, T. and Islam, Md. N. (2021) 'Climate Change-Induced Natural Hazard: Population Displacement, Settlement Relocation, and Livelihood Change Due to Riverbank Erosion in Bangladesh', in Md. N. Islam and A. van Amstel (eds.) *Bangladesh II: Climate Change Impacts, Mitigation and Adaptation in Developing Countries*. Cham: Springer International Publishing, pp. 193–210. https://doi.org/10.1007/978-3-030-71950-0_6.

Maldonado, J. K., Shearer, C., Bronen, R., Peterson, K. and Lazrus, H. (2013) 'The impact of climate change on tribal communities in the US: displacement, relocation, and human rights', *Climatic Change*, 120(3), pp. 601–614. https://doi.org/10.1007/s10584-013-0746-z.

Malecha, M. L., Woodruff, S. C. and Berke, P. R. (2021) 'Planning to exacerbate flooding: evaluating a Houston, Texas, network of plans in place during Hurricane Harvey using a plan integration for resilience scorecard', *Natural Hazards Review*, 22(4), p. 04021030. https://doi.org/10.1061/(ASCE)NH.1527-6996.0000470.

Mallick, B., Best, K., Carrico, A., Ghosh, T., Priodarshini, R., Sultana, Z. and Samanta, G. (2023) 'How do migration decisions and drivers differ against extreme environmental events?', *Environmental Hazards*, 22(5), pp. 475–497. https://doi.org/10.1080/17477891.2023.2195152.

Manly, M. (2007) 'The spirit of Geneva – traditional and new actors in the field of statelessness', *Refugee Survey Quarterly*, 26(4), pp. 255–261.

Mann, M. E., Zhang, Z., Hughes, M. K., Bradley, R. S., Miller, S. K., Rutherford, S. and Ni, F. (2008) 'Proxy-based reconstructions of hemispheric and global surface temperature variations over the past two millennia', *Proceedings of the National Academy of Sciences*, 105(36), pp. 13252–13257. https://doi.org/10.1073/pnas.0805721105.

Manzano, P., Burgas, D., Cadahía, L., Eronen, J. T., Fernández-Llamazares, Á., Bencherif, S., Holand, Ø., Seitsonen, O., Byambaa, B., Fortelius, M., Fernández-Giménez, M. E., Galvin, K. A., Cabeza, M. and Stenseth, N. Chr. (2021) 'Toward a holistic understanding of pastoralism', *One Earth*, 4(5), pp. 651–665. https://doi.org/10.1016/j.oneear.2021.04.012.

Manzoor, Z., Ehsan, M., Khan, M. B., Manzoor, A., Akhter, M. M., Sohail, M. T., Hussain, A., Shafi, A., Abu-Alam, T. and Abioui, M. (2022) 'Floods and flood management and its socio-economic impact on Pakistan: a review of the empirical literature', *Frontiers in Environmental Science*, 10. https://doi.org/10.3389/fenvs.2022.1021862.

Marandi, A. and Main, K. L. (2021) 'Vulnerable City, recipient city, or climate destination? Towards a typology of domestic climate migration impacts in US cities', *Journal of Environmental Studies and Sciences*, 11(3), pp. 465–480. https://doi.org/10.1007/s13412-021-00712-2.

Mardy, T., Uddin, M. N., Sarker, Md. A., Roy, D. and Dunn, E. S. (2018) 'Assessing coping strategies in response to drought: a micro level study in the north-west region of Bangladesh', *Climate*, 6(2). https://doi.org/10.3390/cli6020023.

Marengo, J. A., Alcantara, E., Cunha, A. P., Seluchi, M., Nobre, C. A., Dolif, G., Goncalves, D., Assis Dias, M., Cuartas, L. A., Bender, F., Ramos, A. M., Mantovani, J. R., Alvalá, R. C. and Moraes, O. L. (2023) 'Flash floods and landslides in the city of Recife, Northeast Brazil after heavy rain on May 25–28, 2022: causes, impacts, and disaster preparedness', *Weather and Climate Extremes*, 39, p. 100545. https://doi.org/10.1016/j.wace.2022.100545.

Marengo, J. A., Cunha, A. P. M. A., Nobre, C. A., Ribeiro Neto, G. G., Magalhaes, A. R., Torres, R. R., Sampaio, G., Alexandre, F., Alves, L. M., Cuartas, L.A., Deusdará, K. R. L. and Álvala, R. C. S. (2020) 'Assessing drought in the drylands of northeast Brazil under regional warming exceeding 4°C', *Natural Hazards*, 103(2), pp. 2589–2611. https://doi.org/10.1007/s11069-020-04097-3.

Marengo, J. A., Galdos, M. V., Challinor, A., Cunha, A. P., Marin, F. R., Vianna, M. dos S., Alvala, R. C. S., Alves, L. M., Moraes, O. L. and Bender, F. (2022) 'Drought in Northeast Brazil: a review of agricultural and policy adaptation options for food security', *Climate Resilience and Sustainability*, 1(1), p. e17. https://doi.org/10.1002/cli2.17.

Mariano, C. and Marino, M. (2022) 'Urban planning for climate change: a toolkit of actions for an integrated strategy of adaptation to heavy rains, river floods, and sea level rise', *Urban Science*, 6(3), p. 63. https://doi.org/10.3390/urbansci6030063.

Marxuach, S. M. (2021) 'Taking Stock of Puerto Rico's Reconstruction Process [Policy Brief]'. Center for a New Economy (CNE). https://grupocne.org/wp-content/uploads/2021/05/2021.05.27-Taking-stock-of-Puerto-Ricos-Reconstruction-Process.pdf.

Marzuoli, A. and Liu, F. (2018) 'A data-driven impact evaluation of Hurricane Harvey from mobile phone data', in *2018 IEEE International Conference on Big Data (Big Data). 2018 IEEE International Conference on Big Data (Big Data)*, pp. 3442–3451. https://doi.org/10.1109/BigData.2018.8622641.

Masrur, A., Dewan, A., Botje, D., Kiselev, G. and Murshed, Md. M. (2022) 'Dynamics of human presence and flood-exposure risk in close proximity to Bangladesh's river network: an evaluation with multitemporal satellite imagery', *Geocarto International*, 37(27), pp. 14946–14962. https://doi.org/10.1080/10106049.2022.2093410.

Masselink, G., Beetham, E. and Kench, P. (2020) 'Coral reef islands can accrete vertically in response to sea level rise', *Science Advances*, 6(24), p. eaay3656. https://doi.org/10.1126/sciadv.aay3656.

Massey, D. S., Arango, J., Hugo, G., Kouaouci, A., Pellegrino, A. and Taylor, J. E. (1993) 'Theories of international migration: a review and appraisal', *Population and Development Review*, 19(3), pp. 431–466. https://doi.org/10.2307/2938462.

Massey, D. S. and Espinosa, K. E. (1997) 'What's driving Mexico-U.S. migration? A theoretical, empirical, and policy analysis', *American Journal of Sociology*, 102(4), pp. 939–999. https://doi.org/10.1086/231037.

Masson-Delmotte, V., Zhai, P., Pörtner, H.-O., Roberts, D., Skea, J., Shukla, P. R., Pirani, A., Moufouma-Okia, W., Péan, C., Pidcock, R., Connors, S., Matthews, J. B. R., Chen, Y., Zhou, X., Gomis, M. I., Lonnoy, E., Maycock, T., Tignor, M. and Waterfield, T. (eds.) (2018) 'Summary for Policymakers', in IPCC, *Global Warming of 1.5°C: IPCC Special Report on Impacts of Global Warming of 1.5°C above Pre-industrial Levels in Context of Strengthening Response to Climate Change, Sustainable Development, and Efforts to Eradicate Poverty*. 1st ed. Cambridge, UK and New York, NY, USA: Cambridge University Press, pp. 3–24. https://doi .org/10.1017/9781009157940.

Matthew, R. A. and Gaulin, T. (2001) 'Conflict or cooperation? The social and political impacts of resource scarcity on small island states', *Global Environmental Politics*, 1(2), pp. 48–70. https://doi.org/10.1162/152638001750336596.

Mavhura, E. (2020) 'Dam-induced displacement and resettlement: Reflections from Tokwe-Mukorsi flood disaster, Zimbabwe', *International Journal of Disaster Risk Reduction*, 44, p. 101407. https://doi.org/10.1016/j.ijdrr.2019.101407.

May, C. (2020) 'Rising groundwater and sea-level rise', *Nature Climate Change*, 10(10), pp. 889–890. https://doi.org/10.1038/s41558-020-0886-x.

Mayer, A., Castro-Diaz, L., Lopez, M. C., Leturcq, G. and Moran, E. (2020) 'Is hydropower worth it? Exploring Amazonian resettlement, human development and environmental costs with the Belo Monte project in Brazil', *Energy Research & Social Science*, 78, p. 102129. https://doi.org/10.1016/j.erss.2021.102129.

Mayer, J., Moradi, S., Nejat, A., Ghosh, S., Cong, Z. and Liang, D. (2020) 'Drivers of post-disaster relocations: the case of Moore and Hattiesburg tornados', *International Journal of Disaster Risk Reduction*, 49, p. 101643. https://doi.org/10.1016/j .ijdrr.2020.101643.

McAdam, J. (2011) 'Environmental Migration', in A. Betts (ed.) *Global Migration Governance*. Oxford University Press, pp. 153–188. https://doi.org/10.1093/acprof :oso/9780199600458.003.0007.

McConnell, K., Whitaker, S., Fussell, E., DeWaard, J., Price, K. and Curtis, K. (2021) 'Effects of wildfire destruction on migration, consumer credit, and financial distress'. Rochester, NY. https://doi.org/10.2139/ssrn.3995455.

McFerrin, R., Norman, S. and Wills, D. (2012) 'Determinants of homestead claims and the expansion of Western settlement', *Applied Economics Letters*, 19(18), pp. 1927–1932. https://doi.org/10.1080/13504851.2012.671920.

McGinlay, J., Jones, N., Clark, J. and Maguire-Rajpaul, V. A. (2021) 'Retreating coastline, retreating government? Managing sea level rise in an age of austerity', *Ocean & Coastal Management*, 204, p. 105458. https://doi.org/10.1016/j.ocecoaman.2020.105458.

McGranahan, G., Balk, D. and Anderson, B. (2007) 'The rising tide: assessing the risks of climate change and human settlements in low elevation coastal zones', *Environment and Urbanization*, 19(1), pp. 17–37. https://doi .org/10.1177/0956247807076960.

McGranahan, G., Balk, D., Colenbrander, S., Engin, H. and MacManus, K. (2023) 'Is rapid urbanization of low-elevation deltas undermining adaptation to climate change? A global review', *Environment and Urbanization*, 35(2), pp. 527–559. https://doi .org/10.1177/09562478231192176.

McGregor, H. E., Madden, B., Higgins, M. and Ostertag, J. (2018). Braiding designs for decolonizing research methodologies: theory, practice, ethics. *Reconceptualizing Educational Research Methodology*, 9(2). https://doi.org/10.7577/rerm.2781

McLeman, R. (2006) 'Migration out of 1930s rural Eastern Oklahoma: insights for climate change research', *Great Plains Quarterly*, 26(1), pp. 27–40.

McLeman, R. (2018) 'Thresholds in climate migration', *Population and Environment*, 39(4), pp. 319–338. https://doi.org/10.1007/s11111-017-0290-2.

McLeman, R. (2019) 'International migration and climate adaptation in an era of hardening borders', *Nature Climate Change*, 9(12), pp. 911–918. https://doi.org/10.1038/s41558-019-0634-2.

McLeman, R. (2020a) *How Will International Migration Policy and Sustainable Development Affect Future Climate-Related Migration?* Washington, DC: Migration Policy Institute, p. 39. www.migrationpolicy.org/sites/default/files/publications/tcm-climate-migration-mcleman_final.pdf.

McLeman, R. (2020b) 'Perception of climate migrants', *Nature Climate Change*, 10(7), pp. 600–601. https://doi.org/10.1038/s41558-020-0803-3.

McLeman, R., Fontanella, F., Greig, C., Heath, G. and Robertson, C. (2022a) 'Population responses to the 1976 South Dakota drought: insights for wider drought migration research', *Population, Space and Place*, 28(2), p. e2465. https://doi.org/10.1002/psp.2465.

McLeman, R., Grieg, C., Heath, G. and Robertson, C. (2022b) 'A machine learning analysis of drought and rural population change on the North American Great Plains since the 1970s', *Population and Environment*, 43(4), pp. 500–529. https://doi.org/10.1007/s11111-022-00399-9.

McLeman, R., Herold, S., Reljic, Z., Sawada, M. and McKenney, D. (2010) 'GIS-based modeling of drought and historical population change on the Canadian Prairies', *Journal of Historical Geography*, 36(1), pp. 43–56. https://doi.org/10.1016/j.jhg.2009.04.003.

McLeman, R., Mayo, D., Strebeck, E. and Smit, B. (2008) 'Drought adaptation in rural eastern Oklahoma in the 1930s: lessons for climate change adaptation research', *Mitigation and Adaptation Strategies for Global Change*, 13(4), pp. 379–400. https://doi.org/10.1007/s11027-007-9118-1.

McLeman, R. and Smit, B. (2006) 'Migration as an adaptation to climate change', *Climatic Change*, 76(1), pp. 31–53. https://doi.org/10.1007/s10584-005-9000-7.

McLeman, R. A. (2011) 'Settlement abandonment in the context of global environmental change', *Global Environmental Change*, 21, pp. S108–S120. https://doi.org/10.1016/j.gloenvcha.2011.08.004.

McLeman, R. A. (2013) *Climate and Human Migration: Past Experiences, Future Challenges*. Cambridge: Cambridge University Press. https://doi.org/10.1017/CBO9781139136938.

McLeman, R. A., Dupre, J., Berrang Ford, L., Ford, J., Gajewski, K. and Marchildon, G. (2014) 'What we learned from the Dust Bowl: lessons in science, policy, and adaptation', *Population and Environment*, 35(4), pp. 417–440. https://doi.org/10.1007/s11111-013-0190-z.

McLeman, R. A. and Ploeger, S. K. (2012) 'Soil and its influence on rural drought migration: insights from depression-era Southwestern Saskatchewan, Canada', *Population and Environment*, 33(4), pp. 304–332. https://doi.org/10.1007/s11111-011-0148-y.

McInerney, E., Saxon, J. and Ashley, L. (2022) '*Migration as a climate adaptation strategy: challenges & opportunities for USAID programming [Discussion paper]*', *in*. USAID. www.climatelinks.org/sites/default/files/asset/document/2023-01/FTF1537_USAID_Climate%20Migration%20Strategy_012723.pdf.

McMichael, C. (2015) 'Climate change-related migration and infectious disease', *Virulence*, 6(6), pp. 548–553. https://doi.org/10.1080/21505594.2015.1021539.

McMichael, C., Barnett, J. and McMichael, A. J. (2012) 'An ill wind? Climate change, migration, and health', *Environmental Health Perspectives*, 120(5), pp. 646–654. https://doi.org/10.1289/ehp.1104375.

McMichael, C., Dasgupta, S., Ayeb-Karlsson, S. and Kelman, I. (2020) 'A review of estimating population exposure to sea-level rise and the relevance for migration', *Environmental Research Letters*, 15(12), p. 123005. https://doi.org/10.1088/1748-9326/abb398.

McMichael, C., Farbotko, C., Piggott-McKellar, A., Powell, T. and Kitara, M. (2021) 'Rising seas, immobilities, and translocality in small island states: case studies from Fiji and Tuvalu', *Population and Environment*, 43(1), pp. 82–107. https://doi.org/10.1007/s11111-021-00378-6.

McMichael, C. and Katonivualiku, M. (2020) 'Thick temporalities of planned relocation in Fiji', *Geoforum*, 108, pp. 286–294. https://doi.org/10.1016/j.geoforum.2019.06.012.

McMichael, C., Schwerdtle, P. N. and Ayeb-Karlsson, S. (2023). 'Waiting for the wave, but missing the tide: Case studies of climate-related (im)mobility and health'. *Journal of Migration and Health*, 7, p. 100147. https://doi.org/10.1016/j.jmh.2022.100147.

Menéndez, P., Losada, I. J., Torres-Ortega, S., Narayan, S. and Beck, M. W. (2020) 'The global flood protection benefits of mangroves', *Scientific Reports*, 10(1), p. 4404. https://doi.org/10.1038/s41598-020-61136-6.

Mengis, N., Martin, T., Keller, D. P. and Oschlies, A. (2016) 'Assessing climate impacts and risks of ocean albedo modification in the Arctic', *Journal of Geophysical Research: Oceans*, 121(5), pp. 3044–3057. https://doi.org/10.1002/2015JC011433.

Mercer, L., Whalen, D., Lim, M., Cockney, K., Cormier, S., Irish, C. and Mann, P. J. (2023) 'Towards more inclusive and solution orientated community-based environmental monitoring', *Environmental Research Letters*, 18(6), p. 064003. https://doi.org/10.1088/1748-9326/accfb0.

Methane | Vital Signs (2024) *Climate Change: Vital Signs of the Planet*. https://climate.nasa.gov/vital-signs/methane?intent=121 (Accessed: 27 May 2024).

Met Eireann (2021), The Irish Meteorological Service, Annual Climate Statement for 2021. www.met.ie/annual-climate-statement-for-2021#:~:text=2021%20is%20provisionally%20the%208th,1.22%C2%B0C%20above%20LTA.

Meza, I., Siebert, S., Döll, P., Kusche, J., Herbert, C., Eyshi Rezaei, E., Nouri, H., Gerdener, H., Popat, E., Frischen, J., Naumann, G., Vogt, J. V., Walz, Y., Sebesvari, Z. and Hagenlocher, M. (2020) 'Global-scale drought risk assessment for agricultural systems', *Natural Hazards and Earth System Sciences*, 20(2), pp. 695–712. https://doi.org/10.5194/nhess-20-695-2020.

Meze-Hausken, E. (2008) 'On the (im-)possibilities of defining human climate thresholds', *Climatic Change*, 89(3), pp. 299–324. https://doi.org/10.1007/s10584-007-9392-7.

Migration Data Portal (2024) 'Gender and Migration: Key Trends'. Global Migration Data Analysis Centre (GMDAC) and International Organization for Migration (IOM). www.migrationdataportal.org/themes/gender-and-migration.

Millennium Ecosystem Assessment (2005) *Ecosystems and Human Well-Being: Our Human Planet, Summary for Decision Makers* (4 vol). Washington, DC: Island Press.

Miller, J. (2024) 'Historic Storm Rains Misery on L.A'.s Unhoused', *LAmag – Culture, Food, Fashion, News & Los Angeles*, 7 February. https://lamag.com/news/atmospheric-river-storm-l-a-unhoused (Accessed: 22 April 2024).

Ministerio del Interior de Argentina/Migraciones and PDD (2022) 'Argentina: leading initiatives to address displacement in the context of disasters and climate change [Policy brief]', *Disaster Displacement*. https://disasterdisplacement.org/portfolio-item/policy-brief-leading-initiatives-by-argentina-to-address-displacement-in-the-context-of-disasters-and-climate-change/ (Accessed: 27 May 2024).

Ministry of Foreign Affairs and Trade (MFAT) (2018) *'Pacific climate change-related displacement and migration'*. New Zealand Ministry of Foreign Affairs and Trade. https://apo.org.au/sites/default/files/resource-files/2018-05/apo-nid213946.pdf.

Mitchell, D. (1996) *The Lie of the Land: Migrant Workers and the California Landscape*. University of Minnesota Press, Minneapolis. www.scopus.com/inward/record.url?scp=0029730366&partnerID=8YFLogxK (Accessed: 27 May 2024).

Mitchell, D. C. (1997) *Sold American: The Story of Alaska Natives and Their Land, 1867–1959: The Army to Statehood*. 1st ed. Hanover, New Hampshire: University Press of New England.

Mobarak, A. M. and Reimão, M. E. (2020) 'Seasonal poverty and seasonal migration in Asia', *Asian Development Review*, 37(1), pp. 1–42. https://doi.org/10.1162/adev_a_00139.

Mock, C. J. (2000) 'Rainfall in the garden of the United States Great Plains, 1870–1889', *Climatic Change*, 44(1), pp. 173–195. https://doi.org/10.1023/A:1005570827123.

Mojid, M.A. (2020) 'Climate change-induced challenges to sustainable development in Bangladesh', *IOP Conference Series: Earth and Environmental Science*, 423(1), p. 012001. https://doi.org/10.1088/1755-1315/423/1/012001.

Mokhnacheva, D. (2022) *Baseline mapping of the implementation of commitments related to addressing human mobility challenges in the context of disasters, climate change and environmental degradation under the Global Compact for safe, orderly and regular Migration (GCM)*. Geneva: Platform on Disaster Displacement, p. 88. https://disasterdisplacement.org/wp-content/uploads/2022/05/17052022_PDD_Baseline_Mapping_Report_final_compressed.pdf.

Molina, M. J. and Rowland, F. S. (1974) 'Predicted present stratospheric abundances of chlorine species from photodissociation of carbon tetrachloride', *Geophysical Research Letters*, 1(7), pp. 309–312. https://doi.org/10.1029/GL001i007p00309.

Mondal, Md. S. H., Murayama, T. and Nishikizawa, S. (2020) 'Assessing the flood risk of riverine households: a case study from the right bank of the Teesta River, Bangladesh', *International Journal of Disaster Risk Reduction*, 51, p. 101758. https://doi.org/10.1016/j.ijdrr.2020.101758.

Monirul Qader Mirza, M. (2002) 'Global warming and changes in the probability of occurrence of floods in Bangladesh and implications', *Global Environmental Change*, 12(2), pp. 127–138. https://doi.org/10.1016/S0959-3780(02)00002-X.

Moniruzzaman, M. (2022) 'The impact of remittances on household food security: evidence from a survey in Bangladesh', *Migration and Development*, 11(3), pp. 352–371. https://doi.org/10.1080/21632324.2020.1787097.

de Montjoye, Y.-A., Gambs, S., Blondel, V., Canright, G., de Cordes, N., Deletaille, S., Engø-Monsen, K., Garcia-Herranz, M., Kendall, J., Kerry, C., Krings, G., Letouzé, E., Luengo-Oroz, M., Oliver, N., Rocher, L., Rutherford, A., Smoreda, Z., Steele, J., Wetter, E., Pentland, A. "Sandy" and Bengtsson, L. (2018) 'On the privacy-conscientious use of mobile phone data', *Scientific Data*, 5(1), pp. 1–6. https://doi.org/10.1038/sdata.2018.286.

Motsholapheko, M. R., Kgathi, D. L. and Vanderpost, C. (2011) 'Rural livelihoods and household adaptation to extreme flooding in the Okavango Delta, Botswana', *Physics and Chemistry of the Earth, Parts A/B/C*, 36(14), pp. 984–995. https://doi.org/10.1016/j.pce.2011.08.004.

MPI (2019) *Global refugee resettlement: What do the statistics tell us? Migration data portal*. www.migrationdataportal.org/blog/global-refugee-resettlement-what-do-statistics-tell-us (Accessed: 27 May 2024).

Mueller, V., Gray, C. and Hopping, D. (2020) 'Climate-induced migration and unemployment in middle-income Africa', *Global Environmental Change*, 65, p. 102183. https://doi.org/10.1016/j.gloenvcha.2020.102183.

Mueller, V., Gray, C. and Kosec, K. (2014) 'Heat stress increases long-term human migration in rural Pakistan', *Nature Climate Change*, 4(3), pp. 182–185. https://doi.org/10.1038/nclimate2103.

Mueller, V., Sheriff, G., Dou, X. and Gray, C. (2020) 'Temporary migration and climate variation in Eastern Africa', *World Development*, 126, p. 104704. https://doi.org/10.1016/j.worlddev.2019.104704.

Murakami, H., Delworth, T. L., Cooke, W. F., Zhao, M., Xiang, B. and Hsu, P.-C. (2020) 'Detected climatic change in global distribution of tropical cyclones', *Proceedings of the National Academy of Sciences*, 117(20), pp. 10706–10714. https://doi.org/10.1073/pnas.1922500117.

Murfin, J. and Spiegel, M. (2020) 'Is the risk of sea level rise capitalized in residential real estate?', *The Review of Financial Studies*, 33(3), pp. 1217–1255. https://doi.org/10.1093/rfs/hhz134.

Murphy, A. and Strobl, E. (2009) *The Impact of Hurricanes on Housing Prices: Evidence from US Coastal Cities.* https://mpra.ub.uni-muenchen.de/19360/ (Accessed: 11 April 2024).

Murray-Tortarolo, G. N. and Salgado, M. M. (2021) 'Drought as a driver of Mexico-US migration', *Climatic Change*, 164(3), p. 48. https://doi.org/10.1007/s10584-021-03030-2.

Mustafa, D. and Wrathall, D. (2010) 'Indus basin floods of 2010: souring of a Faustian bargain?', 4(1).

Myers, N. (1995) 'Environmental unknowns', *Science*, 269(5222), pp. 358–360. https://doi.org/10.1126/science.269.5222.358.

Naivinit, W., Le Page, C., Trébuil, G. and Gajaseni, N. (2010) 'Participatory agent-based modeling and simulation of rice production and labor migrations in Northeast Thailand', *Environmental Modelling & Software*, 25(11), pp. 1345–1358. https://doi.org/10.1016/j.envsoft.2010.01.012.

Nanditha, J. S., Kushwaha, A. P., Singh, R., Malik, I., Solanki, H., Chuphal, D. S., Dangar, S., Mahto, S. S., Vegad, U. and Mishra, V. (2023) 'The Pakistan Flood of August 2022: causes and implications', *Earth's Future*, 11(3), p. e2022EF003230. https://doi.org/10.1029/2022EF003230.

Nansen Initiative (2015) Agenda for the Protection of Cross-Border Displaced Persons in the Context of Disasters and Climate Change: Volume I. The Nansen Initiative: Disaster Induced Cross-Border Displacement. *International Journal of Refugee Law*, 28(1), March 2016, pp. 156–162, https://disasterdisplacement.org/wp-content/uploads/2014/08/EN_Protection_Agenda_Volume_I_-low_res.pdf. https://doi.org/10.1093/ijrl/eew004.

National Ocean Service US Department of Commerce and National Oceanic and Atmospheric Administration (NOAA) (2024) *How Do Coral Reefs Form: Corals Tutorial.* https://oceanservice.noaa.gov/education/tutorial_corals/coral04_reefs.html (Accessed: 28 April 2024).

NASA (2019) 'What Is the Sun's Role in Climate Change? – NASA Science', 6 September. https://science.nasa.gov/earth/climate-change/what-is-the-suns-role-in-climate-change/ (Accessed: 27 May 2024).

NASA (2020) 'Milankovitch (orbital) cycles and their role in Earth's climate – NASA science', 27 February. https://science.nasa.gov/science-research/earth-science/milankovitch-orbital-cycles-and-their-role-in-earths-climate/ (Accessed: 27 May 2024).

NASA Earth Data (2024) *How long have sea levels been rising? How does recent sea-level rise compare to that over the previous centuries?* https://sealevel.nasa.gov/faq/13/how-long-have-sea-levels-been-rising-how-does-recent-sea-level-rise-compare-to-that-over-the-previous (Accessed: 28 April 2024).

NASA Global Climate Change (n.d.) *Carbon Dioxide Concentration | NASA Global Climate Change, Climate Change: Vital Signs of the Planet.* https://climate.nasa.gov/vital-signs/carbon-dioxide?intent=121 (Accessed: 27 May 2024).

NASA Jet Propulsion Laboratory, California Institute of Technology (2024) *GIA & Trends | Data Portal, GRACE Tellus: Gravity Recovery and Climate Experiment.* https://grace.jpl.nasa.gov/data/get-data/gia-trends (Accessed: 28 April 2024).

Nash, S. (2015) 'Towards an 'Environmental Migration Management' Discourse a Discursive Turn in Environmental Migration Advocacy?', in K. Rosenow-Williams, F. Gemenne, I. Chirisa, and A. Pilath (eds.) *Organizational perspectives on environmental migration.* 1st ed. London New York: Routledge, Taylor & Francis Group (Routledge studies in development, mobilities and migration).

National Centers for Environmental Information (NCEI) (n.d.). www.ncei.noaa.gov/ (Accessed: 11 April 2024).

National Drought Mitigation Center (2024) Types of Drought. https://drought.unl.edu/Education/DroughtIn-depth/TypesofDrought.aspx (Accessed: 23 April 2024).

National Oceanic and Atmospheric Administration and US Department of Commerce (n.d.) *What is a mangrove forest?* https://oceanservice.noaa.gov/facts/mangroves.html (Accessed: 28 April 2024).

National Oceanic and Atmospheric Administration and US Department of Commerce (n.d.b.) *What is coral bleaching?* https://oceanservice.noaa.gov/facts/coral_bleach.html (Accessed: 28 April 2024).

Natural Resources Canada (2020) *Fire ecology, Government of Canada.* Natural Resources Canada. https://natural-resources.canada.ca/our-natural-resources/forests/wildland-fires-insects-disturbances/forest-fires/fire-ecology/13149 (Accessed: 23 April 2024).

Nawrotzki, R. J. and Bakhtsiyarava, M. (2017) 'International climate migration: evidence for the climate inhibitor mechanism and the agricultural pathway', *Population, Space and Place*, 23(4), p. e2033. https://doi.org/10.1002/psp.2033.

Nawrotzki, R. J., Brenkert-Smith, H., Hunter, L. M. and Champ, P. A. (2014) 'Wildfire-migration dynamics: lessons from Colorado's Fourmile Canyon fire', *Society & Natural Resources*, 27(2), pp. 215–225. https://doi.org/10.1080/08941920.2013.842275.

Nawrotzki, R. J. and DeWaard, J. (2016) 'Climate shocks and the timing of migration from Mexico', *Population and Environment*, 38(1), pp. 72–100. https://doi.org/10.1007/s11111-016-0255-x.

Nawrotzki, R. J., DeWaard, J., Bakhtsiyarava, M. and Ha, J. T. (2017) 'Climate shocks and rural-urban migration in Mexico: exploring nonlinearities and thresholds', *Climatic Change*, 140(2), pp. 243–258. https://doi.org/10.1007/s10584-016-1849-0.

Nawrotzki, R. J., Riosmena, F. and Hunter, L. M. (2013) 'Do rainfall deficits predict U.S.-bound migration from rural Mexico? Evidence from the Mexican census', *Population Research and Policy Review*, 32(1), pp. 129–158. https://doi.org/10.1007/s11113-012-9251-8.

Naylor, A. W. and Ford, J. (2023) 'Vulnerability and loss and damage following the COP27 of the UN Framework Convention on Climate Change', *Regional Environmental Change*, 23(1), p. 38. https://doi.org/10.1007/s10113-023-02033-2.

NBS/NEMA/UNDP (2023) *Nigeria Flood Impact, Recovery and Mitigation Assessment Report 2022–2023.* Final Report. Abuja, Nigeria: National Bureau of Statistics

(NBS), National Emergency Management Agency (NEMA) and United Nations Development Programme (UNDP), p. 84. www.undp.org/sites/g/files/zskgke326/files/2023-12/nigeriafloodimpactrecoverymitigationassessmentreport2023.pdf.

Nebie, E. K. and West, C. T. (2019) 'Migration and land-use and land-cover change in Burkina Faso: a comparative case study', *Journal of Political Ecology*, 26(1), pp. 614–632. https://doi.org/10.2458/v26i1.23070.

Nee, V. and Sanders, J. (2001) 'Understanding the diversity of immigrant incorporation: a forms-of-capital model', *Ethnic and Racial Studies*, 24(3), pp. 386–411. https://doi.org/10.1080/01419870020036710.

Neef, A. and Benge, L. (2022) 'Shifting responsibility and denying justice: New Zealand's contentious approach to Pacific climate mobilities', *Regional Environmental Change*, 22(3), p. 94. https://doi.org/10.1007/s10113-022-01951-x.

Nejat, A., Binder, S. B., Greer, A. and Jamali, M. (2018) 'Demographics and the dynamics of recovery: a latent class analysis of disaster recovery priorities after the 2013 Moore, Oklahoma tornado', *International Journal of Mass Emergencies & Disasters*, 36(1), pp. 23–51. https://doi.org/10.1177/028072701803600102.

Nepal, S., Tripathi, S. and Adhikari, H. (2021) 'Geospatial approach to the risk assessment of climate-induced disasters (drought and erosion) and impacts on out-migration in Nepal', *International Journal of Disaster Risk Reduction*, 59, p. 102241. https://doi.org/10.1016/j.ijdrr.2021.102241.

Neumann, B., Vafeidis, A. T., Zimmermann, J. and Nicholls, R. J. (2015) 'Future coastal population growth and exposure to sea-level rise and coastal flooding – a global assessment', *PLOS ONE*, 10(3), p. e0118571. https://doi.org/10.1371/journal.pone.0118571.

Neumann, K., Sietz, D., Hilderink, H., Janssen, P., Kok, M. and van Dijk, H. (2015) 'Environmental drivers of human migration in drylands – a spatial picture', *Applied Geography*, 56, pp. 116–126. https://doi.org/10.1016/j.apgeog.2014.11.021.

Nguyen, T. P. L. and Sean, C. (2021) 'Do climate uncertainties trigger farmers' out-migration in the Lower Mekong Region?', *Current Research in Environmental Sustainability*, 3, p. 100087. https://doi.org/10.1016/j.crsust.2021.100087.

Nicholls, R. J. and Cazenave, A. (2010) 'Sea-level rise and its impact on coastal zones', *Science*, 328(5985), pp. 1517–1520. https://doi.org/10.1126/science.1185782.

Nicholls, R. J., Lincke, D., Hinkel, J., Brown, S., Vafeidis, A. T., Meyssignac, B., Hanson, S. E., Merkens, J.-L. and Fang, J. (2021) 'A global analysis of subsidence, relative sea-level change and coastal flood exposure', *Nature Climate Change*, 11(4), pp. 338–342. https://doi.org/10.1038/s41558-021-00993-z.

Nicoletti, A., Melde, S., Guadagno, L. and Bilsborrow, R. E. (2023) 'Human mobility in the context of environmental and climate change: recent data collection tools from the International Organization for Migration to address key methodological and conceptual issues', *International Migration*, 61(5), pp. 47–59. https://doi.org/10.1111/imig.13043.

Nigg, J. M., Barnshaw, J. and Torres, M. R. (2006) 'Hurricane Katrina and the Flooding of New Orleans: Emergent Issues in Sheltering and Temporary Housing', *The ANNALS of the American Academy of Political and Social Science*, 604(1), pp. 113–128. https://doi.org/10.1177/0002716205285889.

NOAA (2023) 'Tropical Cyclone Classification'. *National Oceanic and Atmospheric Administration* (NOAA). www.noaa.gov/jetstream/tropical/tropical-cyclone-introduction/tropical-cyclone-classification (Accessed: 12 April 2024).

NOAA (2024) 'Monthly Global Climate Report for Annual 2023: Global Temperatures'. *National Centers for Environmental Information* (NOAA). www.ncei.noaa.gov/access/monitoring/monthly-report/global/202313#gtemp (Accessed: 27 May 2024).

NOAA and US Department of Commerce (n.d.) *Moore, Oklahoma Tornadoes (1890-Present)*. NOAA's National Weather Service. www.weather.gov/oun/tornadodata-city-ok-moore (Accessed: 22 April 2024).

NOAA and GFDL (n.d.) *Climate Modeling – Geophysical Fluid Dynamics Laboratory*. www.gfdl.noaa.gov/climate-modeling/ (Accessed: 2 May 2024).

Nowotarski, C. J., Spotts, J., Edwards, R., Overpeck, S. and Woodall, G.R. (2021) 'Tornadoes in Hurricane Harvey', *Weather and Forecasting*, 36(5), pp. 1589–1609. https://doi.org/10.1175/WAF-D-20-0196.1.

Noy, I. (2020) 'Paying a price of climate change: who pays for managed retreats?', *Current Climate Change Reports*, 6(1), pp. 17–23. https://doi.org/10.1007/s40641-020-00155-x.

Ober, K. (2021) 'At a climate change crossroads: how a Biden-Harris administration can support and protect communities displaced by climate change [Issue brief]', *Refugees International*, 11 February. www.refugeesinternational.org/reports-briefs/at-a-climate-change-crossroads-how-a-biden-harris-administration-can-support-and-protect-communities-displaced-by-climate-change/.

Ober, K. and Sakdapolrak, P. (2017) 'How do social practices shape policy? Analysing the field of "migration as adaptation" with Bourdieu's "Theory of Practice"', *The Geographical Journal*, 183(4), pp. 359–369. https://doi.org/10.1111/geoj.12225.

Obokata, R. and Veronis, L. (2018) 'Transnational approaches to remittances, risk reduction, and disaster relief: evidence from post-typhoon Haiyan experiences of Filipino immigrants in Canada', in *Routledge Handbook of Environmental Displacement and Migration*. Routledge.

O'Connor, A., Riva, B. and Cazabat, C. (2023) *25 years of progress on internal displacement 1998–2023*. Internal Displacement Index 2023. Geneva, Switzerland: IDMC, p. 44. https://api.internal-displacement.org/sites/default/files/publications/documents/IDMC_2023_25_years_of_progress_on_internal_displacement_report.pdf.

OECD (2022) 'Educational Attainment of Migrants', in *OECD Regions and Cities at a Glance 2022*. Paris: OECD Publishing, pp. 82–85. https://doi.org/10.1787/a0fc61dd-en.

OHCHR (2020) '*Historic UN Human Rights case opens door to climate change asylum claims [Press Release]*'. United Nations Office of the High Commissioner for Human Rights. www.ohchr.org/en/press-releases/2020/01/historic-un-human-rights-case-opens-door-climate-change-asylum-claims.

Ohenhen, L. O., Shirzaei, M., Ojha, C., Sherpa, S. F. and Nicholls, R. J. (2024) 'Disappearing cities on US coasts', *Nature*, 627(8002), pp. 108–115. https://doi.org/10.1038/s41586-024-07038-3.

Okaka, F. O. and Odhiambo, B. D. O. (2019) 'Households' perception of flood risk and health impact of exposure to flooding in flood-prone informal settlements in the coastal city of Mombasa', *International Journal of Climate Change Strategies and Management*, 11(4), pp. 592–606. https://doi.org/10.1108/IJCCSM-03-2018-0026.

Okmyung Bin, Thomas W. Crawford, Jamie B. Kruse, and Craig E. Landry (2008) 'Viewscapes and flood hazard: coastal housing market response to amenities and risk', *Land Economics*, 84(3), p. 434. https://doi.org/10.3368/le.84.3.434.

Oliveira, J. and Pereda, P. (2020) 'The impact of climate change on internal migration in Brazil', *Journal of Environmental Economics and Management*, 103, p. 102340. https://doi.org/10.1016/j.jeem.2020.102340.

Oppenheimer, M., Glavovic, B. C., Hinkel, J., van de Wal, R., Magnan, A. K., Abd-Elgawad, A., Cai, R., Cifuentes-Jara, R. M., DeConto, T., Ghosh, J., Hay, F., Isla, B., Marzeion, B., Meyssignac, B. and Sebesvari, Z. (2019) 'Sea Level Rise and Implications for Low-Lying Islands, Coasts and Communities', in Pörtner, H.-O., Roberts, D.C.,

Masson-Delmotte, V., Zhai, P., Tignor, M., Poloczanska, E., Mintenbeck, K., Alegría, A., Nicolai, M., Okem, A., Petzold, J., Rama, B., Weyer, N.M. (eds.). IPCC Special Report on the Ocean and Cryosphere in a Changing Climate. In press.

Oppenheimer, M., Glavovic, B. C., Hinkel, J., van de Wal, R., Magnan, A. K., Abd-Elgawad, A., Cai, R., Cifuentes-Jara, R. M., DeConto, T., Ghosh, J., Hay, F., Isla, B., Marzeion, B., Meyssignac, B. and Sebesvari, Z. (2022) 'Sea Level Rise and Implications for Low-Lying Islands, Coasts and Communities', in IPCC, *The Ocean and Cryosphere in a Changing Climate: Special Report of the Intergovernmental Panel on Climate Change*. Cambridge, UK and New York, NY, USA: Cambridge University Press, pp. 321–446. https://doi.org/10.1017/9781009157964.006.

Opperman, J. J., Galloway, G. E., Fargione, J., Mount, J. F., Richter, B. D. and Secchi, S. (2009) 'Sustainable floodplains through large-scale reconnection to rivers', *Science*, 326(5959), pp. 1487–1488. https://doi.org/10.1126/science.1178256.

Orbe, C., Rind, D., Miller, R. L., Nazarenko, L. S., Romanou, A., Jonas, J., Russell, G. L., Kelley, M. and Schmidt, G. A. (2023) 'Atmospheric response to a collapse of the North Atlantic circulation under a mid-range future climate scenario: a regime shift in northern hemisphere dynamics', *Journal of Climate*, 36(19), pp. 6669–6693. https://doi.org/10.1175/JCLI-D-22-0841.1.

Orlove, B. (2005) 'Human adaptation to climate change: a review of three historical cases and some general perspectives', *Environmental Science & Policy*, 8(6), pp. 589–600. https://doi.org/10.1016/j.envsci.2005.06.009.

Orr, S. A., Richards, J. and Fatorić, S. (2021) 'Climate change and cultural heritage: a systematic literature review (2016–2020)', *The Historic Environment: Policy & Practice*, 12(3–4), pp. 434–477. https://doi.org/10.1080/17567505.2021.1957264.

Otsuyama, K., Dunn, E., Bell, C., and Maki, N. (2021) 'Typology of human mobility and immobility for disaster risk reduction: exploratory case study in Hillsborough County, Florida', *Natural Hazards Review*, 22(4), p. 05021010. https://doi.org/10.1061/(ASCE)NH.1527-6996.0000498.

Otto, F. E. L., Zachariah, M., Saeed, F., Siddiqi, A., Kamil, S., Mushtaq, H., Arulalan, T., AchutaRao, K., Chaithra, S. T., Barnes, C., Philip, S., Kew, S., Vautard, R., Koren, G., Pinto, I., Wolski, P., Vahlberg, M., Singh, R., Arrighi, J., Aalst, M. van, Thalheimer, L., Raju, E., Li, S., Yang, W., Harrington, L. J. and Clarke, B. (2023) 'Climate change increased extreme monsoon rainfall, flooding highly vulnerable communities in Pakistan', *Environmental Research: Climate*, 2(2), p. 025001. https://doi.org/10.1088/2752-5295/acbfd5.

Oulahen, G. and Doberstein, B. (2012) 'Citizen participation in post-disaster flood hazard mitigation planning in Peterborough, Ontario, Canada', *Risk, Hazards & Crisis in Public Policy*, 3(1), pp. 1–26. https://doi.org/10.1515/1944-4079.1098.

Özdemir, I. (2023) 'The climate refugee crisis is landing on Europe's shores – and we are far from ready', *POLITICO*, 20 February. www.politico.eu/article/climate-refugee-crisis-europe-policy/ (Accessed: 23 April 2024).

Özerdem, A. (2010) 'The "responsibility to protect" in natural disasters: another excuse for interventionism? Nargis Cyclone, Myanmar', *Conflict, Security & Development*, 10(5), pp. 693–713. https://doi.org/10.1080/14678802.2010.511511.

Paik, S., Min, S., Zhang, X., Donat, M. G., King, A. D., & Sun, Q. (2020). 'Determining the Anthropogenic Greenhouse Gas Contribution to the Observed Intensification of Extreme Precipitation'. *Geophysical Research Letters*, 47(12), p. e2019GL086875. https://doi.org/10.1029/2019GL086875.

Pajuelo, Ó. (2020) 'Between disbelief and indifference: are the "environmentally displaced" refugees under the extended definition of the Cartagena Declaration?', *Internacia: Revista de relaciones internacionales*, (1), pp. 1–12.

Palagi, E., Coronese, M., Lamperti, F. and Roventini, A. (2022) 'Climate change and the nonlinear impact of precipitation anomalies on income inequality', *Proceedings of the National Academy of Sciences*, 119(43), p. e2203595119. https://doi.org/10.1073/pnas.2203595119.

Palloni, A., Massey, D. S., Ceballos, M., Espinosa, K. and Spittel, M. (2001) 'Social capital and international migration: a test using information on family networks', *American Journal of Sociology*, 106(5), pp. 1262–1298. https://doi.org/10.1086/320817.

Pandolfi, J. M., Connolly, S. R., Marshall, D. J. and Cohen, A. L. (2011) 'Projecting coral reef futures under global warming and ocean acidification', *Science*, 333(6041), pp. 418–422. https://doi.org/10.1126/science.1204794.

du Parc, E. and Yasukawa, L. (2020) *The 2019–2020 Australian Bushfires: from temporary evacuation to longer-term displacement*. Geneva, Switzerland: The Internal Displacement Monitoring Centre, p. 25.

Parisien, M.-A., Barber, Q. E., Bourbonnais, M. L., Daniels, L. D., Flannigan, M. D., Gray, R. W., Hoffman, K. M., Jain, P., Stephens, S. L., Taylor, S. W. and Whitman, E. (2023) 'Abrupt, climate-induced increase in wildfires in British Columbia since the mid-2000s', *Communications Earth & Environment*, **4**, p. 309. https://doi.org/10.1038/s43247-023-00977-1.

Parisien, M.-A., Barber, Q. E., Hirsch, K. G., Stockdale, C. A., Erni, S., Wang, X., Arseneault, D. and Parks, S. A. (2020) 'Fire deficit increases wildfire risk for many communities in the Canadian boreal forest', *Nature Communications*, 11(1), p. 2121. https://doi.org/10.1038/s41467-020-15961-y.

Pathak, S., Panta, H. K., Bhandari, T. and Paudel, K. P. (2020) 'Flood vulnerability and its influencing factors', *Natural Hazards*, 104(3), pp. 2175–2196. https://doi.org/10.1007/s11069-020-04267-3.

Patricola, C. M. and Wehner, M. F. (2018) 'Anthropogenic influences on major tropical cyclone events', *Nature*, 563(7731), pp. 339–346. https://doi.org/10.1038/s41586-018-0673-2.

Pattyn, F. and Morlighem, M. (2020) 'The uncertain future of the Antarctic Ice Sheet', *Science*, 367(6484), pp. 1331–1335. https://doi.org/10.1126/science.aaz5487.

Paul, B. K. (1998) 'Coping with the 1996 tornado in Tangail, Bangladesh: an analysis of field data', *The Professional Geographer*, 50(3), pp. 287–301. https://doi.org/10.1111/0033-0124.00121.

Paul, B. K. (2009) 'Why relatively fewer people died? The case of Bangladesh's Cyclone Sidr', *Natural Hazards*, 50(2), pp. 289–304. https://doi.org/10.1007/s11069-008-9340-5.

Pauli, B.J. (2020) 'The Flint water crisis', *WIREs Water*, 7(3), p. e1420. https://doi.org/10.1002/wat2.1420.

Pavanello, F., De Cian, E., Davide, M., Mistry, M., Cruz, T., Bezerra, P., Jagu, D., Renner, S., Schaeffer, R. and Lucena, A. F. P. (2021) 'Air-conditioning and the adaptation cooling deficit in emerging economies', *Nature Communications*, 12(1), p. 6460. https://doi.org/10.1038/s41467-021-26592-2.

Pavel, T., Hasan, S., Halim, N. and Mozumder, P. (2023) 'Impacts of transient and permanent environmental shocks on internal migration', *Applied Economics*, 55(4), pp. 333–356. https://doi.org/10.1080/00036846.2022.2087859.

Payne, A. E., Demory, M.-E., Leung, L. R., Ramos, A. M., Shields, C. A., Rutz, J. J., Siler, N., Villarini, G., Hall, A. and Ralph, F. M. (2020) 'Responses and impacts of atmospheric rivers to climate change', *Nature Reviews Earth & Environment*, 1(3), pp. 143–157. https://doi.org/10.1038/s43017-020-0030-5.

PDD (2020) *Internal Displacement in the Context of Disasters and the Adverse Effects of Climate Change: Submission to the High-Level Panel on Internal Displacement by the Envoy of the Chair of the Platform on Disaster Displacement*. Geneva: Platform on Disaster Displacement (PDD), p. 52. https://disasterdisplacement.org/wp-content/uploads/2020/09/27052020_HLP_submission_SCREEN_compressed.pdf.

Pearce, T. (2018) 'Incorporating Indigenous Knowledge in Research', in R. McLeman and F. Gemenne (eds.) *Routledge Handbook of Environmental Displacement and Migration*. 1st ed. University of the Sunshine Coast, Queensland; Sustainability Research Centre: Routledge, pp. 125–134. www.taylorfrancis.com/chapters/edit/10.4324/9781315638843-10/incorporating-indigenous-knowledge-research-tristan-pearce.

Pearce, T., Ford, J., Willox, A. C. and Smit, B. (2015) 'Inuit traditional ecological knowledge (TEK), subsistence hunting and adaptation to climate change in the Canadian arctic', *Arctic*, 68(2), pp. 233–245.

Pearse, R. (2017) 'Gender and climate change', *WIREs Climate Change*, 8(2), p. e451. https://doi.org/10.1002/wcc.451.

Peek, L. (2019) 'Review of *Weathering Katrina*: culture and recovery among Vietnamese Americans', *Contemporary Sociology*, 48(2), pp. 225–227.

Pena, L. D. and Goldstein, S. L. (2014) 'Thermohaline circulation crisis and impacts during the mid-Pleistocene transition', *Science*, 345(6194), pp. 318–322. https://doi.org/10.1126/science.1249770.

Penu, D. A. K. and Paalo, S. A. (2021) 'Institutions and pastoralist conflicts in Africa: a conceptual framework', *Journal of Peacebuilding & Development*, 16(2), pp. 224–241. https://doi.org/10.1177/15423166211995733.

Pessar, P. R. and Mahler, S. J. (2003) 'Transnational migration: bringing gender in', *International Migration Review*, 37(3), pp. 812–846. https://doi.org/10.1111/j.1747-7379.2003.tb00159.x.

Pham, E. O., Emrich, C. T., Li, Z., Mitchem, J. and Cutter, S. L. (2020) 'Evacuation departure timing during Hurricane Matthew', *Weather, Climate, and Society*, 12(2), pp. 235–248. https://doi.org/10.1175/WCAS-D-19-0030.1.

Pielke, R. A., Rubiera, J., Landsea, C., Fernández, M. L. and Klein, R. (2003) 'Hurricane vulnerability in Latin America and the Caribbean: normalized damage and loss potentials', *Natural Hazards Review*, 4(3), pp. 101–114. https://doi.org/10.1061/(ASCE)1527-6988(2003)4:3(101).

Piggott-McKellar, A., McMichael, C. and Powell, T. (2021) 'Generational retreat: locally driven adaption to coastal hazard risk in two Indigenous communities in Fiji', *Regional Environmental Change*, 21(2), p. 49. https://doi.org/10.1007/s10113-021-01780-4.

Piguet, E. (2010) 'Linking climate change, environmental degradation, and migration: a methodological overview', *WIREs Climate Change*, 1(4), pp. 517–524. https://doi.org/10.1002/wcc.54.

Piguet, E. (2022) 'Linking climate change, environmental degradation, and migration: an update after 10 years', *WIREs Climate Change*, 13(1), p. e746. https://doi.org/10.1002/wcc.746.

Piguet, E., Kaenzig, R. and Guélat, J. (2018) 'The uneven geography of research on "environmental migration"', *Population and Environment*, 39(4), pp. 357–383. https://doi.org/10.1007/s11111-018-0296-4.

Pinheiro de Castro Simão, H. (2024) 'The Cartagena "Spirit" as a third world human rights alternative to refugee protection: lessons to learn from Brazil's approach to Venezuelan socio-economic refugee', *The International Journal of Human Rights*, 28(4), pp. 529–554. https://doi.org/10.1080/13642987.2023.2277743.

Plane, D. A., Henrie, C.J. and Perry, M. J. (2005) 'Migration up and down the urban hierarchy and across the life course', *Proceedings of the National Academy of Sciences*, 102(43), pp. 15313–15318. https://doi.org/10.1073/pnas.0507312102.

Pokhrel, R., Cos, S. del, Montoya Rincon, J. P., Glenn, E. and González, J. E. (2021) 'Observation and modeling of Hurricane Maria for damage assessment', *Weather and Climate Extremes*, 33, p. 100331. https://doi.org/10.1016/j.wace.2021.100331.

Pompeani, D. P., Bird, B. W., Wilson, J. J., Gilhooly, W. P., Hillman, A. L., Finkenbinder, M.S. and Abbott, M. B. (2021) 'Severe Little Ice Age drought in the midcontinental United States during the Mississippian abandonment of Cahokia', *Scientific Reports*, 11(1), p. 13829. https://doi.org/10.1038/s41598-021-92900-x.

Pörtner, H.-O., Roberts, D. C., Masson-Delmotte, V., Zhai, P., Poloczanska, E., Mintenbeck, K., Tignor, M., Algeria, A., Nicolai, M., Okem, J., Petzold, J., Rama, B. and Weyer, N. M. (eds.) (2019) 'Technical Summary', in IPCC, *IPCC Special Report on the Ocean and Cryosphere in a Changing Climate*. Cambridge, UK and New York, NY, USA: Cambridge University Press, pp. 39–70. https://doi.org/10.1017/9781009157964.002.

Potvin, C. K., Broyles, C., Skinner, P. S. and Brooks, H. E. (2022) 'Improving Estimates of U.S. Tornado Frequency by Accounting for Unreported and Underrated Tornadoes', *Journal of Applied Meteorology and Climatology*, 61(7), pp. 909–930. https://doi.org/10.1175/JAMC-D-21-0225.1.

Pragathi, M. S. and Anitha, M. (2019) 'Coping strategies during drought situation: a case of dry land farmers in Rayalaseema region of Andhra Pradesh, India', *Indian Journal of Economics and Development*, pp. 1–14.

Pralle, S. (2019) 'Drawing lines: FEMA and the politics of mapping flood zones', *Climatic Change*, 152(2), pp. 227–237. https://doi.org/10.1007/s10584-018-2287-y.

Protection of Persons Displaced Across Borders in the Context of Disasters and the Adverse Effects of Climate Change [Policy Brief] (2023). Platform on Disaster Displacement (PDD) and The UN Refugee Agency (UNHCR), p. 14. https://disasterdisplacement.org/wp-content/uploads/2023/12/PDD-UNHCR_GRF_Policy_Brief.pdf.

Public Safety Canada (2024) 'Canadian Disaster Database entry'. https://cdd.publicsafety.gc.ca/dtprnt-eng.aspx?cultureCode=en-Ca&eventTypes=%27WF%27&normalizedCostYear=1&dynamic=false&eventId=1135&prnt=both.

Puente, G.B., Perez, F. and Gitter, R. J. (2016) 'The effect of rainfall on migration from Mexico to the United States', *International Migration Review*, 50(4), pp. 890–909. https://doi.org/10.1111/imre.12116.

Qiang, Y. (2019) 'Disparities of population exposed to flood hazards in the United States', *Journal of Environmental Management*, 232, pp. 295–304. https://doi.org/10.1016/j.jenvman.2018.11.039.

Quiñones, E. J., Liebenehm, S. and Sharma, R. (2021) 'Left home high and dry-reduced migration in response to repeated droughts in Thailand and Vietnam', *Population and Environment*, 42(4), pp. 579–621. https://doi.org/10.1007/s11111-021-00374-w.

Rademacher-Schulz, C., Schraven, B. and Mahama, E. S. (2014) 'Time matters: shifting seasonal migration in Northern Ghana in response to rainfall variability and food insecurity', *Climate and Development*, 6(1), pp. 46–52. https://doi.org/10.1080/17565529.2013.830955.

Radford, J. and Connor, P. (2019) 'Canada now leads the world in refugee resettlement, surpassing the U.S.', *Pew Research Center*, 19 June. www.pewresearch.org/short-reads/2019/06/19/canada-now-leads-the-world-in-refugee-resettlement-surpassing-the-u-s/ (Accessed: 27 May 2024).

Raff, J. L., Goodbred, S. L., Pickering, J. L., Sincavage, R. S., Ayers, J. C., Hossain, M. S., Wilson, C. A., Paola, C., Steckler, M. S., Mondal, D. R., Grimaud, J.-L., Grall, C. J., Rogers, K. G., Ahmed, K. M., Akhter, S. H., Carlson, B. N., Chamberlain, E. L., Dejter, M., Gilligan, J. M., Hale, R. P., Khan, M. R., Muktadir, M. G., Rahman, M. M.

and Williams, L. A. (2023) 'Sediment delivery to sustain the Ganges-Brahmaputra delta under climate change and anthropogenic impacts', *Nature Communications*, 14(1), p. 2429. https://doi.org/10.1038/s41467-023-38057-9.

Rahman, Md. S. (2013) 'Climate change, disaster and gender vulnerability: a study on two divisions of Bangladesh', *American Journal of Human Ecology*, 2(2), pp. 72–82. https://doi.org/10.11634/216796221504315.

Railsback, S. F. and Grimm, V. (2012) *Agent-based and individual-based modeling: a practical introduction*. Princeton: Princeton University Press.

Rain, D. (1999) *Eaters of The Dry Season: Circular Labor Migration In The West African Sahel*. Boulder, Colorado: Westview Press. https://doi.org/10.4324/9780429500886.

Raker, E. J. (2020) 'Natural hazards, disasters, and demographic change: the case of severe tornadoes in the United States, 1980–2010', *Demography*, 57(2), pp. 653–674. https://doi.org/10.1007/s13524-020-00862-y.

Raleigh, C. and Kniveton, D. (2012) 'Come rain or shine: an analysis of conflict and climate variability in East Africa', *Journal of Peace Research*, 49(1), pp. 51–64. https://doi.org/10.1177/0022343311427754.

Ramseyer, C., Holliday, L. and Floyd, R. (2015) 'Enhanced residential building code for tornado safety', *Journal of Performance of Constructed Facilities*, 30(4), p. 04015084. https://doi.org/10.1061/(ASCE)CF.1943-5509.0000832.

Randall, A. (2014) Don't call them "refugees": Why climate-change victims need a different label. *The Guardian*. www.theguardian.com/vital-signs/2014/sep/18/refugee-camps-climate-change-victims-migration-pacific-islands. (online, *18 September 2014*).

Rao, S., Doherty, F. C., Teixeira, S., Takeuchi, D. T. and Pandey, S. (2023) 'Social and structural vulnerabilities: associations with disaster readiness', *Global Environmental Change*, 78, p. 102638. https://doi.org/10.1016/j.gloenvcha.2023.102638.

Ratha, D., Chandra, V., Eung Ju, K., Plaza, S. and Shaw, W. (2023) *Migration and Development Brief 39: Leveraging Diaspora Finances for Private Capital Mobilization*. Washington, D.C.: World Bank, p. 62. https://knomad.org/sites/default/files/publication-doc/migration_development_brief_39.pdf.

Raymond, C., Matthews, T. and Horton, R. M. (2020) 'The emergence of heat and humidity too severe for human tolerance', *Science Advances*, 6(19), p. eaaw1838. https://doi.org/10.1126/sciadv.aaw1838.

Refugee Council of Australia (2020) *Australia's offshore processing regime: The facts*. www.refugeecouncil.org.au/offshore-processing-facts/ (Accessed: 26 May 2024).

Refugees International (2021) Task Force Report to the President on the Climate Crisis and Global Migration: A Pathway to Protection for People on the Move. www.refugeesinternational.org/reports-briefs/task-force-report-to-the-president-on-the-climate-crisis-and-global-migration-a-pathway-to-protection-for-people-on-the-move/#endnotes.

Reichman, D. R. (2022) 'Putting climate-induced migration in context: the case of Honduran migration to the USA', *Regional Environmental Change*, 22(3), p. 91. https://doi.org/10.1007/s10113-022-01946-8.

Rentschler, J. and Salhab, M. (2020) *People in Harm's Way: Flood Exposure and Poverty in 189 Countries*. The World Bank (Policy Research Working Papers). https://doi.org/10.1596/1813-9450-9447.

Rentschler, J., Avner, P., Marconcini, M., Su, R., Strano, E., Vousdoukas, M. and Hallegatte, S. (2023) 'Global evidence of rapid urban growth in flood zones since 1985'. *Nature*, 622, pp. 87–92. https://doi.org/10.1038/s41586-023-06468-9.

Rhoades, J. L., Gruber, J. S. and Horton, B. (2018) 'Developing an in-depth understanding of elderly adult's vulnerability to climate change', *The Gerontologist*, 58(3), pp. 567–577. https://doi.org/10.1093/geront/gnw167.

Riccio, B. (2001) 'From "ethnic group" to "transnational community"? Senegalese migrants' ambivalent experiences and multiple trajectories', *Journal of Ethnic and Migration Studies*, 27(4), pp. 583–599. https://doi.org/10.1080/13691830120090395.

Rigaud, K. K., de Sherbinin, A., Jones, B., Bergmann, J., Clement, V., Ober, K., Schewe, J., Adamo, S., McCusker, B., Heuser, S. and Midgley, A. (2018) *Groundswell: Preparing for Internal Climate Migration. Publications & Research*. Washington, DC: World Bank. https://doi.org/10.1596/29461.

Rigg, J., Grundy-Warr, C., Law, L. and Tan-Mullins, M. (2008) 'Grounding a natural disaster: Thailand and the 2004 tsunami', *Asia Pacific Viewpoint*, 49(2), pp. 137–154. https://doi.org/10.1111/j.1467-8373.2008.00366.x.

Riosmena, F., Nawrotzki, R. and Hunter, L. (2018) 'Climate migration at the height and end of the great Mexican emigration era', *Population and development review*, 44(3), pp. 455–488. https://doi.org/10.1111/padr.12158.

Ristroph, E. B. (2021) 'Navigating climate change adaptation assistance for communities: a case study of Newtok Village, Alaska', *Journal of Environmental Studies and Sciences*, 11(3), pp. 329–340. https://doi.org/10.1007/s13412-021-00711-3.

Rivera-Batiz, F. L. (1999) 'Undocumented workers in the labor market: an analysis of the earnings of legal and illegal Mexican immigrants in the United States', *Journal of Population Economics*, 12(1), pp. 91–116. https://doi.org/10.1007/s001480050092.

Robinson, P. J. (2005) 'Ice and snow in paintings of Little Ice Age winters', *Weather*, 60(2), pp. 37–41. https://doi.org/10.1256/wea.164.03.

Robinson, S. and Butchart, C. (2022) 'Planning for climate change in small island developing states: can Dominica's climate resilience and recovery plan be a model for transformation in the Caribbean?', *Sustainability*, 14(9). https://doi.org/10.3390/su14095089.

Robinson, S., Vega Troncoso, A., Roberts, J. T. and Peck, M. (2023) '"We are a people": sovereignty and disposability in the context of Puerto Rico's post-Hurricane Maria experience', *The Geographical Journal*, 189(4), pp. 575–583. https://doi.org/10.1111/geoj.12472.

Roderick, M. L., Greve, P. and Farquhar, G. D. (2015) 'On the assessment of aridity with changes in atmospheric CO_2', *Water Resources Research*, 51(7), pp. 5450–5463. https://doi.org/10.1002/2015WR017031.

Rodina, L. (2019) 'Water resilience lessons from Cape Town's water crisis', *WIREs Water*, 6(6), p. e1376. https://doi.org/10.1002/wat2.1376.

Rodriguez-Díaz, C. E. and Lewellen-Williams, C. (2020) 'Race and racism as structural determinants for emergency and recovery response in the aftermath of Hurricanes Irma and Maria in Puerto Rico', *Health Equity*, 4(1), pp. 232–238. https://doi.org/10.1089/heq.2019.0103.

Romanello, M., Napoli, C. D., Green, C., Kennard, H., Lampard, P., Scamman, D., Walawender, M., Ali, Z., Ameli, N., Ayeb-Karlsson, S., Beggs, P. J., Belesova, K., Berrang Ford, L., Bowen, K., Cai, W., Callaghan, M., Campbell-Lendrum, D., Chambers, J., Cross, T. J., Costello, A. (2023). 'The 2023 report of the Lancet Countdown on health and climate change: The imperative for a health-centred response in a world facing irreversible harms'. *The Lancet*, 402(10419), pp. 2346–2394. https://doi.org/10.1016/S0140-6736(23)01859-7

Romero, L. A. (2023) 'Malleable detention: the restructuring of carceral space within U.S. immigration detention', *Punishment & Society*, 25(4), pp. 848–866. https://doi.org/10.1177/14624745221109539.

Roos, N., Kovats, S., Hajat, S., Filippi, V., Chersich, M., Luchters, S., Scorgie, F., Nakstad, B., Stephansson, O. and Consortium, C. (2021) 'Maternal and newborn health risks of climate change: a call for awareness and global action', *Acta Obstetricia et Gynecologica Scandinavica*, 100(4), pp. 566–570. https://doi.org/10.1111/aogs.14124.

Rosencrants, T. D. and Ashley, W. S. (2015) 'Spatiotemporal analysis of tornado exposure in five US metropolitan areas', *Natural Hazards*, 78(1), pp. 121–140. https://doi.org/10.1007/s11069-015-1704-z.

Rosengärtner, S. K., De Sherbinin, A. M. and Stojanov, R. (2023) 'Supporting the agency of cities as climate migration destinations', *International Migration*, 61(5), pp. 98–115. https://doi.org/10.1111/imig.13024.

Rosenow-Williams, K. and Gemenne, F. (eds.) (2016) *Organizational Perspectives on Environmental Migration*. 1st ed. London and New York: Routledge (Routledge Studies in Development, Mobilities and Migration).

Rude, B., Niederhöfer, B. and Ferrara, F. (2021) 'Deforestation and migration', *CESifo Forum*, 22(01), pp. 49–57.

Rumbach, A., Sullivan, E. and Makarewicz, C. (2020) 'Mobile home parks and disasters: understanding risk to the third housing type in the United States', *Natural Hazards Review*, 21(2), p. 05020001. https://doi.org/10.1061/(ASCE)NH.1527-6996.0000357.

Sadiq, K. and Tsourapas, G. (2021) 'The postcolonial migration state', *European Journal of International Relations*, 27(3), pp. 884–912. https://doi.org/10.1177/13540661211000114.

Saintilan, N., Khan, N. S., Ashe, E., Kelleway, J. J., Rogers, K., Woodroffe, C. D. and Horton, B. P. (2020) 'Thresholds of mangrove survival under rapid sea level rise', *Science*, 368(6495), pp. 1118–1121. https://doi.org/10.1126/science.aba2656.

Sakib, M. S., Alam, S., Shampa, Murshed, S. B., Kirtunia, R., Mondal, M. S. and Chowdhury, A. I. (2023) 'Impact of urbanization on pluvial flooding: insights from a fast growing megacity, Dhaka', *Water*, 15(21). https://doi.org/10.3390/w15213834.

Salt, J. and Stein, J. (1997) 'Migration as a business: the case of trafficking', *International Migration*, 35(4), pp. 467–494. https://doi.org/10.1111/1468-2435.00023.

Samal, P., Babu, S. C., Mondal, B. and Mishra, S. N. (2022) 'The global rice agriculture towards 2050: an inter-continental perspective', *Outlook on Agriculture*, 51(2), pp. 164–172. https://doi.org/10.1177/00307270221088338.

Sana, M. (2005) 'Buying membership in the transnational community: migrant remittances, social status, and assimilation', *Population Research and Policy Review*, 24(3), pp. 231–261. https://doi.org/10.1007/s11113-005-4080-7.

Sanders, B. F., Schubert, J. E., Goodrich, K. A., Houston, D., Feldman, D. L., Basolo, V., Luke, A., Boudreau, D., Karlin, B., Cheung, W., Contreras, S., Reyes, A., Eguiarte, A., Serrano, K., Allaire, M., Moftakhari, H., AghaKouchak, A. and Matthew, R. A. (2020) 'Collaborative modeling with fine-resolution data enhances flood awareness, minimizes differences in flood perception, and produces actionable flood maps', *Earth's Future*, 8(1), p. e2019EF001391. https://doi.org/10.1029/2019EF001391.

Santos-Lozada, A. R., Kaneshiro, M., McCarter, C. and Marazzi-Santiago, M. (2020) 'Puerto Rico exodus: long-term economic headwinds prove stronger than Hurricane Maria', *Population and Environment*, 42(1), pp. 43–56. https://doi.org/10.1007/s11111-020-00355-5.

Sarkar, B., Dutta, S. and Singh, P. K. (2022) 'Drought and temporary migration in rural India: a comparative study across different socio-economic groups with a cross-sectional nationally representative dataset', *PLOS ONE*, 17(10), p. e0275449. https://doi.org/10.1371/journal.pone.0275449.

Sassen, S. (2001) 'Global Cities and Global City-Regions: A Comparison', in A.J. Scott (ed.) *Global City-Regions Trends, Theory, Policy*. Oxford University Press, pp. 78–95. https://doi.org/10.1093/oso/9780198297994.003.0007.

Scheffran, J., Marmer, E. and Sow, P. (2012) 'Migration as a contribution to resilience and innovation in climate adaptation: social networks and co-development in Northwest Africa', *Applied Geography*, 33, pp. 119–127. https://doi.org/10.1016/j.apgeog.2011.10.002.

Schewel, K. (2020) 'Understanding immobility: moving beyond the mobility bias in migration studies', *International Migration Review*, 54(2), pp. 328–355. https://doi.org/10.1177/0197918319831952.

Schewel, K. and Asmamaw, L. B. (2021) 'Migration and development in Ethiopia: exploring the mechanisms behind an emerging mobility transition', *Migration Studies*, 9(4), pp. 1673–1707. https://doi.org/10.1093/migration/mnab036.

Schewel, K., Dickerson, S., Madson, B. and Nagle Alverio, G. (2024) 'How well can we predict climate migration? A review of forecasting models', *Frontiers in Climate*, 5. https://doi.org/10.3389/fclim.2023.1189125.

Schlingmann, A., Graham, S., Benyei, P., Corbera, E., Martinez Sanesteban, I., Marelle, A., Soleymani-Fard, R. and Reyes-García, V. (2021) 'Global patterns of adaptation to climate change by Indigenous Peoples and local communities. A systematic review', *Current Opinion in Environmental Sustainability*, 51, pp. 55–64. https://doi.org/10.1016/j.cosust.2021.03.002.

Schmidlin, T. W., Hammer, B. O., Ono, Y. and King, P. S. (2009) 'Tornado shelter-seeking behavior and tornado shelter options among mobile home residents in the United States', *Natural Hazards*, 48(2), pp. 191–201. https://doi.org/10.1007/s11069-008-9257-z.

Schmidtke, R. and Ober, K. (2023) *Two years after eta and iota: displaced and forgotten in Guatemala*. Refugees International. www.refugeesinternational.org/reports-briefs/two-years-after-eta-and-iota-displaced-and-forgotten-in-guatemala/.

Schraven, B. (2011) 'Action approaches for environmentally induced migration (a state of the art report)', *International Conference*, p. 17.

Schwerdtle, P., Bowen, K. and McMichael, C. (2018) 'The health impacts of climate-related migration', *BMC Medicine*, 16(1), p. 1. https://doi.org/10.1186/s12916-017-0981-7.

Schwerdtle, P. N., McMichael, C., Mank, I., Sauerborn, R., Danquah, I. and Bowen, K. J. (2020). 'Health and migration in the context of a changing climate: a systematic literature assessment'. *Environmental Research Letters*, 15(10), p. 103006. https://doi.org/10.1088/1748-9326/ab9ece.

Scissa, C. (2022) 'The climate changes, should EU migration law change as well? Insights from Italy', *European Journal of Legal Studies*, 14(1), pp. 5–23. https://doi.org/10.2924/EJLS.2022.011.

Scott, M. (2018) *Hurricane Maria's devastation of Puerto Rico | NOAA Climate.gov, Understanding Climate*. www.climate.gov/news-features/understanding-climate/hurricane-marias devastation-puerto-rico (Accessed: 21 April 2024).

Seddon, N., Chausson, A., Berry, P., Girardin, C. A. J., Smith, A. and Turner, B. (2020) 'Understanding the value and limits of nature-based solutions to climate change and other global challenges', *Philosophical Transactions of the Royal Society B: Biological Sciences*, 375(1794), p. 20190120. https://doi.org/10.1098/rstb.2019.0120.

Šedová, B., Čizmaziová, L. and Cook, A. (2021) 'A meta-analysis of climate migration literature', *CEPA Discussion Papers* [Preprint], (29). https://doi.org/10.25932/publishup-49982.

Sedova, B. and Kalkuhl, M. (2020) 'Who are the climate migrants and where do they go? Evidence from rural India', *World Development*, 129, p. 104848. https://doi.org/10.1016/j.worlddev.2019.104848.

Selby, J., Dahi, O. S., Fröhlich, C. and Hulme, M. (2017) 'Climate change and the Syrian civil war revisited', *Political Geography*, 60, pp. 232–244. https://doi.org/10.1016/j.polgeo.2017.05.007.

Seneviratne, S.I., Zhang, X., Muhammad, A., Badi, W., Dereczynski, C., Di Luca, A., Ghosh, S., Iskandar, J., Kossin, S., Lewis, F., Otto, I., Pinto, M., Satoh, S. M., Vicente-Serrano, M., Wehner, M. and Zhou, B. (2021) 'Weather and Climate Extreme Events in a Changing Climate', in V. Masson-Delmotte, P. Zhai, A. Pirani, S. L. Connors, C. Péan, S. Berger, N. Caud, L. Goldfarb, M. I. Gomis, K. Huang, E. Leitzell, E. Lonnoy, J. B. R. Matthews, T. K. Maycock, O. Waterfield, R. Yelekçi, R. Yu, and B. Zhou (eds.) *Climate Change 2021: The Physical Science Basis. Working Group I Contribution to the Sixth Assessment Report of the Intergovernmental Panel on Climate Change.* Cambridge, United Kingdom and New York, NY, USA: Cambridge University Press, pp. 1513–1766. https://doi.org/10.1017/9781009157896.013.

Serdeczny, O. (2017) What does it mean to 'address displacement' under the UNFCCC? An analysis of the negotiations process and the role of research [Discussion paper]. Bonn: German Development Institute/Deutsches Institut für Entwicklungspolitik (DIE).

Sesana, E., Gagnon, A.S., Ciantelli, C., Cassar, J. and Hughes, J. J. (2021) 'Climate change impacts on cultural heritage: a literature review', *WIREs Climate Change*, 12(4), p. e710. https://doi.org/10.1002/wcc.710.

Shahi, A. (2019) 'Drought: the Achilles heel of the Islamic Republic of Iran', *Asian Affairs*, 50(1), pp. 18–39. https://doi.org/10.1080/03068374.2019.1567100.

Sharygin, E. (2021) 'Estimating migration impacts of wildfire: California's 2017 North Bay fires', in D. Karácsonyi, A. Taylor, and D. Bird (eds.) *The Demography of Disasters: Impacts for Population and Place.* Cham: Springer International Publishing, pp. 49–70. https://doi.org/10.1007/978-3-030-49920-4_3.

Shaw, J. M. (2003) 'Climate change and deforestation: implications for the Maya collapse', *Ancient Mesoamerica*, 14(1), pp. 157–167. https://doi.org/10.1017/S0956536103132063.

Shehzad, K. (2023) 'Extreme flood in Pakistan: is Pakistan paying the cost of climate change? A short communication', *Science of The Total Environment*, 880, p. 162973. https://doi.org/10.1016/j.scitotenv.2023.162973.

de Sherbinin, A., Grace, K., McDermid, S., van der Geest, K., Puma, M. J. and Bell, A. (2022) 'Migration theory in climate mobility research', *Frontiers in Climate*, 4. https://doi.org/10.3389/fclim.2022.882343.

Sherwood, S. C. and Huber, M. (2010) 'An adaptability limit to climate change due to heat stress', *Proceedings of the National Academy of Sciences*, 107(21), pp. 9552–9555. https://doi.org/10.1073/pnas.0913352107.

Shi, J.-R., Santer, B. D., Kwon, Y.-O. and Wijffels, S. E. (2024) 'The emerging human influence on the seasonal cycle of sea surface temperature', *Nature Climate Change*, 14(4), pp. 364–372. https://doi.org/10.1038/s41558-024-01958-8.

Shi, K. and Touge, Y. (2023) 'Identifying the shift in global wildfire weather conditions over the past four decades: an analysis based on change-points and long-term trends', *Geoscience Letters*, 10(1), p. 3. https://doi.org/10.1186/s40562-022-00255-6.

Shi, L. and Varuzzo, A.M. (2020) 'Surging seas, rising fiscal stress: exploring municipal fiscal vulnerability to climate change', *Cities*, 100, p. 102658. https://doi.org/10.1016/j.cities.2020.102658.

Shivakoti, R., Henderson, S. and Withers, M. (2021) 'The migration ban policy cycle: a comparative analysis of restrictions on the emigration of women domestic workers', *Comparative Migration Studies*, 9(1), p. 36. https://doi.org/10.1186/s40878-021-00250-4.

Shultz, J. M., Rechkemmer, A., Rai, A. and McManus, K. T. (2019) 'Public health and mental health implications of environmentally induced forced migration', *Disaster Medicine and Public Health Preparedness*, 13(2), pp. 116–122. https://doi.org/10.1017/dmp.2018.27.

Shvidenko, A. Z. and Schepaschenko, D. G. (2013) 'Climate change and wildfires in Russia', *Contemporary Problems of Ecology*, 6(7), pp. 683–692. https://doi.org/10.1134/S199542551307010X.

Shvidenko, A. Z. and Schepaschenko, D. G. (2014). Carbon budget of Russian forests. *Siberian Journal of Forest Science*, 1, 69–92.

Siebeneck, L., Schumann, R., Kuenanz, B.-J., Lee, S., Benedict, B. C., Jarvis, C. M. and Ukkusuri, S. V. (2020) 'Returning home after Superstorm Sandy: phases in the return-entry process', *Natural Hazards*, 101(1), pp. 195–215. https://doi.org/10.1007/s11069-020-03869-1.

Siegert, M., Alley, R.B., Rignot, E., Englander, J. and Corell, R. (2020) 'Twenty-first century sea-level rise could exceed IPCC projections for strong-warming futures', *One Earth*, 3(6), pp. 691–703. https://doi.org/10.1016/j.oneear.2020.11.002.

Simms, J. R. Z., Waller, H. L., Brunet, C. and Jenkins, P. (2021) 'The long goodbye on a disappearing, ancestral island: a just retreat from Isle de Jean Charles', *Journal of Environmental Studies and Sciences*, 11(3), pp. 316–328. https://doi.org/10.1007/s13412-021-00682-5.

Singer, A. and Massey, D. S. (1998) 'The social process of undocumented border crossing among Mexican migrants', *International Migration Review*, 32(3), pp. 561–592. https://doi.org/10.1177/019791839803200301.

Singer, A. and Wilson, J. H. (2010) *The Impact of the Great Recession on Metropolitan Immigration Trends. Research.* Washington, D.C.: The Brookings Institution. www.brookings.edu/articles/the-impact-of-the-great-recession-on-metropolitan-immigration-trends/.

Sjaastad, L. A. (1962) 'The costs and returns of human migration', *Journal of Political Economy*, 70(5, Part 2), pp. 80–93. https://doi.org/10.1086/258726.

Slater, L., Villarini, G., Archfield, S., Faulkner, D., Lamb, R., Khouakhi, A. and Yin, J. (2021) 'Global changes in 20-year, 50-year, and 100-year river floods', *Geophysical Research Letters*, 48(6), p. e2020GL091824. https://doi.org/10.1029/2020GL091824.

Smirnov, O., Lahav, G., Orbell, J., Zhang, M., and Xiao, T. (2023). 'Climate change, drought, and potential environmental migration flows under different policy scenarios'. *International Migration Review*, 57(1), pp. 36–67. https://doi.org/10.1177/01979183221079850.

Smith, G. and Vila, O. (2020) 'A national evaluation of state and territory roles in hazard mitigation: building local capacity to implement FEMA hazard mitigation assistance grants', *Sustainability*, 12(23), p. 10013. https://doi.org/10.3390/su122310013.

Smith, S. K. and House, M. (2006) 'Snowbirds, sunbirds, and stayers: seasonal migration of elderly adults in Florida', *The Journals of Gerontology: Series B*, 61(5), pp. S232–S239. https://doi.org/10.1093/geronb/61.5.S232.

Smith, C. D. (2014). 'Modelling migration futures: Development and testing of the Rainfalls Agent-Based Migration Model – Tanzania'. *Climate and Development*, 6(1), pp. 77–91. https://doi.org/10.1080/17565529.2013.872593.

Snead, R. E. (2010) 'Bangladesh', in E.C.F. Bird (ed.) *Encyclopedia of the World's Coastal Landforms*. Dordrecht: Springer Netherlands, pp. 1077–1080. https://doi.org/10.1007/978-1-4020-8639-7_201.

Sobczak-Szelc, K. and Fekih, N. (2020) 'Migration as one of several adaptation strategies for environmental limitations in Tunisia: evidence from El Faouar', *Comparative Migration Studies*, 8(1), p. 8. https://doi.org/10.1186/s40878-019-0163-1.

     References

Song, J., Pan, R., Yi, W., Wei, Q., Qin, W., Song, S., Tang, C., He, Y., Liu, X., Cheng, J. and Su, H. (2021) 'Ambient high temperature exposure and global disease burden during 1990–2019: an analysis of the Global Burden of Disease Study 2019', *Science of The Total Environment*, 787, p. 147540. https://doi.org/10.1016/j.scitotenv.2021.147540.

Speare, A. (1974) 'Residential satisfaction as an intervening variable in residential mobility', *Demography*, 11(2), pp. 173–188. https://doi.org/10.2307/2060556.

Spilker, G., Nguyen, Q., Koubi, V. and Böhmelt, T. (2020) 'Attitudes of urban residents towards environmental migration in Kenya and Vietnam', *Nature Climate Change*, 10(7), pp. 622–627. https://doi.org/10.1038/s41558-020-0805-1.

Stark, O. and Bloom, D. E. (1985) 'The new economics of labor migration', *The American Economic Review*, 75(2), pp. 173–178.

Stavi, I., Roque de Pinho, J., Paschalidou, A. K., Adamo, S. B., Galvin, K., de Sherbinin, A., Even, T., Heaviside, C. and van der Geest, K. (2022) 'Food security among dryland pastoralists and agropastoralists: the climate, land-use change, and population dynamics nexus', *The Anthropocene Review*, 9(3), pp. 299–323. https://doi.org/10.1177/20530196211007512.

Stevens-Rumann, C. S., Kemp, K. B., Higuera, P. E., Harvey, B. J., Rother, M. T., Donato, D. C., Morgan, P. and Veblen, T. T. (2018) 'Evidence for declining forest resilience to wildfires under climate change', *Ecology Letters*, 21(2), pp. 243–252. https://doi.org/10.1111/ele.12889.

Stoler, J., Brewis, A., Kangmennang, J., Keough, S. B., Pearson, A. L., Rosinger, A. Y., Stauber, C. and Stevenson, E. G. (2021) 'Connecting the dots between climate change, household water insecurity, and migration', *Current Opinion in Environmental Sustainability*, 51, pp. 36–41. https://doi.org/10.1016/j.cosust.2021.02.008.

Stolte, T. R., Moel, H. de, Koks, E. E., Wens, M. L. K., Veldhoven, F. van, Garg, S., Farhad, N. and Ward, P. J. (2023) 'Global drought risk in cities: present and future urban hotspots', *Environmental Research Communications*, 5(11), p. 115008. https://doi.org/10.1088/2515-7620/ad0210.

Storlazzi, C. D., Gingerich, S. B., van Dongeren, A., Cheriton, O. M., Swarzenski, P. W., Quataert, E., Voss, C. I., Field, D. W., Annamalai, H., Piniak, G. A. and McCall, R. (2018) 'Most atolls will be uninhabitable by the mid-21st century because of sea-level rise exacerbating wave-driven flooding', *Science Advances*, 4(4), p. eaap9741. https://doi.org/10.1126/sciadv.aap9741.

Strader, S. M., Ash, K., Wagner, E. and Sherrod, C. (2019) 'Mobile home resident evacuation vulnerability and emergency medical service access during tornado events in the Southeast United States', *International Journal of Disaster Risk Reduction*, 38, p. 101210. https://doi.org/10.1016/j.ijdrr.2019.101210.

Streets, D. G. and Glantz, M. (2000) 'Exploring the concept of climate surprise', *Global Environmental Change*, 10(2), pp. 97–107. https://doi.org/10.1016/S0959-3780(00)00015-7.

Strode, M. and Khanna, M. (2021) 'Improving official statistics on stateless people: challenges, solutions, and the road ahead', *Statistical Journal of the IAOS*, 37(4), pp. 1087–1101. https://doi.org/10.3233/SJI-210878.

Suckall, N., Fraser, E. and Forster, P. (2017) 'Reduced migration under climate change: evidence from Malawi using an aspirations and capabilities framework', *Climate and Development*, 9(4), pp. 298–312. https://doi.org/10.1080/17565529.2016.1149441.

Sue, C. A., Riosmena, F. and LePree, J. (2019) 'The influence of social networks, social capital, and the ethnic community on the U.S. destination choices of Mexican migrant men', *Journal of Ethnic and Migration Studies*, 45(13), pp. 2468–2488. https://doi.org/10.1080/1369183X.2018.1447364.

Suksomboon, P. (2008) 'Remittances and "Social Remittances": Their Impact on Livelihoods of Thai Women in the Netherlands and Non-migrants in Thailand', *Gender, Technology and Development*, 12(3), pp. 461–482. https://doi.org/10.1177/097185240901200309.

Sulikowska, A. and Wypych, A. (2020) 'Summer temperature extremes in Europe: how does the definition affect the results?', *Theoretical and Applied Climatology*, 141(1), pp. 19–30. https://doi.org/10.1007/s00704-020-03166-8.

Summers, J. K., Lamper, A., McMillion, C. and Harwell, L. C. (2022) 'Observed changes in the frequency, intensity, and spatial patterns of nine natural hazards in the United States from 2000 to 2019', *Sustainability*, 14(7). https://doi.org/10.3390/su14074158.

Sunam, R., Barney, K. and McCarthy, J. F. (2021) 'Transnational labour migration and livelihoods in rural Asia: tracing patterns of agrarian and forest change', *Geoforum*, 118, pp. 1–13. https://doi.org/10.1016/j.geoforum.2020.11.004.

Sutter, D. and Simmons, K. M. (2010) 'Tornado fatalities and mobile homes in the United States', *Natural Hazards*, 53(1), pp. 125–137. https://doi.org/10.1007/s11069-009-9416-x.

Swapan, M. S. H. and Sadeque, S. (2021) 'Place attachment in natural hazard-prone areas and decision to relocate: Research review and agenda for developing countries', *International Journal of Disaster Risk Reduction*, 52, p. 101937. https://doi.org/10.1016/j.ijdrr.2020.101937.

Sweet, W. V., Hamlington, B. D., Kopp, R. E., Weaver, C. P., Barnard, P. L., Bekaert, D., Brooks, W., Craghan, M., Dusek, G., Frederikse, T., Garner, G., Genz, A. S., Krasting, J. P., Larour, E., Marcy, D., Marra, J. J., Obeysekera, J., Osler, M., Pendleton, M., Roman, D., Schmied, L., Veatch, W., White, K. D. and Zuzak, C. (2022) *Global and Regional Sea Level Rise Scenarios for the United States: Updated Mean Projections and Extreme Water Level Probabilities Along U.S. Coastlines.* NOAA Technical Report NOS 01. Silver Spring, MD: National Oceanic and Atmospheric Administration (NOAA), U.S. Department of Commerce, National Ocean Service, p. 111. https://oceanservice.noaa.gov/hazards/sealevelrise/noaa-nostechrpt01-global-regional-SLR-scenarios-US.pdf.

Syvitski, J. P. M., Kettner, A. J., Overeem, I., Hutton, E. W. H., Hannon, M. T., Brakenridge, G. R., Day, J., Vörösmarty, C., Saito, Y., Giosan, L., and Nicholls, R. J. (2009). 'Sinking deltas due to human activities'. *Nature Geoscience*, 2(10), pp. 681–686. https://doi.org/10.1038/ngeo629.

Tabari, H. (2020) 'Climate change impact on flood and extreme precipitation increases with water availability'. *Scientific Reports,* 10, p. 13768. https://doi.org/10.1038/s41598-020-70816-2

Tabari, H., Hosseinzadehtalaei, P., Thiery, W. and Willems, P. (2021) 'Amplified drought and flood risk under future socioeconomic and climatic change', *Earth's Future*, 9(10), p. e2021EF002295. https://doi.org/10.1029/2021EF002295.

Taherkhani, M., Vitousek, S., Barnard, P. L., Frazer, N., Anderson, T. R. and Fletcher, C. H. (2020) 'Sea-level rise exponentially increases coastal flood frequency', *Scientific Reports*, 10(1), p. 6466. https://doi.org/10.1038/s41598-020-62188-4.

Tanim, S. H., Wiernik, B. M., Reader, S., and Hu, Y. (2022). 'Predictors of hurricane evacuation decisions: a meta-analysis', *Journal of Environmental Psychology*', 79, p. 101742. https://doi.org/10.1016/j.jenvp.2021.101742.

Tate, E., Rahman, M. A., Emrich, C. T. and Sampson, C. C. (2021) 'Flood exposure and social vulnerability in the United States', *Natural Hazards*, 106(1), pp. 435–457. https://doi.org/10.1007/s11069-020-04470-2.

Taylor, J. E., Poleacovschi, C. and Perez, M. (2023) 'Evaluating the climate change adaptation barriers of critical infrastructure in rural Alaska', *Climate and Development*, 15(7), pp. 553–564. https://doi.org/10.1080/17565529.2022.2123698.

Teicher, H. M. and Marchman, P. (2024) 'Integration as adaptation: advancing research and practice for inclusive climate receiving communities', *Journal of the American Planning Association*, 90(1), pp. 30–49. https://doi.org/10.1080/01944363.2023.2188242.

Tellman, B., Sullivan, J. A., Kuhn, C., Kettner, A. J., Doyle, C. S., Brakenridge, G. R., Erickson, T. A. and Slayback, D. A. (2021) 'Satellite imaging reveals increased proportion of population exposed to floods', *Nature*, 596(7870), pp. 80–86. https://doi.org/10.1038/s41586-021-03695-w.

Tesselaar, M., Botzen, W. J. W., Haer, T., Hudson, P., Tiggeloven, T., and Aerts, J. C. J. H. (2020). 'Regional Inequalities in Flood Insurance Affordability and Uptake under Climate Change'. *Sustainability*, 12(20), p. 8734. https://doi.org/10.3390/su12208734.

Terry, J. P. (ed.) (2007) 'Tropical Cyclogenesis', in *Tropical Cyclones: Climatology and Impacts in the South Pacific*. New York, NY: Springer, pp. 15–25. https://doi.org/10.1007/978-0-387-71543-8_2.

Thalheimer, L. and Oh, W. S. (2023) 'An inventory tool to assess displacement data in the context of weather and climate-related events', *Climate Risk Management*, 40, p. 100509. https://doi.org/10.1016/j.crm.2023.100509.

Thalheimer, L., Williams, D. S., van der Geest, K. and Otto, F. E. L. (2021) 'Advancing the evidence base of future warming impacts on human mobility in African drylands', *Earth's Future*, 9(10), p. e2020EF001958. https://doi.org/10.1029/2020EF001958.

Thanh Thi Pham, N., Nong, D., Raghavan Sathyan, A. and Garschagen, M. (2020) 'Vulnerability assessment of households to flash floods and landslides in the poor upland regions of Vietnam', *Climate Risk Management*, 28, p. 100215. https://doi.org/10.1016/j.crm.2020.100215.

The Causes of Climate Change: Human activities are driving the global warming trend observed since the mid-20th century – NASA Science (2024). https://science.nasa.gov/climate-change/causes/ (Accessed: 27 May 2024).

The Guardian (2005) 'The first climate change refugees', 2 December. www.theguardian.com/theguardian/2005/dec/02/guardianweekly.guardianweekly11 (Accessed: 30 April 2024).

The Maldives Journal (2023) 'Male' area population on the rise: census reveals increase in residents', *The Maldives Journal: Fourth pillar of democracy at work*. Dhivehi Edition, 1 April. https://themaldivesjournal.com/48882 (Accessed: 29 April 2024).

The National Institute of Demographic Studies (INED) (2023) 'Everything you need to know about the population: figures: all countries in the world'. www.ined.fr/fr/tout-savoir-population/chiffres/tous-les-pays-du-monde/#r151.

The New Humanitarian (2008) 'The world's first climate change "refugees"', *The New Humanitarian*, 8 June. www.thenewhumanitarian.org/feature/2008/06/08.

Thiede, B., Gray, C. and Mueller, V. (2016) 'Climate variability and inter-provincial migration in South America, 1970–2011', *Global Environmental Change*, 41, pp. 228–240. https://doi.org/10.1016/j.gloenvcha.2016.10.005.

Thiede, B. C., Robinson, A. and Gray, C. (2024) 'Climatic variability and internal migration in Asia: evidence from big microdata', *Population and Development Review*, n/a(n/a). https://doi.org/10.1111/padr.12612.

Thober, J., Schwarz, N. and Hermans, K. (2018) 'Agent-based modeling of environment-migration linkages: a review', *Ecology and Society*, 23(2). https://doi.org/10.5751/ES-10200-230241.

Thomas, K., Hardy, R. D., Lazrus, H., Mendez, M., Orlove, B., Rivera-Collazo, I., Roberts, J. T., Rockman, M., Warner, B. P. and Winthrop, R. (2019) 'Explaining differential vulnerability to climate change: a social science review', *WIREs Climate Change*, 10(2), p. e565. https://doi.org/10.1002/wcc.565.

Thompson, R. R., Garfin, D.R. and Silver, R. C. (2017) 'Evacuation from natural disasters: a systematic review of the literature', *Risk Analysis*, 37(4), pp. 812–839. https://doi .org/10.1111/risa.12654.

Thorpe, H. (2017) '"The endless winter": transnational mobilities of skilled snow sport workers', *Journal of Ethnic and Migration Studies*, 43(3), pp. 528–545. https://doi .org/10.1080/1369183X.2016.1190638.

Tietjen, B. and Gopalakrishnan, T. (2023) 'Loss and damage funding in the UN climate nego-tiations: from dialogue to reality', *Environment: Science and Policy for Sustainable Development*, 65(3), pp. 18–28. https://doi.org/10.1080/00139157.2023.2180268.

Tiggeloven, T., de Moel, H., Winsemius, H. C., Eilander, D., Erkens, G., Gebremedhin, E., Diaz Loaiza, A., Kuzma, S., Luo, T., Iceland, C., Bouwman, A., van Huijstee, J., Ligtvoet, W. and Ward, P. J. (2020) 'Global-scale benefit–cost analysis of coastal flood adaptation to different flood risk drivers using structural measures', *Natural Hazards and Earth System Sciences*, 20(4), pp. 1025–1044. https://doi.org/10.5194/ nhess-20-1025-2020.

Titus, J. G. (2023) 'Population in floodplains or close to sea level increased in US but declined in some counties – especially among Black residents', *Environmental Research Letters*, 18(3), p. 034001. https://doi.org/10.1088/1748-9326/acadf5.

Tockner, K. and Stanford, J. A. (2002) 'Riverine flood plains: present state and future trends', *Environmental Conservation*. 2002/11/13 ed., 29(3), pp. 308–330. https:// doi.org/10.1017/S037689290200022X.

Todaro, M. P. (1969) 'A model of labor migration and urban unemployment in less devel-oped countries', *The American Economic Review*, 59(1), pp. 138–148.

Tornadoes | National Oceanic and Atmospheric Administration (n.d.) National Oceanic and Atmospheric Administration. www.noaa.gov/education/resource-collections/ weather-atmosphere/tornadoes (Accessed: 21 April 2024).

Torres, J. M. and Casey, J. A. (2017) 'The centrality of social ties to climate migration and mental health', *BMC Public Health*, 17(1), p. 600. https://doi.org/10.1186/ s12889-017-4508-0.

Tory, K. J. and Frank, W. M. (2010) 'Tropical Cyclone Formation', in *Global Perspectives on Tropical Cyclones*. WORLD SCIENTIFIC (World Scientific Series on Asia-Pacific Weather and Climate, Volume 4), pp. 55–91. https://doi.org/10.1142/97898 14293488_0002.

Toth, L. T., Storlazzi, C. D., Kuffner, I. B., Quataert, E., Reyns, J., McCall, R., Stathakopoulos, A., Hillis-Starr, Z., Holloway, N. H., Ewen, K. A., Pollock, C. G., Code, T. and Aronson, R. B. (2023) 'The potential for coral reef restoration to mit-igate coastal flooding as sea levels rise', *Nature Communications*, 14(1), p. 2313. https://doi.org/10.1038/s41467-023-37858-2.

Tran, D. D., Dang, M. M., Du Duong, B., Sea, W. and Vo, T. T. (2021) 'Livelihood vul-nerability and adaptability of coastal communities to extreme drought and salinity intrusion in the Vietnamese Mekong Delta', *International Journal of Disaster Risk Reduction*, 57, p. 102183. https://doi.org/10.1016/j.ijdrr.2021.102183.

Triandafyllidou, A. (2022) 'Migrant Smuggling', in *Routledge Handbook of Immigration and Refugee Studies*. 2nd ed. London: Routledge.

Tropical Cyclone Classification | National Oceanic and Atmospheric Administration (2023) *NOAA.* www.noaa.gov/jetstream/tropical/tropical-cyclone-introduction/ tropical-cyclone-classification (Accessed: 12 April 2024).

Tubridy, F., Lennon, M. and Scott, M. (2022) 'Managed retreat and coastal climate change adaptation: the environmental justice implications and value of a coproduction approach', *Land Use Policy*, 114, p. 105960. https://doi.org/10.1016/j.landusepol.2021.105960.

Tuholske, C., Caylor, K., Funk, C., Verdin, A., Sweeney, S., Grace, K., Peterson, P. and Evans, T. (2021) 'Global urban population exposure to extreme heat', *Proceedings of the National Academy of Sciences*, 118(41), p. e2024792118. https://doi.org/10.1073/pnas.2024792118.

Twinomuhangi, R., Sseviiri, H. and Kato, A. M. (2023) 'Contextualising environmental and climate change migration in Uganda', *Local Environment*, 28(5), pp. 580–601. https://doi.org/10.1080/13549839.2023.2165641.

Ty Miller, J. and Thai Vu, A. (2021) 'Emerging research methods in environmental displacement and forced migration research', *Geography Compass*, 15(4), p. e12558. https://doi.org/10.1111/gec3.12558.

Tyler, J., Sadiq, A.A., and Noonan, D.S. (2019) 'A review of the community flood risk management literature in the USA: lessons for improving community resilience to floods', *Nat Hazards*, 96, pp. 1223–1248. https://doi.org/10.1007/s11069-019-03606-3.

Udmale, P., Ichikawa, Y., Ning, S., Shrestha, S. and Pal, I. (2020) 'A statistical approach towards defining national-scale meteorological droughts in India using crop data', *Environmental Research Letters*, 15(9), p. 094090. https://doi.org/10.1088/1748-9326/abacfa.

von Uexkull, N., Croicu, M., Fjelde, H. and Buhaug, H. (2016) 'Civil conflict sensitivity to growing-season drought', *Proceedings of the National Academy of Sciences*, 113(44), pp. 12391–12396. https://doi.org/10.1073/pnas.1607542113.

UCAR (2023) *What Are Monsoons and Why Do They Happen? UCAR Center for Science Education.* https://scied.ucar.edu/learning-zone/storms/monsoons (Accessed: 22 April 2024).

UK Environment Agency (2021) 'Thames Barrier closed for 200th time: Barrier being closed to protect London from a high tide as a result of low pressure and northerly winds coinciding with spring tides [Press Release]'. GOV.UK. www.gov.uk/government/news/thames-barrier-closed-for-200th-time.

UN Climate Change (n.d. a) *Cancun Agreements | UNFCCC.* https://unfccc.int/process/conferences/pastconferences/cancun-climate-change-conference-november-2010/statements-and-resources/Agreements?gad_source=1 (Accessed: 23 May 2024).

UN Climate Change (n.d. b) *Fund for responding to loss and damage | UNFCCC.* https://unfccc.int/loss-and-damage-fund-joint-interim-secretariat (Accessed: 23 May 2024).

UN Climate Change (n.d. c) *Task Force on Displacement | UNFCCC.* https://unfccc.int/process/bodies/constituted-bodies/WIMExCom/TFD#Phase-3- (Accessed: 23 May 2024).

UN General Assembly (2023) *The Report of the Main findings and recommendations of the Midterm Review of the implementation of the Sendai Framework for Disaster Risk Reduction 2015–2030. UN resolutions and reports.* United Nations General Assembly, p. 20. https://documents.un.org/doc/undoc/gen/n22/764/26/pdf/n2276426.pdf?token=kWsHFV5scZPyn9sGU5&fe=true.

UN High Commissioner for Refugees (UNHCR) (2020) *Legal considerations regarding claims for international protection made in the context of the adverse effects of climate change and disasters.* UNHCR, p. 11. www.refworld.org/policy/legalguidance/unhcr/2020/en/123356.

UNDRR (2023) *What is the Sendai Framework for Disaster Risk Reduction? | UNDRR.* www.undrr.org/implementing-sendai-framework/what-sendai-framework (Accessed: 23 May 2024).

UNEP (2008) *Sudan Post-Conflict Environmental Assessment.* Assessment. Nairobi, Kenya: United Nations Environment Programme (UNEP), p. 354. www.unep.org/resources/assessment/sudan-post-conflict-environmental-assessment (Accessed: 23 April 2024).

UNFCCC (2023) 'Decision -/CP.27 -/CMA.4: Funding arrangements for responding to loss and damage associated with the adverse effects of climate change, including a focus on addressing loss and damage'. https://unfccc.int/sites/default/files/resource/cma4_auv_8f.pdf.

UNHCR (2024) 'Figures at a glance: 108.4 million people worldwide were forcibly displaced'. *The UN Refugee Agency* (UNHCR). www.unhcr.org/about-unhcr/who-we-are/figures-glance.

United Nations Convention to Combat Desertification (UNCCD) (2017) *The Global Land Outlook, 1st ed.* Bonn, Germany. https://digitallibrary.un.org/record/3972346?ln=en&v=pdf.

United Nations Department of Economic and Social Affairs (2021) *Handbook on the Management of Population and Housing Censuses: Revision 2.* United Nations (Studies in Methods (Ser. F)). https://doi.org/10.18356/9789210601467.

United Nations Department of Economic and Social Affairs (2023) *The Sustainable Development Goals Report 2023: Special Edition.* United Nations (The Sustainable Development Goals *Report).* https://doi.org/10.18356/9789210024914.

United Nations Development Programme (UNDP) (2022) *Socio-Economic Impact Assessment of Hurricane Dorian and the COVID-19 Pandemic on MSMEs in The Bahamas: Building Resilience for MSMEs in the face of unprecedented crisis in The Bahamas.* Multi Country Office in Jamaica, p. 181. www.undp.org/jamaica/publications/socio-economic-impact-assessment-hurricane-dorian-and-covid-19-pandemic-msmes-bahamas.

United Nations Environment Programme (2023) *Emissions Gap Report 2023: Broken Record – Temperatures hit new highs, yet world fails to cut emissions (again).* United Nations Environment Programme. https://doi.org/10.59117/20.500.11822/43922.

United Nations (2024) 'Fast Facts – What is Sustainable Development?' United Nations Sustainable Development Goals, 8 August. www.un.org/sustainabledevelopment/blog/2023/08/what-is-sustainable-development/ (Accessed: 23 May 2024).

United States Government Accountability Office (2022) *2017 Hurricanes: Update on FEMA's Disaster Recovery Efforts in Puerto Rico and the U.S. Virgin Islands | U.S. GAO.* Testimony Before the Subcommittee on Economic Development, Public Buildings, and Emergency Management, Committee on Transportation and Infrastructure, House of Representatives. GAO, p. 34. www.gao.gov/assets/gao-22-106211.pdf (Accessed: 21 April 2024).

US Department of Commerce and National Oceanic and Atmospheric Administration (NOAA) (n.d. a) *Tornadoes FAQ, National Weather Service.* NOAA's National Weather Service. www.weather.gov/lmk/tornadoesfaq (Accessed: 21 April 2024).

US Department of Commerce and National Oceanic and Atmospheric Administration (NOAA) (n.d. c) *How do hurricanes form?* https://oceanservice.noaa.gov/facts/how-hurricanes-form.html (Accessed: 21 April 2024).

US Department of Commerce and National Oceanic and Atmospheric Administration (NOAA) (2024) *The Global Conveyor Belt – Currents: NOAA's National Ocean Service Education.* https://oceanservice.noaa.gov/education/tutorial_currents/05conveyor2.html (Accessed: 16 May 2024).

US Department of Commerce and National Oceanic and Atmospheric Administration (NOAA) (n.d.) *What are El Nino and La Nina?* https://oceanservice.noaa.gov/facts/ninonina.html (Accessed: 27 May 2024).

US EPA and OAR (2016) 'Climate Change Indicators: Sea Surface Temperature'. www
.epa.gov/climate-indicators/climate-change-indicators-sea-surface-temperature
(Accessed: 21 April 2024).

USAID (2020) *Honduras: Climate Change, Food Security, and Migration.* United States
Agency of *International Development* (USAID), p. 4. https://pdf.usaid.gov/pdf_docs/
PA00XXBJ.pdf.

Van Coppenolle, R. and Temmerman, S. (2020) 'Identifying ecosystem surface areas
available for nature-based flood risk mitigation in coastal cities around the
world', *Estuaries and Coasts*, 43(6), pp. 1335–1344. https://doi.org/10.1007/
s12237-020-00718-z.

Van Hear, N. (2004) '"I went as far as my money would take me": conflict, forced migra-
tion and class'. University of Oxford: Centre on Migration, Policy and Society
(COMPAS) (Working Paper No. 6). www.compas.ox.ac.uk/wp-content/uploads/
WP-2004-006-VanHear_Forced_Migration_Class.pdf.

VanLandingham, M. J. and Bankston, C. L. (2017) *Weathering Katrina: Culture and
Recovery among Vietnamese Americans.* Russell Sage Foundation. www.jstor.org/
stable/10.7758/9781610448642 (Accessed: 11 April 2024).

Vanos, J., Guzman-Echavarria, G., Baldwin, J. W., Bongers, C., Ebi, K. L. and Jay, O.
(2023) 'A physiological approach for assessing human survivability and liveability
to heat in a changing climate', *Nature Communications*, 14(1), p. 7653. https://doi
.org/10.1038/s41467-023-43121-5.

van de Ven, D.-J., Mittal, S., Gambhir, A., Lamboll, R. D., Doukas, H., Giarola, S.,
Hawkes, A., Koasidis, K., Köberle, A. C., McJeon, H., Perdana, S., Peters, G. P.,
Rogelj, J., Sognnaes, I., Vielle, M. and Nikas, A. (2023) 'A multimodel analysis of
post-Glasgow climate targets and feasibility challenges', *Nature Climate Change*,
13(6), pp. 570–578. https://doi.org/10.1038/s41558-023-01661-0.

Vicedo-Cabrera, A. M., Scovronick, N., Sera, F., Royé, D., Schneider, R., Tobias, A.,
Astrom, C., Guo, Y., Honda, Y., Hondula, D. M., Abrutzky, R., Tong, S., Coelho,
M. de S. Z. S., Saldiva, P. H. N., Lavigne, E., Correa, P.M., Ortega, N. V., Kan, H.,
Osorio, S., Kyselý, J., Urban, A., Orru, H., Indermitte, E., Jaakkola, J.J.K., Ryti, N.,
Pascal, M., Schneider, A., Katsouyanni, K., Samoli, E., Mayvaneh, F., Entezari, A.,
Goodman, P., Zeka, A., Michelozzi, P., de'Donato, F., Hashizume, M., Alahmad,
B., Diaz, M. H., Valencia, C. D. L. C., Overcenco, A., Houthuijs, D., Ameling, C.,
Rao, S., Di Ruscio, F., Carrasco-Escobar, G., Seposo, X., Silva, S., Madureira, J.,
Holobaca, I. H., Fratianni, S., Acquaotta, F., Kim, H., Lee, W., Iniguez, C., Forsberg,
B., Ragettli, M. S., Guo, Y. L. L., Chen, B. Y., Li, S., Armstrong, B., Aleman, A.,
Zanobetti, A., Schwartz, J., Dang, T. N., Dung, D. V., Gillett, N., Haines, A., Mengel,
M., Huber, V. and Gasparrini, A. (2021) 'The burden of heat-related mortality attrib-
utable to recent human-induced climate change', *Nature Climate Change*, 11(6), pp.
492–500. https://doi.org/10.1038/s41558-021-01058-x.

Vicente-Serrano, S. M., Beguería, S. and López-Moreno, J. I. (2010) 'A multisca-
lar drought index sensitive to global warming: the standardized precipitation
evapotranspiration index', *Journal of Climate*, 23(7), pp. 1696–1718. https://doi
.org/10.1175/2009JCLI2909.1.

Vigil, S. (2022) *Land Grabbing and Migration in a Changing Climate: Comparative
Perspectives from Senegal and Cambodia.* 1st ed. London: Routledge. https://doi
.org/10.4324/9781003193142.

Villarreal, A. (2014) 'Explaining the decline in Mexico-U.S. migration: the effect of
the great recession', *Demography*, 51(6), pp. 2203–2228. https://doi.org/10.1007/
s13524-014-0351-4.

Vinke, K., Rottmann, S., Gornott, C., Zabre, P., Nayna Schwerdtle, P. and Sauerborn, R. (2022) 'Is migration an effective adaptation to climate-related agricultural distress in sub-Saharan Africa?', *Population and Environment*, 43(3), pp. 319–345. https://doi.org/10.1007/s11111-021-00393-7.

Voices from the Frontlines – Africa Climate Mobility Initiative (2022) Africa Climate Mobility Initiative. https://africa.climatemobility.org/ (Accessed: 2 May 2024).

Vose, R. S., Huang, B., Yin, X., Arndt, D., Easterling, D. R., Lawrimore, J. H., Menne, M. J., Sanchez-Lugo, A. and Zhang, H. M. (2021) 'Implementing Full Spatial Coverage in NOAA's Global Temperature Analysis', *Geophysical Research Letters*, 48(4), p. e2020GL090873. https://doi.org/10.1029/2020GL090873.

Vousdoukas, M. I., Mentaschi, L., Hinkel, J., Ward, P. J., Mongelli, I., Ciscar, J.-C. and Feyen, L. (2020) 'Economic motivation for raising coastal flood defenses in Europe', *Nature Communications*, 11(1), p. 2119. https://doi.org/10.1038/s41467-020-15665-3.

Vousdoukas, M. I., Ranasinghe, R., Mentaschi, L., Plomaritis, T. A., Athanasiou, P., Luijendijk, A. and Feyen, L. (2020) 'Sandy coastlines under threat of erosion', *Nature Climate Change*, 10(3), pp. 260–263. https://doi.org/10.1038/s41558-020-0697-0.

Wachinger, G., Renn, O., Begg, C. and Kuhlicke, C. (2013) 'The risk perception paradox – implications for governance and communication of natural hazards', *Risk Analysis*, 33(6), pp. 1049–1065. https://doi.org/10.1111/j.1539-6924.2012.01942.x.

Wahl, D., Byrne, R., Schreiner, T. and Hansen, R. (2007) 'Palaeolimnological evidence of late-Holocene settlement and abandonment in the Mirador Basin, Peten, Guatemala', *The Holocene*, 17(6), pp. 813–820. https://doi.org/10.1177/0959683607080522.

van de Wal, R. S. W., Nicholls, R. J., Behar, D., McInnes, K., Stammer, D., Lowe, J. A., Church, J. A., DeConto, R., Fettweis, X., Goelzer, H., Haasnoot, M., Haigh, I. D., Hinkel, J., Horton, B. P., James, T. S., Jenkins, A., LeCozannet, G., Levermann, A., Lipscomb, W. H., Marzeion, B., Pattyn, F., Payne, A. J., Pfeffer, W. T., Price, S. F., Seroussi, H., Sun, S., Veatch, W. and White, K. (2022) 'A high-end estimate of sea level rise for practitioners', *Earth's Future*, 10(11), p. e2022EF002751. https://doi.org/10.1029/2022EF002751.

Ward, R. C. (1978) *Floods: A Geographical Perspective*. Macmillan Press, London.

Warner, K. (2012) 'Human migration and displacement in the context of adaptation to climate change: the Cancun adaptation framework and potential for future action', *Environment and Planning C: Government and Policy*, 30(6), pp. 1061–1077. https://doi.org/10.1068/c1209j.

Warner, K. (2018) 'Coordinated approaches to large-scale movements of people: contributions of the Paris Agreement and the Global Compacts for migration and on refugees', *Population and Environment*, 39(4), pp. 384–401.

Warner, K. and Afifi, T. (2014) 'Where the rain falls: evidence from 8 countries on how vulnerable households use migration to manage the risk of rainfall variability and food insecurity', *Climate and Development*, 6(1), pp. 1–17. https://doi.org/10.1080/17565529.2013.835707.

Waseem, H. B. and Rana, I. A. (2023) 'Floods in Pakistan: a state-of-the-art review', *Natural Hazards Research*, 3(3), pp. 359–373. https://doi.org/10.1016/j.nhres.2023.06.005.

Wasko, C., Nathan, R., Stein, L. and O'Shea, D. (2021) 'Evidence of shorter more extreme rainfalls and increased flood variability under climate change', *Journal of Hydrology*, 603, p. 126994. https://doi.org/10.1016/j.jhydrol.2021.126994.

Weerasinghe, S. (2020) *Refugee Law in a Time of Climate Change, Disaster and Conflict*. UN High Commissioner for Refugees (UNHCR), p. 112. www.refworld.org/reference/research/unhcr/2020/en/123415.

Weinkle, J., Maue, R. and Pielke, R. (2012) 'Historical global tropical cyclone landfalls', *Journal of Climate*, 25(13), pp. 4729–4735. https://doi.org/10.1175/JCLI-D-11-00719.1.

Werner, A. D., Sharp, H. K., Galvis, S. C., Post, V. E. A. and Sinclair, P. (2017) 'Hydrogeology and management of freshwater lenses on atoll islands: review of current knowledge and research needs', *Journal of Hydrology*, 551, pp. 819–844. https://doi.org/10.1016/j.jhydrol.2017.02.047.

van Westen, R. M., Kliphuis, M. and Dijkstra, H. A. (2024) 'Physics-based early warning signal shows that AMOC is on tipping course', *Science Advances*, 10(6), p. eadk1189. https://doi.org/10.1126/sciadv.adk1189.

Whalen, D., Forbes, D. L., Kostylev, V., Lim, M., Fraser, P., Nedimović, M. R. and Stuckey, S. (2022) 'Mechanisms, volumetric assessment, and prognosis for rapid coastal erosion of Tuktoyaktuk Island, an important natural barrier for the harbour and community', *Canadian Journal of Earth Sciences*, 59(11), pp. 945–960. https://doi.org/10.1139/cjes-2021-0101.

What are atmospheric rivers? | National Oceanic and Atmospheric Administration (n.d.). www.noaa.gov/stories/what-are-atmospheric-rivers (Accessed: 22 April 2024).

What are drylands? | Dryland Forestry | Food and Agriculture Organization of the United Nations (n.d.). www.fao.org/dryland-forestry/background/what-are-drylands/en/ (Accessed: 23 April 2024).

Wheater, H. and Evans, E. (2009) 'Land use, water management and future flood risk', *Land Use Policy*, 26, pp. S251–S264. https://doi.org/10.1016/j.landusepol.2009.08.019.

White, K. D. (2013) 'Nature–society linkages in the Aral Aea region', *Journal of Eurasian Studies*, 4(1), pp. 18–33. https://doi.org/10.1016/j.euras.2012.10.003.

White House (2021) *Report on the Impact of Climate Change on Migration*. Washington, DC: The White House, p. 37. www.whitehouse.gov/wp-content/uploads/2021/10/Report-on-the-Impact-of-Climate-Change-on-Migration.pdf.

Whitley Binder, L. and Jurjevich, J. R. (2016) Winds of Change? Exploring Climate Change-Driven Migration and Related Impacts in the Pacific Northwest: Symposium Summary. Portland, Oregon and Seattle, Washington: Portland State University Population Research Center and the University of Washington Climate Impacts Group, p. 34. https://pdxscholar.library.pdx.edu/cgi/viewcontent.cgi?article=1037&context=prc_pub.

Wiegel, H. (2023) 'Complicating the tale of "first climate migrants": resource-dependent livelihoods, drought and labour mobilies in semi-arid Chile', *Geoforum*, 138, p. 103663. https://doi.org/10.1016/j.geoforum.2022.11.005.

Williams, N. E. and Gray, C. (2020) 'Spatial and temporal dimensions of weather shocks and migration in Nepal', *Population and Environment*, 41(3), pp. 286–305. https://doi.org/10.1007/s11111-019-00334-5.

Willison, C. E., Singer, P. M., Creary, M. S. and Greer, S. L. (2019) 'Quantifying inequities in US federal response to hurricane disaster in Texas and Florida compared with Puerto Rico', *BMJ Global Health*, 4(1), p. e001191. https://doi.org/10.1136/bmjgh-2018-001191.

Wilmsen, B., Van Hulten, A., Duan, Y. (2021). Leaving the Three Gorges After Resettlement: Who Left, Why Did They Leave, and Where Did They Go?. La Trobe. https://doi.org/10.26181/6041959f422c6

Wilmsen, B. and Rogers, S. (2023) 'Putting "rupture" to work at the Three Gorges Dam', *Dialogues in Human Geography*, 13(2), pp. 211–215. https://doi.org/10.1177/20438206231168882.

Wilson, C. (2023) 'Indian Census 2021: Impact of COVID-19 on Its Operation & Implementation', *International Journal of Law Management & Humanities*, 6(2), p. 1318.

Wing, O.E.J., Lehman, W., Bates, P.D., Sampson, C.C., Quinn, N., Smith, A.M., Neal, J.C., Porter, J.R. and Kousky, C. (2022) 'Inequitable patterns of US flood risk in the Anthropocene', *Nature Climate Change*, 12(2), pp. 156–162. https://doi.org/10.1038/s41558-021-01265-6.

Wink Junior, M. V., dos Santos, L. G., Ribeiro, F. G. and da Trindade, C. S. (2023) 'Natural disasters and poverty: evidence from a flash flood in Brazil', *Environment, Development and Sustainability* [Preprint]. https://doi.org/10.1007/s10668-023-03623-0.

Winkler, R. L. and Rouleau, M. D. (2021) 'Amenities or disamenities? Estimating the impacts of extreme heat and wildfire on domestic US migration', *Population and Environment*, 42(4), pp. 622–648. https://doi.org/10.1007/s11111-020-00364-4.

Wischnath, G. and Buhaug, H. (2014) 'On climate variability and civil war in Asia', *Climatic Change*, 122(4), pp. 709–721. https://doi.org/10.1007/s10584-013-1004-0.

WOLA (2022) *Migration, country by country, at the U.S.-Mexico border.* www.wola.org/2022/11/migration-country-by-country-at-the-u-s-mexico-border/ (Accessed: 23 April 2024).

Wolde, S. G., D'Odorico, P. and Rulli, M. C. (2023) 'Environmental drivers of human migration in sub-Saharan Africa', *Global Sustainability*, 6, p. e9. https://doi.org/10.1017/sus.2023.5.

Wolpert, J. (1966) 'Migration as an adjustment to environmental stress', *Journal of Social Issues*, 22(4), pp. 92–102. https://doi.org/10.1111/j.1540-4560.1966.tb00552.x.

Woodroffe, C.D. (2008) 'Reef-island topography and the vulnerability of atolls to sea-level rise', *Global and Planetary Change*, 62(1), pp. 77–96. https://doi.org/10.1016/j.gloplacha.2007.11.001.

World Bank (2023) 'Data: Net migration – Dominica'. https://data.worldbank.org/indicator/SM.POP.NETM?end=2021&locations=DM&start=1960&view=chart (Accessed: 11 January 2024).

World Bank (2024) 'Bangladesh'. https://data.worldbank.org/country/BD.

World Bank Open Data Portal (2024) 'GDP (current US$) [Data File]'. https://data.worldbank.org/indicator/NY.GDP.MKTP.CD (Accessed: 29 April 2024).

World Commission on Environment and Development (WCED) and Bruntland, G.H. (1987) *Our Common Future: Report of the World Commission on Environment and Development.* United Nations: Oxford University Press.

World Meteorological Organization (WMO) (2024) 'El Niño weakens but impacts continue [Press Release]'. https://wmo.int/news/media-centre/el-nino-weakens-impacts-continue.

World Meteorological Organization (2022) 'Early Warning systems must protect everyone within five years [Press Release]'. https://wmo.int/news/media-centre/early-warning-systems-must-protect-everyone-within-five-years (Accessed: 11 April 2024).

World Meterological Organization (2023) 'Tropical cyclone naming [Fact Sheet]'. https://wmo.int/content/tropical-cyclone-naming (Accessed: 12 April 2024).

Worster, D. (1979) *Dust Bowl: The Southern Plains in the 1930s.* New York: Oxford University Press.

Wu, L., Zhao, H., Wang, C., Cao, J. and Liang, J. (2022) 'Understanding of the effect of climate change on tropical cyclone intensity: a review', *Advances in Atmospheric Sciences*, 39(2), pp. 205–221. https://doi.org/10.1007/s00376-021-1026-x.

Xiao, T., Oppenheimer, M., He, X. and Mastrorillo, M. (2022) 'Complex climate and network effects on internal migration in South Africa revealed by a network model', *Population and Environment*, 43(3), pp. 289–318. https://doi.org/10.1007/s11111-021-00392-8.

Xu, L. and Famiglietti, J. S. (2023) 'Global patterns of water-driven human migration', *WIREs Water*, 10(4), p. e1647. https://doi.org/10.1002/wat2.1647.

Xue, P., Malanotte-Rizzoli, P., Wei, J. and Eltahir, E. A. B. (2020) 'Coupled ocean-atmosphere modeling over the Maritime continent: a review', *Journal of Geophysical Research: Oceans*, 125(6), p. e2019JC014978. https://doi.org/10.1029/2019JC014978.

Yang, D. and Choi, H. (2007) 'Are remittances insurance? Evidence from rainfall shocks in the Philippines', *The World Bank Economic Review*, 21(2), pp. 219–248. https://doi.org/10.1093/wber/lhm003.

Ye, Q. and Glantz, M. H. (2005) 'The 1998 Yangtze floods: the use of short-term forecasts in the context of seasonal to interannual water resource management', *Mitigation and Adaptation Strategies for Global Change*, 10(1), pp. 159–182. https://doi.org/10.1007/s11027-005-7838-7.

Yee, M., Piggott-McKellar, A. E., McMichael, C. and McNamara, K. E. (2022) 'Climate change, voluntary immobility, and place-belongingness: insights from Togoru, Fiji', *Climate*, 10(3), p. 46. https://doi.org/10.3390/cli10030046.

Yeoh, B. (2016) 'Migration and gender politics in Southeast Asia', *Migration, Mobility, & Displacement*, 2(1). https://doi.org/10.18357/mmd21201615022.

Yeung, W.-J. J. and Mu, Z. (2020) 'Migration and marriage in Asian contexts', *Journal of Ethnic and Migration Studies*, 46(14), pp. 2863–2879. https://doi.org/10.1080/1369183X.2019.1585005.

Yoo, S. H. and Agadjanian, V. (2024) 'Drought and migration: a case study of rural Mozambique', *Population and Environment*, 46(1), p. 3. https://doi.org/10.1007/s11111-023-00444-1.

Younes, H., Darzi, A. and Zhang, L. (2021) 'How effective are evacuation orders? An analysis of decision making among vulnerable populations in Florida during Hurricane Irma', *Travel Behaviour and Society*, 25, pp. 144–152. https://doi.org/10.1016/j.tbs.2021.07.006.

Yu, F., Chen, Z., Ren, X. and Yang, G. (2009) 'Analysis of historical floods on the Yangtze River, China: characteristics and explanations', *Geomorphology*, 113(3), pp. 210–216. https://doi.org/10.1016/j.geomorph.2009.03.008.

Zander, K. K. and Garnett, S. (2020) 'Risk and experience drive the importance of natural hazards for peoples' mobility decisions', *Climatic Change*, 162(3), pp. 1639–1654. https://doi.org/10.1007/s10584-020-02846-8.

Zander, K. K., Wilson, T. and Garnett, S. T. (2020) 'Understanding the role of natural hazards in internal labour mobility in Australia', *Weather and Climate Extremes*, 29, p. 100261. https://doi.org/10.1016/j.wace.2020.100261.

Zazyki, M. A., da Silva, W. V., de Moura, G.L., Kaczam, F. and da Veiga, C.P. (2022) 'Property rights in informal settlements', *Cities*, 122, p. 103540. https://doi.org/10.1016/j.cities.2021.103540.

Zemp, M., Huss, M., Eckert, N., Thibert, E., Paul, F., Nussbaumer, S. U. and Gärtner-Roer, I. (2020) 'Brief communication: ad hoc estimation of glacier contributions to sea-level rise from the latest glaciological observations', *The Cryosphere*, 14(3), pp. 1043–1050. https://doi.org/10.5194/tc-14-1043-2020.

Zhang, K. and Leatherman, S. (2011) 'Barrier Island Population along the U.S. Atlantic and Gulf Coasts', *Journal of Coastal Research*, 27(2), pp. 356–363. https://doi.org/10.2112/JCOASTRES-D-10-00126.1.

Zhang, X., Chen, N., Sheng, H., Ip, C., Yang, L., Chen, Y., Sang, Z., Tadesse, T., Lim, T. P. Y., Rajabifard, A., Bueti, C., Zeng, L., Wardlow, B., Wang, S., Tang, S., Xiong, Z., Li, D. and Niyogi, D. (2019) 'Urban drought challenge to 2030 sustainable development goals', *Science of The Total Environment*, 693, p. 133536. https://doi.org/10.1016/j.scitotenv.2019.07.342.

Zhu, M., Zhang, Z., Zhu, B., Kong, R., Zhang, F., Tian, J. and Jiang, T. (2020) 'Population and economic projections in the Yangtze river basin based on shared socioeconomic pathways', *Sustainability*, 12(10), p. 4202. https://doi.org/10.3390/su12104202.

Zou, Y., Rasch, P.J., Wang, H., Xie, Z. and Zhang, R. (2021) 'Increasing large wildfires over the western United States linked to diminishing sea ice in the Arctic', *Nature Communications*, 12(1), p. 6048. https://doi.org/10.1038/s41467-021-26232-9.

Zscheischler, J., Martius, O., Westra, S., Bevacqua, E., Raymond, C., Horton, R. M., van den Hurk, B., AghaKouchak, A., Jézéquel, A., Mahecha, M. D., Maraun, D., Ramos, A. M., Ridder, N. N., Thiery, W. and Vignotto, E. (2020) 'A typology of compound weather and climate events', *Nature Reviews Earth & Environment*, 1(7), pp. 333–347. https://doi.org/10.1038/s43017-020-0060-z.

Zúñiga, E., Magaña, V. and Piña, V. (2020) 'Effect of urban development in risk of floods in Veracruz, Mexico', *Geosciences*, 10(10). https://doi.org/10.3390/geosciences10100402.

Zuo, J., Pullen, S., Palmer, J., Bennetts, H., Chileshe, N. and Ma, T. (2015) 'Impacts of heat waves and corresponding measures: a review', *Journal of Cleaner Production*, 92, pp. 1–12. https://doi.org/10.1016/j.jclepro.2014.12.078.

Index

Page numbers in bold refer to keywords appearing in boxes; page numbers in italics refer to keywords appearing in figures.

Printed by Integrated Books International,
United States of America